Lasers in Polymer Science and Technology: Applications

Volume IV

Editors

Jean-Pierre Fouassier, Ph.D.

Professor
Laboratory of General Photochemistry
Ecole Nationale Superieure de Chimie
University of Haute-Alsace
Mulhouse, France

Jan F. Rabek, Ph.D.

Department of Polymer Technology
The Royal Institute of Technology
Stockholm, Sweden

CRC Press, Inc.
Boca Raton, Florida

Library of Congress Cataloging-in-Publication Data

Lasers in polymer science and technology:applications / editors, Jean
 -Pierre Fouassier and Jan F. Rabek.
   p.   cm.
  Bibliography: p.
  Includes index.
  ISBN 0-8493-4844-7 (v. 1)
  1. Polymers--Analysis.   2. Laser spectroscopy.   I. Fouassier,
Jean-Pierre, 1947- II. Rabek, J. F.
TP1140.L37   1990
668.9--dc20   89-9822
                                                                      CIP

Direct all inquiries to CRC Press, Inc., 2000 Corporate Blvd., N.W., Boca Raton, Florida, 33431.

© 1990 by CRC Press, Inc.

International Standard Book Number 0-8493-4844-7 (v. 1)
International Standard Book Number 0-8493-4845-5 (v. 2)
International Standard Book Number 0-8493-4846-3 (v. 3)
International Standard Book Number 0-8493-4847-1 (v. 4)

Library of Congress Number 89-9822
Printed in the United States

DEDICATION

To our wives, partners through life
Geneviève — Ewelina
and our children
Patrick, Laurence, and Yann — Dominika
for their patience and understanding.

PREFACE

Laser spectroscopy and laser technology have been growing ever since the first laser was developed in 1960 and cover now a wide range of applications. Among them, three groups came into prominence as regards polymer science and technology: molecular gas lasers (notably CO_2 lasers) in the IR region, gas, solid, and dye lasers in the visible and near IR region, and the relatively new group of UV excimer lasers. Lasers are unique sources of light. Many recent advances in science are dependent on the application of their uniqueness to specific problems. Lasers can produce the most spectrally pure light available, enabling atomic and molecular energy levels to be studied in greater detail than ever before. Certain types of laser can give rise to the shortest pulses of light available from any light source, thus providing a means for measuring some of the fastest processes in nature.

Measurements of luminescence (fluorescence and phosphorescence) provide some of the most sensitive and selective methods of spectroscopy. In addition, luminescence measurements provide important information about the properties of excited states, because the emitted light originates from electronically excited states. The measurement of luminescence intensities makes it possible to monitor the changes in concentration of the emitting chemical species as a function of time, whereas the wavelength distribution of the luminescence provides information on the nature and energy of the emitting species.

Such areas as laser luminescence spectroscopy, pico- and nanosecond absorption spectroscopy, CIDNP and CIDEP laser flash photolysis, holographic spectroscopy, and time-resolved diffuse reflectance laser spectroscopy, have evolved from esoteric research specialities into standard procedures, and in some cases routinely applied in a number of laboratories all over the world.

Application of Rayleigh, Brillouin, and Raman laser spectroscopy in polymer science gives information about local polymer chain motion, large-scale diffusion, relaxation behavior, phase transitions, and ordered states of macromolecules.

During the last decade the photochemistry and photophysics of polymers have grown into an important and pervasive branch of polymer science. Great strides have been made in the theory of photoreactions, energy transfer processes, the utilization of photoreactions in polymerization, grafting, curing, degradation, and stabilization of polymers. The progress of powerful laser techniques has not been limited to spectroscopical studies in polymer matrix, colloids, dyed fabrics, photoinitiators, photosensitizers, photoresists, materials for solar energy conversion, or biological molecules and macromolecules; it has also found a number of practical and even industrial applications.

One of the most important applications of lasers is the use of a high intensity beam for material processing in polymers. In these materials, the laser beam can be employed for drilling, cutting, and welding. Lasers can produce holes at very high speeds and dimensions, unobtainable by other processing methods.

Lasers can be successfully used to study surface processes and surface modification of polymeric materials, such as molecular beam scattering, oxidation, etching, annealing, phase transitions, surface mobility, and thin films and vapor phase deposition.

UV laser radiation causes the breakup and spontaneous removal of material from the surface of organic polymers (ablative photodecomposition). The surface of the solid is etched away to a depth of a few tenths of a micron, and the products are expelled at supersonic velocity. This method has found practical applications in photolithography, optics, electronics, and the aerospace industries.

The newest process includes stereolithography, which involves building three-dimensional plastic prototypes (models) from computer-aided designs. Stereolithography is actually a combination of four technologies: photochemistry, computer-aided, laser light, and laser-image formation. The device (which consists in a mechanically scanned, computer driven three-dimensional solid pattern generator) builds parts by creating, under the laser exposure, cross sections of the part out of a liquid photopolymer, then "fusing" the sections together until a complete model is formed.

Another new development is technology of micromachines such as gears, turbines, and motors which are 100 to 200 μm in diameter which can be used in a space technology, microrobots, or missile-guidance systems. These micromachines are made by a process of etching patterns on silicon chips. Beside making such micromachines, microscopic tools on a catheter, inserted through a blood vessel, would enable surgeons to do "closed heart" surgery. Developing of micromachine technology would not be possible without photopolymers and UV lasers.

The editors went to great lengths in order to secure the cooperation of the most outstanding specialists to complete this monography. A number of invited authorities were not able to accept our invitation, due to other commitments, but all authors who presented their contributions "poured their hearts out" in this endeavour. We would like to thank them for their efforts and cooperation. This monography strongly favors the inclusion of experimental details, apparatus, and techniques, thus allowing the neophyte to learn the "tricks of the trade" from the experts. This is an effort to show, in compact form, the bulk of information available on applications of lasers to polymer science and technology. The editors are pleased to submit to the readers the state-of-the art in this field.

J.-P. Fouassier and J. R. Rabek

CONTRIBUTORS

Angelika Anders, Dr. rer. nat.
Professor
Institute of Biophysics
University of Hannover
Hannover, West Germany

Robert S. Brown, Ph.D.
Assistant Professor
Department of Chemistry
Colorado State University
Fort Collins, Colorado

Jean-Marc Haudin, D.Sc.
Professor
Ecole des Mines de Paris
CEMEF
Valbonne, France

Gary R. Holtom, Ph.D.
Head of Laser Operations
Department of Chemistry
Regional Laser and Biotechnology
 Laboratories
University of Pennsylvania
Philadelphia, Pennsylvania

Marita Knälmann, Dr. rer. nat.
Institute of Biophysics
University of Hannover
Hannover, West Germany

Marian Kryszewski, Ph.D.
Professor
Centre of Molecular and Macromolecular
 Studies
Polish Academy of Sciences
Lodz, Poland

Yukio Kubota, Ph.D.
Professor
Department of Chemistry
Yamaguchi University
Yoshida, Yamaguchi, Japan

Patrick Navard, Ph.D.
Ecole des Mines de Paris
CEMEF
Valbonne, France

T. J. Schaafsma, Ph.D.
Professor
Department of Molecular Physics
Agricultural University
NL-6703 Wageningen
The Netherlands

G. F. W. Searle, Ph.D.
Department of Molecular Physics
Agricultural University
P.O. Box 8091
NL-6700 Wageningen
The Netherlands

Robert F. Steiner, Ph.D.
Professor
Department of Chemistry
University of Maryland
Catonsville, Maryland

A. van Hoek
Department of Molecular Physics
Agricultural University
De Dreijen 11
NL-6703 Wageningen
The Netherlands

Charles L. Wilkins, Ph.D.
Professor
Department of Chemistry
University of California
Riverside, California

SERIES TABLE OF CONTENTS

TABLE OF CONTENTS

Chapter 1

LASER MASS SPECTROMETRY: APPLICATION TO POLYMER ANALYSIS

Robert S. Brown and Charles L. Wilkins

TABLE OF CONTENTS

I. INTRODUCTION

Mass spectrometry (MS) is concerned with the generation and analysis of ions produced from a variety of sample types and methods. This review will cover application of MS to polymer characterization and, in particular, the use of lasers as ionization and volatilization sources for MS. In addition, the use of lasers for subsequent further fragmentation of ions (photodissociation) as well as for multiphoton ionization of laser desorbed neutrals will be discussed briefly. Potential applications of lasers to polymer characterization by MS also will be considered.

Various mass spectrometric techniques which permit direct study of molecules in the 10,000- to 20,000-amu range currently exist. These methods can directly cover low molecular weight commercial polymers. However, direct mass spectrometric examination of polymers at the other extreme, where average molecular weights of 1 million or higher is common, is not practical. Nevertheless, degradation methods can be utilized to indirectly study higher molecular weight polymers. In this way, MS can be used for higher molecular weight polymer characterization, although not to obtain molecular weight information. Fingerprint patterns are often generated and the evolved material can be further separated by gas chromatography, followed by mass spectrometric detection. Computerized pattern recognition techniques also have been applied to identify polymers and important structural information can be obtained from the low mass fragment ions produced.

The current major limitation for directly analyzing intact species with higher masses is the lack of suitable ionization techniques to get these large molecules into the vapor phase as ions. As ionization methods have improved, instrumental mass ranges of commercial mass spectrometers have been increased to take advantage of the new methods. As a consequence, modern mass spectrometers often have the capability to examine greater mass ranges, given suitable sources of ions. The unique ability of MS to provide actual molecular weights, in addition to information regarding detailed molecular structure, is a powerful driving force for the utilization of MS as a problem-solving tool. The high resolving power of even low-resolution MS, when compared with other separation methods such as size exclusion chromatography, allows the analyst to discern minor components or impurities differing only slightly in molecular structure. Examples of direct analysis of polymers with average masses in the 1,000- to 10,000-amu range are numerous and applications of mass spectrometry to the analysis of prepolymers are also numerous.

II. TYPES OF MASS SPECTROMETERS AND ION SOURCES

To better understand the capabilities of MS for polymer characterization and provide some background on both the development of MS for nonvolatile sample analysis and the problems which the mass spectroscopist faces with such analyses, it is appropriate to discuss briefly the various forms of MS, along with their advantages and disadvantages. A review of the basic ionization methods, other than those employing lasers, currently available to the mass spectroscopist will be presented. The current state of laser ionization sources will be covered in detail in Section III. The capabilities of nonlaser ionization techniques for polymer characterization will be discussed, as will examples of polymers studied to date by these techniques.

A. TYPES OF MASS SPECTROMETERS

Although there is a large variety of mass spectrometers the most common utilize four basic designs, which can be further subdivided into two types: scanning and pulsed mass spectrometers. The two basic types of scanning mass spectrometers are quadrupole and sector mass spectrometers. Important pulsed mass spectrometers are time-of-flight (TOF) and Fourier transform mass spectrometers (FTMSs). These categories are distinguished by their requirements for continuous or pulsed sources of ions. Although scanning spectrometers are more common, pulsed spectrometers appear to be better suited for polymer analysis, particularly when laser ionization techniques are employed.

Sector instruments generally include two ion separation regions, called sectors. One sector effects ion separation based upon applied magnetic fields (the magnetic sector), while the other utilizes applied potential fields (the electrostatic sector). These two sectors are used to separate ions by mass to charge ratio (m/z) and mass spectrometers using them are called double sector mass spectrometers. Single sector mass spectrometers are generally not used due to their low resolution. The low resolving power of a single magnetic sector instrument results from the distribution of the kinetic energy of ions of a given m/z, which broadens the ion beam reaching the detector. This causes a corresponding loss in mass spectral resolution. Ions produced from the sample are selected by varying the applied magnetic field which, in turn, focuses ions of different m/z onto the detector. By introduction of a second electrostatic sector, this small kinetic energy spread can be reduced by only focusing upon the detector ions within a given kinetic energy range. This combination of sectors can produce mass spectrometers with high resolving power. Typical resolutions for such instruments (m/Δm) are between 25,000 and 100,000. A practical upper mass limit in excess of 10,000 amu is obtainable, but sensitivity can suffer at higher masses. A good overview of the tradeoffs for sector instruments with respect to mass range and resolution has been published.[1]

The second type of scanning mass spectrometer is the quadrupole mass analyzer. This spectrometer utilizes four metal rods arranged symmetrically. A continuous beam of ions is accelerated between the rods which have DC and radio-frequency (rf) AC potentials applied to them. The effect is to produce a stable path to the detector only for ions with a particular m/z. By changing the potentials applied to the rods, ions with various m/z are focused sequentially onto the detector. Quadrupole mass analyzers are relatively inexpensive but suffer from low resolving power and have an upper mass limit which is typically about 1000 amu, although some commercial quadrupole analyzers can go as high as 2000 amu.

The two pulsed mass spectrometers (TOF and Fourier transform) have several features which make them well suited for polymer analysis. Both have very high upper mass limits and are compatible with pulsed ionization techniques which the scanning spectrometers cannot effectively employ. The TOF mass analyzer is based on the difference in the time it takes for ions of different mass produced in the source to travel a fixed distance down a field-free drift tube and strike a detector. The ions are accelerated out of the ion source by an electric field and, because they all have roughly the same kinetic energy distribution, their relative velocities in the drift tube are inversely proportional to their masses. Thus, heavier ions reach the detector after lighter ions. Resolution is relatively low due to the initial kinetic energy distribution of the ions being analyzed (mass resolution of 500 to 600 is typical, although more recent instruments employing reflectron mass analysis can obtain mass resolution approaching 10,000). Mass range is extremely high since there are no focusing optics which, in scanned instruments, set upper mass transmission limits. A recent issue of *Analytical Instrumentation* was devoted entirely to TOF instruments and applications.[2]

The second type of pulsed spectrometer is the FTMS, sometimes referred to as a Fourier transform ion cyclotron resonance mass spectrometer because it is based upon the cyclotron resonance phenomenon. A comprehensive review of instrumental developments and applications has been published.[3] In the FTMS, pulsed ion sources, similar to those used in TOF spectrometers, are practical. However, the ions formed are trapped in an electrostatic ion trap which is contained within a high magnetic field (typically 1.5 to 7.2 T). Ions are constrained to move along the lines of force of the magnetic field and are effectively trapped by the applied electric fields of the trapped ion cell which produces a potential well from which the ions cannot escape. Ions can be trapped in this way for extremely long times. Ion cyclotron motion is a circular path about the magnetic field axis. If rf energy is applied to the ions, they increase their radii from the center of the magnetic field because they are constrained to maintain constant angular frequency (proportional to m/z and the magnetic field strength, B). As a result of the rf excitation, a signal, which is the summation of the cyclotron frequencies of all ions present, is produced. This time-domain signal is sorted into its constituent frequencies by the Fourier transform technique and converted by use of the cyclotron equation (described qualitatively above) into m/z values. As in the TOF spectrometer, the upper mass limit is high due to the absence of focusing optics, but resolution is much better in the FTMS instrument. Because the major factor limiting resolution in FTMS is damping of the ion signal due to collisions with background neutrals in the vacuum system, pressures must be low in order to obtain the highest resolution (pressures between 10^{-8} and 10^{-9} torr are used). There is an inverse relationship between mass and resolution, so that although extremely high resolution is obtainable at lower masses (>1,000,000 at m/z = 100), this drops off as the mass is increased (e.g., under conditions where 1 million resolution is obtained for m/z 100, the resolution would be 100,000 for m/z 1000). In addition to the high mass range advantage of pulsed-source mass spectrometers, their ability to analyze all ions produced from a rapid ionization event makes both suitable for use with high-power pulsed laser sources.

B. NONLASER IONIZATION SOURCES

In recent years the number of ionization sources for MS has increased dramatically. Most significant are new "soft" ionization techniques developed primarily for the analysis of polar biological species, but which have found use for a wide range of applications, including polymer analysis. Several reviews on the use of these new soft ionization methods for polymer characterization have appeared.[4-7] Soft ionization is distinguished from other ionization methods, such as conventional electron ionization, by its ability to produce a significant number of ions indicative of molecular weight for large molecules, which normally would be considered too nonvolatile for mass spectrometric analysis. Although many of these techniques can be applied, in principle, with either scanning or pulsed mass spectrometers, in practice, some are restricted to one or the other due to certain operating constraints. In particular, several of the pulsed ionization techniques simply cannot be used conveniently with sector or quadrupole mass spectrometers.

1. Field Desorption

One of the first of the new ionization techniques developed was the field desorption (FD) ion source.[8,9] It employs a special emitter onto which the sample is deposited. Emitters have been made of a variety of materials, including silicon and tungsten, and have as their common feature thin microneedles on their surface. The emitter is heated and a high electric field, which causes the sample to be desorbed from the surface as ions, is applied. A large percentage of the desorbed ions are molecular ions. Most applications to date have utilized sector mass spectrometers which are compatible with the high electric fields which must be employed. Furthermore, their superior mass range and resolution over quadrupole spectrometers has made sector instruments predominant for FD studies. Recently, a FD source has been described for use with a FTMS.[10] As with most desorption/ionization methods, the polarity of the sample determines the types of ions formed. Nonpolar molecules tend to produce mainly M^+ or M^- ions, while polar molecules produce ions mainly by attachment reactions with cations (K^+, Na^+) or protons (H^+). These characteristics appear to be similar for all desorption/ionization techniques. A clear difference with the FD technique is its success with nonpolar samples, which generally do not produce intact molecular ions by other desorption/ionization techniques.

FD, being the oldest of the soft ionization sources available for larger molecules, has been widely applied to polymer characterization. Selected examples include the determination of molecular weight distributions for low molecular weight polystyrene samples by FD-MS[11] and the observation of ions from polystyrene with masses in excess of 10,000 amu.[12] Polybutadiene samples were studied using FD-MS by different workers with some reporting observation of proton attachment species[13] and others reporting M^+ species.[14] In these studies, accurate molecular weight distributions were determined for the low molecular weight polymers studied (average molecular weights between 1000 and 3000 amu). Polyglycols have also been studied by several workers.[10,15] In both cases, attachment ion species were observed (either H^+ or Na^+) for these polymers, with average molecular weights between 1000 and 2000 amu. The ion distributions also appeared to track the molecular weight distribution of each sample. Additional examples can be found in the reviews already cited.[4-7]

2. Particle Bombardment — Secondary Ion Mass Spectrometry and Fast Atom Bombardment

Another soft ionization technique is the use of highly energetic particles to cause desorption and ionization. Such methods have become the techniques of choice in recent years, supplanting FD in most cases, especially for polar biological materials. Particle bombardment sources can be grouped into two types, secondary ion mass spectrometry (SIMS) sources,[16] which utilize high energy ion beams (typically 2 to 10 KeV argon or cesium ions), and the fast atom bombardment (FAB) ionization method,[17] which utilizes a high energy beam of neutral atoms (typically 3 to 8 KeV argon or xenon atoms). The two techniques have a great deal in common and many of the distinctions in terminology appear to be somewhat arbitrary. In fact, the desorption mechanisms involved in the two techniques appear to be very similar. The distinction between static and dynamic SIMS adds additional complications to the terminology. Not only does this refer to the intensity of the primary ion beam but also to the state of the sample. In dynamic SIMS, a high particle flux is employed with an associated liquid matrix. The latter prevents major sample decomposition and aids in the desorption/ ionization process. This is also referred to as liquid SIMS because of the liquid matrix (glycerol is the common choice) employed. Liquid SIMS and FAB are essentially the same except that the latter uses neutral bombarding particles. Static SIMS is different in that it utilizes no matrix and a lower primary ion flux. These distinctions are discussed in more detail by Barber and co-authors[17] in the introduction of their MS review and by Burlingame and co-authors[18] in their review.

FAB-MS has been applied in several cases to the study of low molecular weight polymers. These have been limited to polar polymers due to the need for solubility in the liquid matrix employed in FAB-MS. In particular, polyglycols have been examined extensively[19,20] and the results were compared[21] with previous FD-MS results. In contrast to the earlier FD studies, the FAB-MS technique did not provide quantitative data; the observed ion distribution was shifted to lower mass, probably due to excessive fragmentation. Similar results[22] for the closely related but highly cross-linked poly(ethylene imine) (PEI) polymers were observed. Because of the regular repeating units of the polyglycols and the wide range of molecular weights available, it has been proposed that they be used as mass calibration standards[19,20] in MS.

Unlike FAB-MS which has been employed mainly to examine intact molecular ions from low molecular weight polar polymers, static SIMS has been used for both the generation of molecular ion species and for degradation studies of larger polymers. An excellent example of the study of intact polymers of modest polarity has been published by Bletsos, et al.[23] demonstrating static SIMS with a TOF mass spectrometer. In this work they show the SIMS spectra of thin films of polydimethylsiloxane (Figure 1) and polystyrene with ions extending

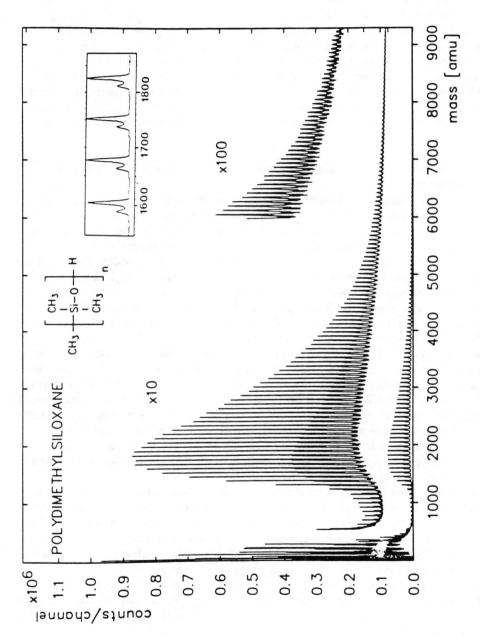

FIGURE 1.   SIMS-TOF mass spectrum of poly(dimethylsiloxane). (From Bletsos, I. V., Hercules, D. M., vanLeyen, D., and Benninghoven, A., *Macromolecules, 20*, 407, 1987. With permission.)

beyond 9000 and 7000 amu, respectively. Several reports of SIMS of higher molecular weight polymers which show characteristic fragment ions have appeared.[24-26] This technique provides results similar to thermal pyrolysis MS of polymers.[27-33]

3. Plasma Desorption

Another soft ionization method common to MS in recent years is plasma desorption ionization. It has been used minimally for polymer characterization to date, but is used extensively in the area of biological MS, with particular success in the ionization of large peptides (with masses up to approximately 20,000 amu).[34] Plasma desorption is also a particle bombardment technique but, in contrast to the previously discussed particle bombardment methods, it utilizes million electronvolts fission fragments (typically from ^{252}Cf radioactive decay). In this respect, it is in a quite different energy regime than any of the previous desorption methods and this is why they have not been grouped together. One of the few plasma desorption applications to polymer characterization involved the study of polyethers.[35] The amount of fragmentation was excessive, even more so than for the FAB-MS of similar polymers. Being such a high-energy technique, this possibly is not surprising and may be why more work in this area has not appeared. Peptides, which are apparently more robust, are amenable to plasma desorption mass spectral analysis and produce molecular ions, sometimes multiply charged. A primary current limitation to this method is that it is employed almost exclusively with low resolution TOF mass analysis, which cannot distinguish individual masses in the mass ranges of interest. Some recently reported work with an FTMS instrument[36] may allow higher resolution spectra to be obtained.

4. Other Ionization Techniques

There exists another class of ionization sources for nonvolatile samples which are different from those previously described. These are the solvent assisted ionization techniques of electrohydrodynamic ionization and thermospray ionization. Electrohydrodynamic mass spectrometry (EH-MS) extracts ions (usually as cationized species) directly from solution using an applied electric field. EH-MS has been used to characterize the molecular weight distributions of low molecular weight poly(ethylene glycols) (PEGs) with molecular weights in the 400- to 3200-amu range.[37,38] However, PEI samples have been shown to consistently produce molecular weight distributions[39] which are low. This appears to be due to the high degree of branching in PEI and solvent-solute interactions. These interactions appear to cause lower molecular weight species to be preferentially desorbed from solution. The thermospray ionization technique has similarities with EH-MS and has been quite successful as an interface between liquid chromatography and MS. Here, ions are thought to be preformed in solution and are desorbed as the solvent is evaporated in the mass spectrometer. Both of these techniques are most useful for the ionization of polar species and the thermospray method has been successfully applied to a wide range of biological materials.

III. LASER MASS SPECTROMETRY

The use of lasers in combination with MS can be traced back to experiments conducted in the early 1960s. One of the first experiments[40] utilized a focused ruby laser to study metals, semiconductors, and insulators with a conventional sector instrument which had a modified source. Another early experiment was concerned with the vaporization of graphite[41] and utilized a single magnetic sector instrument and a ruby laser. Volatilized material was ionized by a conventional electron beam. Most of the early work used scanning or low-resolution TOF spectrometers to characterize the evolved products. Most studies also used electron beam-assisted ionization of the pyrolysis products, but all had in common the study of small molecules or fragment species. It was not until very high-powered lasers were developed and coupled to pulsed spectrometers that laser ionization MS became a more versatile analysis technique. By utilizing high-powered pulsed lasers, direct desorption and ionization of intact large molecules became a reality. This has come to be known as laser desorption (LD). Probably the first true LD-MS experiment was performed in 1970 by Vastola and co-workers,[42] who observed cationized salts and dimers. Other workers utilized a hybrid sector mass spectrometer to study LD spectra of a variety of nonvolatile biological molecules.[43] That thermal mechanisms may play a major role in the LD process was demonstrated by the direct desorption and ionization of glucose by rapid conventional heating of the sample.[44]

A. CURRENT INSTRUMENTATION

The first practical laser mass spectrometer was the laser microprobe, developed in 1975.[45] It utilized a focused beam from a high energy (10^8 to 10^9 W cm^{-2}) pulsed laser coupled with a TOF mass spectrometer. The first use of this new instrument was to study samples of epoxy resin doped with Li, Mg, and Co. Due to its very high spatial resolution, it was used initially to produce surface maps of the elemental composition of samples. Spatial resolution was so good that even single cells could be probed.[46] The commercial introduction of the laser microprobe (LAMMA) instrument in the late 1970s has resulted by far in the most reports of laser MS. Two recent reviews dealing with LAMMA have been published, part 1 detailing the basic principles and instrumentation[47] and part 2, concerned with applications to structural analysis.[48] The demonstration in 1982 by McCreery, Ledford, and Gross[49] of a LD interface coupled to FTMS has led to the development of an instrument not only as well suited to pulsed laser ionization sources and high mass analysis as the TOF instrument but with the capability for much higher resolution mass analysis.

Several reviews have discussed early work in the coupling of lasers to mass spectrometry along with more recent instrumental advances.[50-52] The probable mechanisms involved in the direct desorption of ions from nonvolatile samples[53,54] have also been examined. Accordingly, instrumentation and mechanisms of laser ionization will be discussed only briefly here. Laser desorption and instrumentation basically falls into two categories: LD-TOF and LD-FTMS instrumentations. This is due to the success of high-energy pulsed lasers in inducing desorption and ionization. The actual desorption event is over very quickly, so a pulsed spectrometer is the instrument of choice. As TOF MS and FTMS are the only well-developed pulsed mass spectrometers currently available which have high mass range capabilities, it is not surprising that they dominate the field of laser MS.

B. IONIZATION MECHANISMS

A wide variety of pulsed lasers have been employed for LD, operating at wavelengths from the UV to the mid-IR. The most widely used lasers have been pulsed carbon dioxide lasers operating at 10.6 μm in the IR and Nd:YAG lasers, which operate at 1.06 μm in the near IR or at 265 nm in the UV by employing frequency-quadrupling techniques. The wavelength of light appears in most cases to cause only slight differences in the number and types of ions produced, but, due to use of different instrumentation, it is hard to make direct comparisons. Significant differences have been noted in the spectra of chlorophyll when studied with pulsed carbon dioxide lasers of similar specifications, but employing either a TOF or FTMS instrument for mass analysis.[55] This is probably due to the different time delays between ion production and ion detection for the two analyzers. The laser power level also plays an important part in the number and types of ions produced.[56] Not surprisingly, the amount of fragmentation appears to increase when higher laser power densities are used.

The mechanism for all desorption events appears to be similar, in that the best results are obtained with more polar materials. Different desorption methods often produce very similar mass spectra for the same compound, even though the primary desorption event is caused by greatly different interactions. Although there are many similarities between the different desorption/ionization techniques, there are probably subtle fundamental differences which need to be discovered before a thorough understanding and optimization of these techniques can be realized. For the laser desorption technique, a simple thermal mechanism may explain much of what is observed. When a highly focused laser beam strikes the sample, rapid heating which produces a nonthermal equilibrium condition occurs. In conventional heating, thermal ionization competes with thermal decomposition of the sample. When a laser is employed, it has been proposed that heating is so fast that decomposition reactions are not kinetically favored and that intact species are desorbed preferentially. Although this may be too simple a model, it does explain many of the experimental observations. Certainly, interactions in the gas phase directly above the surface between ions and neutrals must also be taken into account. As many more neutrals than ions are produced (perhaps a ratio of 10,000:1), gas-phase reactions may also be an important mechanism for ion formation. That most intact molecular ion species observed are the result of attachment reactions, either cationization or protonation, lends support to the concept that these interactions are important in the overall ionization mechanism. Fragment ions may or may not be attachment ions, indicating that fragmentation may occur to some extent by direct thermal decomposition pathways. For negative ions, other mechanisms must be considered, such as electron capture reactions of thermal electrons produced from the sample. Although analogous anion capture reactions (e.g., halide attachment ions) are sometimes observed, they are far less common than cation attachment in the corresponding positive ion LD mass spectra. For more detailed discussions, the reader is referred to the previously cited review articles.

IV. LASER PYROLYSIS MASS SPECTROMETRY OF POLYMERS

Laser ionization techniques applied to polymer characterization can be divided into two types of analysis, pyrolysis methods and the direct study of intact ions from low molecular weight polymers. As the majority of polymers of commercial interest are often too large for direct ionization with current mass spectrometric techniques, it is not surprising that the largest number of applications have been in the area of laser pyrolysis. Pyrolysis MS has already been discussed briefly with regard to nonlaser ionization techniques in Section II.B. Kistemaker and co-workers[57] have published a review of the early application of laser pyrolysis MS to polymers which adequately covers work prior to 1976. Some additional references to laser pyrolysis studies of polymers are contained in the previously cited review of LAMMA studies.[50]

Pyrolysis MS can be further subdivided into two additional categories: direct ionization by the laser process (desorption and ionization) and simple volatilization coupled with conventional ionization sources such as electron ionization (EI) or chemical ionization (CI). The latter is restricted to the analysis of more volatile evolved species and is often coupled with gas chromatographic separation prior to mass spectrometric analysis.[58] It is used in a manner similar to conventional thermal pyrolysis. Direct laser ionization methods are also used to generate the fragment ions produced from laser desorption of higher molecular weight polymers. This information can assist in obtaining bulk structural information about higher molecular weight polymers, including backbone and side-chain structures.

One characteristic that all laser pyrolysis mass spectra of commercial polymers have in common is that the majority of ions produced are of relatively low mass (typically a few hundred atomic mass units or less). This suggests either substantial polymer degradation, small molecule expulsion, or cleavage at the ends of the polymer chain occurs upon laser pyrolysis. A question which remains with this type of analysis is how much the observed ions reflect the structure of the native polymer and how much rearrangement and recombination occurs. A pyrolysis MS technique utilizing laser vaporization was described by Lum.[59] It utilizes continuous wave (CW) laser radiation in the visible (argon-ion laser, λ = 514.5 nm) or IR (carbon dioxide laser, λ = 10.6 μm) regions. Power levels in the 10- to 1000-W/cm^2 range were employed by utilizing different degrees of focusing of the laser beam. Volatile products were analyzed by 70-eV impact ionization, followed by mass analysis via a quadrupole mass filter. Thick ($\sim$0.5 mm) polymer samples of polyvinyl chloride (rigid and plasticized), polyoxymethylene, and a polyester elastomer block copolymer were studied. Comparisons with conventional thermal heating pyrolysis spectra were made and a method was presented for distinguishing between fragment ions and parents, both of which are produced by electron impact ionization. Evolution profiles of small molecules as a function of irradiation time were presented.

Later work with this instrument[60] extended the analysis to polymer contaminants in a trioxane-ethylene oxide copolymer which contained stress fractures thought to be caused by volatile impurities. In addition, depth profiling of evolved species was examined. Corrosion products on gold-coated electrical contacts were also studied. A final paper[61] utilizing the same experimental setup was concerned with the laser-induced thermal decomposition of poly(butylene terephthalate). Two major decomposition pathways were presented and activation energies calculated. A similar ionization source for a quadrupole mass spectrometer which utilized either a CW or pulsed carbon dioxide laser to study laser pyrolysis of polymeric material, as well as laser vaporization of graphite, was described.[62] A study of the wavelength dependence of the incident laser light on the pyrolysis products of polymethylmethacrylate has also been published.[63] By the use of a Raman shifter and a quadrupled Nd:YAG laser, several wavelengths in the UV were studied. Approximately 3000 laser pulses were required to record a complete mass spectrum, due to the use of a quadrupole mass spectrometer.

Direct production of ions by interaction of the laser with the sample, thus requiring no additional ionization source, is a second means of studying degradation products from high molecular weight polymers. As has already been mentioned, one of the first applications of the new laser microprobe was the study of an epoxy resin[45] which produced spectra dominated by ions corresponding to different metal dopants. A subsequent paper[64] reported similar experiments on thin samples of epoxy resin doped with Li, Na, K, Rb, and Cs at concentration between 6 and 120 ppm. Background levels of low molecular weight fragment ions were also determined for a series of organic polymer substrates. A discussion of organic ion interference on the quantitative determination of metals in such samples was presented. A study of naturally occurring Na, K, and Ca in frog skeletal muscle fiber which had been embedded in an epoxy resin matrix was also discussed. Much of this work was also covered in a second publication by the same authors.[65]

The majority of the published work concerning the study of ions formed directly from laser pyrolysis of organic polymers was done in the early 1980s by Hercules' group at the University of Pittsburgh and utilized a commercial LAMMA 500 instrument. In one publication,[66] they utilized the LAMMA to study the hydrocarbon backbone of several straight chain polymers [polyethylene, poly(tetrafluoroethylene), and poly(vinyl chloride)] and to examine small changes in the side chains of poly(phenylmethacrylate) and poly(benzylmethacrylate) of unspecified molecular weights (although it is assumed they were large). High laser powers (in the 10^{10}-W/cm^2 range) from a focused frequency-quadrupled Nd:YAG pulsed laser were employed. Both positive and negative low mass fragment ions (in the 20- to 200-amu range) were studied. The major species produced were clusters of C which extended as high as C^+_{17} for poly(phenyl methacrylate) and smaller amounts of C-H clusters. Small changes in the fragmentation patterns were observed when side chains were modified.

In a subsequent paper, the commercial polymers Avcothane and Biomer, which are segmented polyether/polyurethanes used in biomedical applications, were studied.[67] Again,

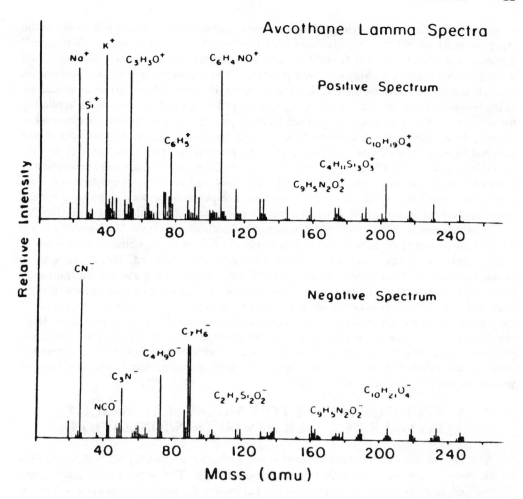

FIGURE 2.   Positive and negative LAMMA laser pyrolysis spectrum of Avcothane. (From Graham, S. W. and Hercules, D. M., *Spectros. Lett.*, 15, 1, 1982. With permission.)

only low mass fragment ions were observed as can be seen in the positive and negative ion LAMMA spectra of Avcothane in Figure 2. Detailed structural information is inferred from the much more complicated fragmentation pattern than was observed for the previous polymers. The presence of teflon and poly(dimethyl siloxane) contaminants was inferred from the fragment ions. Ions corresponding to polyurethane and polyether fragments were also observed. A good summary of this work and the capabilities of this type of analysis have been published.[68]

The LAMMA instrument was employed by other workers to study coal and shale samples,[69] which can be considered as naturally occurring heterogeneous polymers with inclusions of low molecular weight material. Both negative and positive ion spectra with masses in the 100-amu range (i.e., <200 amu) were recorded. Ions observed were attributed to both organic and inorganic (mineral) species. As in most laser desorption mass spectra, cationization is an important ionization mechanism. Interpretation of the results is made difficult by the tendency of the samples to preferentially desorb more volatile, low molecular weight inclusion compounds.

The potential for the use of multiple lasers operating at different wavelengths to first desorb and then ionize neutrals of high molecular weight, such as polymers, will be discussed in more detail later in this review. However, some recent work with multiple lasers in the study of small neutral species volatilized from polymers has started to appear. One paper[70] describes use of a commercial LAMMA instrument coupled with a second tunable dye laser to study metals (Cd, Cu, Mo) in doped polymers. The second laser was used to perform multiphoton ionization of the species desorbed with the first laser. This offers some selectivity in the ionization process and can enhance sensitivity. This procedure can also be applied to the ionization of desorbed organic neutrals from polymer pyrolysis. Such an application has recently been published[71] and utilized an excimer laser operating at 248 nm to volatize material from several commercial polymers (e.g., polystyrene, plexiglas, and teflon), followed by a picosecond UV laser pulse to ionize desorbed neutrals which were subsequently analyzed by a TOF mass spectrometer. The effect of various delay times between laser pulses upon the ions produced was also studied. The potential of such techniques for the direct analysis of intact higher molecular weight oligimers will be discussed in Section V.

Finally, some results which must be classified as laser pyrolysis MS, but which are different in several respects from those already described, have been published. In this work, a laser ionization source coupled to a FTMS instrument was employed. These results differ in that the spectra obtained are characterized by much higher molecular weight fragments than for previous studies. In the LD FTMS instrument, a high-power pulsed carbon dioxide laser which is focused to a spot of about 0.5 to 1.0 mm in diameter is used. In one report[72] the results for several polysaccharides (dextran, cellulose, and starch) were presented. Regular fragmentation series were observed with ions with masses in excess of 1000 amu being produced. Specific cleavage sites within the molecules were proposed. Other work from the same laboratory[73] has examined malto-oligosaccharides, including starch, and the effect of various dopants (NaCl, KBr, and Ag_2O).

V. STUDY OF INTACT LOW MOLECULAR WEIGHT POLYMERS BY LASER DESORPTION/IONIZATION

This section deals with the use of lasers as soft ionization sources for MS with application to the study of species in the 1,000- to 10,000-amu range. This section will include commercial polymers as well as some examples of LD used in the study of biopolymers. Interest in studies of biopolymers has been a significant factor in the continued development of higher mass capabilities for MS.

Although 10,000 Da is the current approximate upper limit for LD MS with present methods, MS has been performed on species up to 20,000 amu using other ionization sources. Some thoughts on improving laser ionization techniques will be explored at the end of this section. Applicability of these techniques to the study of higher mass polymers and nonpolar polymers, which currently do not produce intact molecular ions or even large fragment ions by LD methods, will be discussed.

A. SYNTHETIC POLYMERS

As is the case with the use of lasers for pyrolysis MS, the TOF and FTMS instruments have been utilized almost exclusively for LD MS studies of intact low molecular weight polymers. This is again due to their compatibility with high-power pulsed lasers which are preferred for direct desorption/ionization. Although fewer papers have appeared than for laser pyrolysis, it should be noted that work in this area only started to appear in 1985. Unlike the dominance of the laser microprobe for pyrolysis work, laser FTMS studies have outnumbered laser TOF reports for the study of intact low molecular weight polymers. The instrumental differences between the two techniques have already been discussed and both appear to be well suited for the analysis of polymers in the 1000- to 10,000-amu range. Due to the ionization mechanisms involved, polar polymers have received the most study and those studies have produced the most encouraging results. Polystyrene also can be studied by this technique, but it is one of the few nonpolar polymers to date which has produced molecular ions with masses up to several thousand amu by LD. In contrast to FD-MS, where cation attachment is not observed for nonpolar species, polystyrene showed cation attachment reactions under LD conditions.

In contrast to the laser pyrolysis studies utilizing LAMMA instruments, lower laser power levels must be employed with LD in order to generate intact molecular ion species and avoid pyrolysis. Laser powers on the order of 100 times lower than those typically employed in pyrolysis work have been reported. Two reports of research using a commercial LAMMA system have appeared. The first was by the Hercules and Mattern[74] and dealt with the study of low molecular weight polyglycols. Both PEG and poly(propylene glycol) (PPG) were examined. It was determined that, although reasonable molecular weight distributions were obtained for lower mass species, as the average molecular weight was raised to approximately 1000 amu, the effect of postacceleration of the ions after the flight tube was a critical factor affecting the observed ion intensities of higher mass ions. This is due to the dependence of the electron multiplier response on the velocity of the impacting ion. As ion mass is increased, ion velocity is reduced at constant applied voltage. If this velocity drops below a certain threshold, electron multipliers do not respond linearly. This problem was overcome by increasing the postacceleration voltage from the normal 6000 to 9500 V. With this added postacceleration, good agreement was obtained with the reported nominal average molecular weights for the higher mass polymers of the manufacturer. It should be noted that this problem is increased as masses of ions increase and requires additional increases in postaccelerating voltages. At some point a practical limit where the voltages cannot be increased further is reached. This problem does not occur with FTMS instruments because ions are not detected with electron multipliers. In addition to calculating average molecular weights for the polymers studied, relative affinities for attachment reaction with Na^+, K^+, Li^+, and Cs^+ were studied. These differed in both order and magnitude for the PEG and PPG samples. Both low mass pyrolysis fragmentation and C-O bond cleavage ions, similar to those observed in FAB spectra, were present in some of the mass spectra.

The second paper reports research utilizing a LAMMA instrument to study several industrial polymers.[75] These authors studied poly(dimethylsiloxane), polystyrene, and polyamide-6 samples for which they observed both low mass fragments and higher mass cation attachment species. The poly(dimethylsiloxane) (average molecular weight ~1500) sample produced observable ions with masses up to approximately 3500 amu, but the distribution was skewed to lower mass, probably due to inadequate postacceleration (see Figure 3). The polyamide-6 sample was an extract of low molecular weight oligomers obtained from a higher molecular weight sample. This sample was also subjected to high performance liquid chromatography (HPLC) analysis. Similar distributions were observed by the two methods with oligomers of up to seven repeating units observed. Discussions of the similarity of the laser results with SIMS-TOF were presented.

Cotter and co-workers have taken a different approach by modifying a commercially available TOF mass spectrometer for LD work. This particular instrument does not have the microprobe capabilities of the LAMMA instrument (extremely small spot size) but this feature is not required for LD work and adds to the cost of the instrument. The instrumentation has been described[76,77] elsewhere. It makes use of a pulsed carbon dioxide laser and a CVC model 2000 TOF mass analyser. A block diagram of this system is shown in Figure 4. The mass resolution of this instrument is reported to be ~200 at m/z = 500. Typically, 20 to 50 laser pulses are signal averaged per spectrum. In one of the papers[77] describing the instrumentation, performance is demonstrated for a sample of low molecular weight (average molecular weight 580) polystyrene doped with Li, Na, and K salts.

Another paper reports studies[78] of some additional polymers, including several poly-

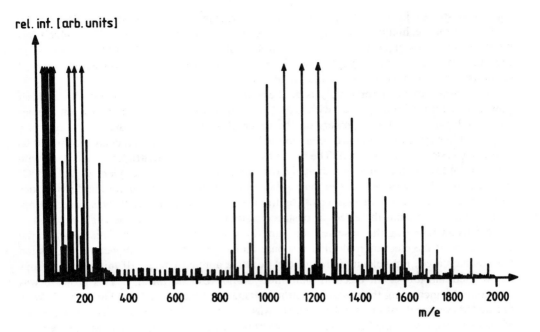

FIGURE 3.   LAMMA 1000 laser desorption mass spectrum of poly(dimethylsiloxane) of average molecular weight 1500. (From Holm, R., Karas, M., and Vogt, H., *Anal. Chem.*, 59, 371, 1987. With permission.)

glycols (PEG, PPG) and PEI, with average molecular weights up to 3000. Two industrial polymers were also examined, Flectol® DS (an antioxidant) and Hycar® CTB (carboxyl terminated *cis*-1,4-polybutadiene). All of the polymers produced cation attachment ions from intact oligomer species. The Hycar® CTB sample had a known molecular weight average of ~3000, but ions were observed only to about 2000 amu, with the maximum of the distribution around 800 amu. There appears to be some preferential desorption of low molecular weight oligomers for this sample. No molecular weight averages were available for the Flectol DS, so comparisons cannot be made for this sample.

In contrast to the Hycar CTB results, the polyglycols studied provided average molecular weight distributions consistent with those determined by more conventional methods. Comparisons of these results with several other mass spectral characterization methods showed that very similar averages were obtained in all cases. The molecular weights of the polymers studied ranged from PPG 790 (nominal molecular weight of 790) to PEG 3350. An interesting difference with this work is that the authors utilized a time-delayed drawout pulse (~15 μs) to improve the results obtained. This was presumably the result of selectively extracting lower energy ions formed after the initial ionization event, which were believed to include more intact oligomer ions and fewer fragmentation products.

The final polymers studied were three PEI samples with average molecular weights of 600, 1200, and 1800. PEIs are another class of polymers which have been studied extensively by MS utilizing different instruments and ionization techniques. These are summarized in Table 1. Unlike the polyglycols, significant discrepancies exist between the nominal molecular weights reported by the manufacturer and most of the different MS values. Indeed, for a variety of reasons, no two mass spectrometric techniques appear to give the same results. This has been explained for the FD and FAB techniques as resulting from excessive fragmentation, skewing the distribution to lower masses. For the EH-MS results, the cause is thought to be preferential desorption of low molecular weight oligomers. The LD results are somewhat puzzling in that the TOF (generally lower than the values of the manufacturer)

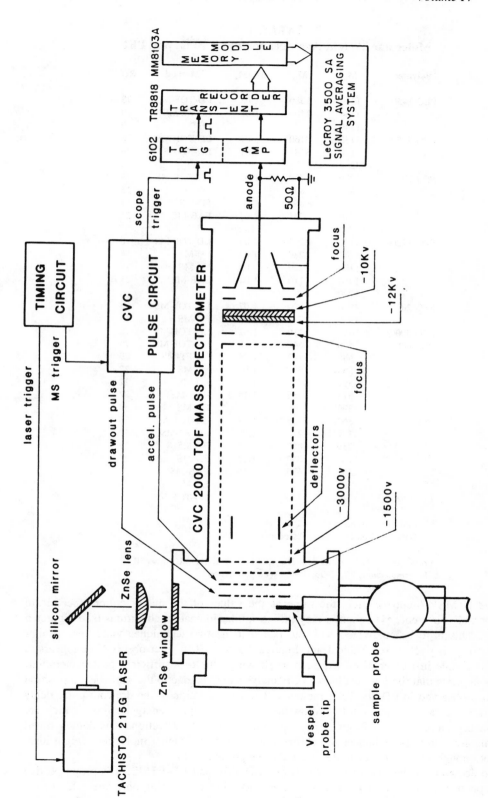

FIGURE 4. Block diagram of LD-TOF mass spectrometer. (From Olthoff, J. K., Lys, I., Demirev, P., and Cotter, R. J., *Anal. Instrum.*, 16, 93, 1987. With permission.)

TABLE 1
Molecular Weight Averages for PEG, PPG, and PEI

Polymer	M_n	M_w	M_w/M_n	Method	Ref.
PEG 1450	1350	1380	1.02	LD-TOF-MS	78
	1384	1404	1.01	LD-FT-MS	79
	1349			EGT[a]	79
PEG 3350	3130	3160	1.01	LD-TOF-MS	78
	3160	3189	1.01	LD-FT-MS	79
	3297			EGT[a]	12
PPG 790	730	750	1.03	LD-TOF-MS	78
	805	835	1.04	FD-MS	15
	819	855	1.04	FD-MS	15
	710	741	1.04	FAB-MS	21
	790[b]		~1.05[b]		
PPG 1220	1170	1295	1.03	LD-TOF-MS	78
	1240	1280	1.03	FD-MS	15
	1190	1220	1.03	FD-MS	15
	1064	1110	1.04	FAB-MS	21
	1220[b]				
PPG 2020	1920	1960	1.02	LD-TOF-MS	78
	1930	1980	1.03	FD-MS	15
PPG 3000	3057	3110	1.02	LD-TOF-MS	78
PEI 600	590	640	1.08	LD-TOF-MS	78
	685	742	1.08	LD-FT-MS	79
	553	583	1.05	FD-MS	22
	557	627	1.13	FD-MS	22
	513	585	1.14	FAB-MS	22
	399			EG-MS	39
PEI 1200	860	925	1.08	LD-TOF-MS	78
	1137	1226	1.08	LD-FT-MS	79
	739	850	1.15	FD-MS	22
	766	860	1.12	FD-MS	22
	551	636	1.15	FAB-MS	22
	505			EH-MS	39
PEI 1800	1010	1150	1.14	LD-TOF-MS	78

[a]   End-group titration.
[b]   Data supplied by Waters Associates.

From Cotter, R. J., Honovich, J. P., Olthoff, J. K., and Lattimer, R. P., *Macromolecules,* 19, 2996, 1986. With permission.

and the FTMS[79] (approximately the same as the values of the manufacturer) spectra are considerably different, even though they utilize similar ionization conditions (see Figures 5 and 6). This is to be contrasted with the fact that the two techniques yield very similar results for polyglycols (see Table 1 and Figures 7 and 8). The latter observation appears to rule out simple instrumental effects such as skewing, due to insufficient postacceleration. Cotter suggests that the low results by TOF analysis may be due to the same decomposition reactions observed in FD and FAB methods. The incorporation of the drawout pulse delay appears to act as an energy filter, allowing some of the more energetic ions (which are presumably more likely to fragment) to be removed prior to detection. It is demonstrated that this drawout pulse enhances the observed abundances of higher molecular weight ions, but not enough to provide accurate molecular weight averages.

To account for the discrepancies between LD-TOF and LD-FTMS, it is proposed that the trapped ion cell in the FTMS acts as an energy filter by trapping only ions which have very low internal and kinetic energies. This, it is argued, would tend to favor observation

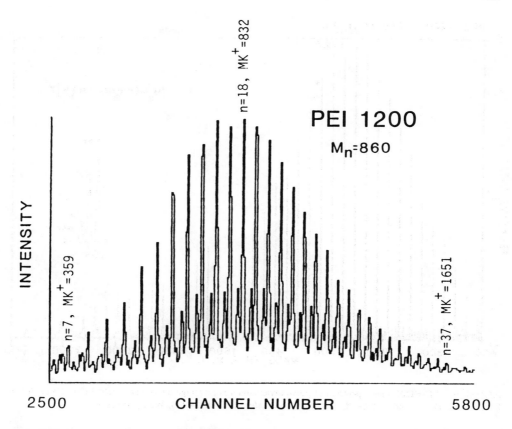

FIGURE 5. LD-TOF mass spectrum of poly(ethyleneimine) of average molecular weight 1200. (From Cotter, R. J., Honovich, J. P., Olthoff, J. K., and Lattimer, R. P., *Macromolecules,* 19, 2996, 1986. With permission.)

of intact molecular ions. Although this argument may have some merit, it should be pointed out that ion trapping in a FTMS instrument is limited only by the kinetic energy of the ion in the z-axis of the magnetic field. This does not insure either low internal energy or even low kinetic energy outside of the z-axis component. If an ion has large kinetic energy components other than the z-axis, it may or may not be trapped. Ions with high kinetic energy components outside of the z-axis may end up far from the center of the trapped ion cell. These may be subsequently lost during the excitation pulse which is applied prior to detection.

An alternative explanation to the one proposed by Cotter and co-workers has been presented for similar discrepancies between the LD-TOF and LD-FTMS spectra of chlorophyll.[55] In that report the authors of the current review proposed that relatively slow reactions, similar to those observed in chemical ionization, occur between desorbed neutrals and the cationic species produced. These presumably take place on a time scale that is too long for the TOF mass spectrometer to observe. The FTMS spectra are typically taken 1 to 3 s after the laser is fired, while the TOF instrument only can observe species with delays in the microsecond time frame. It is also possible that high energy cation attachment ions observed in the microsecond time frame, comprising both molecular and fragment ion species, decompose by expulsion of the cation, producing a neutral organic species at longer time frames and leaving only the charged cation in the trap. Large numbers of inorganic cations are the main low mass species observed in the FTMS spectra of most organics. This could result in observation of only low energy ions formed directly during laser excitation

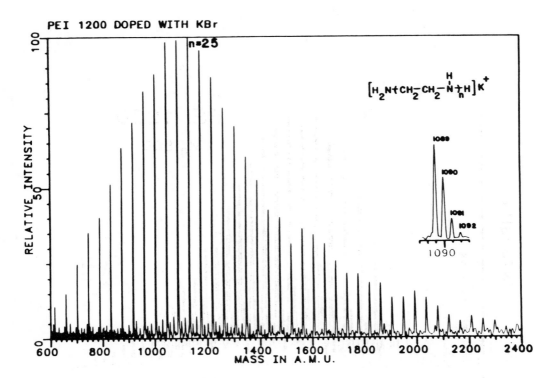

FIGURE 6.   LD-FTMS mass spectrum of poly(ethyleneimine) of average molecular weight 1200. (From Brown, R. S., Weil, D. A., and Wilkins, C. L., *Macromolecules*, 19, 1255, 1986. With permission.)

or ions formed via reactions of cations with background neutrals in the trap, at times long (in MS terms) after the initial laser event. Whatever the actual mechanisms involved, and it is most likely not a single mechanism, the result is that the FTMS spectra are more useful in determining accurate molecular weight distributions for PEI samples than any other mass spectrometric technique to date. It should be pointed out that for a sample[80] of PEI 1800 the LD-FTMS technique produced number average molecular weights of ~1600. This is still well above the values reported by LD-TOF for the same sample and may be due to a variety of causes. It is a distinct possibility that the nominal values of the manufacturer are in error.

LD-FTMS has been used to study intact polymers. The status of FTMS for high mass analysis was reviewed in 1986.[81] The major difference remaining between the LD-TOF and LD-FTMS techniques is the obtainable mass resolution of the two instruments. With a commercial FTMS instrument of early 1980s vintage, unit mass resolution was obtained[82] for a PEG 3350 sample up to a mass of ~4000 amu as demonstrated in Figure 9. Unit mass resolution is the resolution required to resolve adjacent masses (i.e., the difference between ^{12}C and ^{13}C or ^{13}C and $^{12}C^{+}H$). This corresponds to an $m/\Delta m$ resolution of about 8000. This compares with the $m/\Delta m$ resolution of 200 at 500 amu for the LD-TOF system.

An interesting feature of the Fourier transform process is the ability to reduce the number of experimental data points employed in the calculation and obtain a lower resolution spectrum. This is demonstrated for the lower resolution spectrum in Figure 8 of the same PEG 3350 sample shown in Figure 9. Here, resolution is sacrificed for improved signal-to-noise ratio. The higher signal-to-noise level in the lower resolution spectrum facilitates the calculation of more accurate average molecular weight values. It should be pointed out that, with an FTMS instrument, resolution is inversely proportional to mass and that, at 200 amu, the corresponding FTMS resolution would be at least 80,000 and probably much higher. In

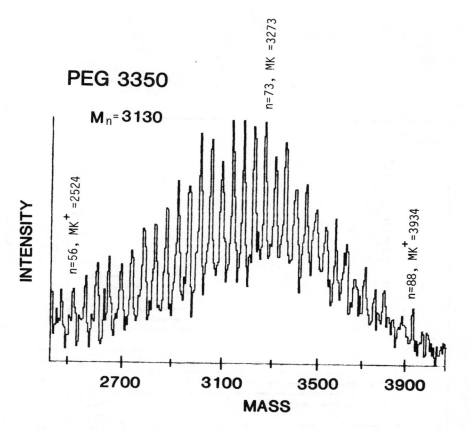

FIGURE 7.   LD-TOF mass spectrum of poly(ethyleneglycol) of average molecular weight 3350. (From Cotter, R. J., Honovich, J. P., Olthoff, J. K., and Lattimer, R. P., *Macromolecules*, 19, 2996, 1986. With permission.)

the same context, the resolution at 8000 amu would be reduced to about 4000. Due to the large number of isotope peaks at such masses, the apparent resolution is much worse due to the overlap of unresolved isotope ions. This is observed for the PEG 6000 spectrum taken from Reference 79 and included as Figure 10 in this Chapter.

Current commercial instruments have improved in design and one can expect some increases in resolution for newer instruments. The incorporation of higher magnetic field strength superconducting magnets above the 3.0-T magnets currently most commonly employed should also produce improved resolution (resolution scales as the ratio of the field strengths, assuming the same field homogeneity). Current instrumentation with 7.0-T magnets should more than double the attainable resolution. Alternatively, lower pressures in the FTMS analyzer cell region also can be used to improve resolution. This may be accomplished by the use of differentially pumped systems, such as that employed by the second generation FTMS-2000 available commercially from Nicolet, or by use of a differentially pumped external ion source. A diagram of the FTMS-2000 with laser interface is shown in Figure 11 and a spectrum of polydimethyl siloxane taken with this instrument is reproduced in Figure 12. Although not the same sample as that used in Benninghoven's SIMS-TOF study (Figure 1), the similarities are striking. The LD-FTMS spectrum appears to have a higher average mass distribution than the reported nominal average molecular weight for this sample (2000). The SIMS-TOF ions extend to higher mass, although with a skewed distribution, which may indicate a higher average molecular weight sample was employed in the SIMS-

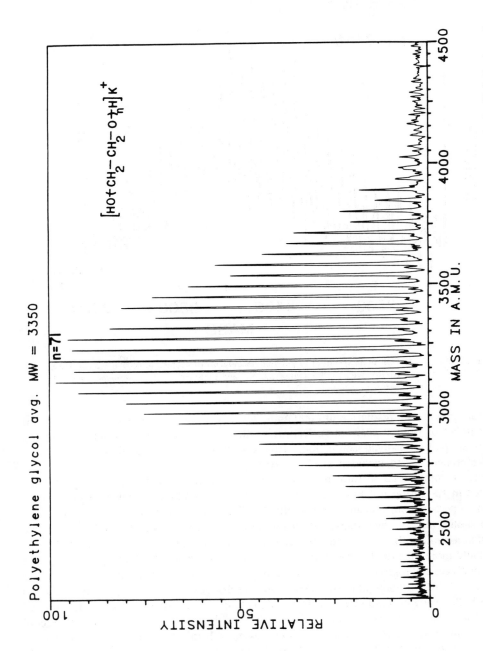

FIGURE 8. LD-FTMS mass spectrum of poly(ethyleneglycol) of average molecular weight 3350. (From Brown, R. S., Weil, D. A., and Wilkins, C. L., *Macromolecules*, 19, 1255, 1986. With permission.)

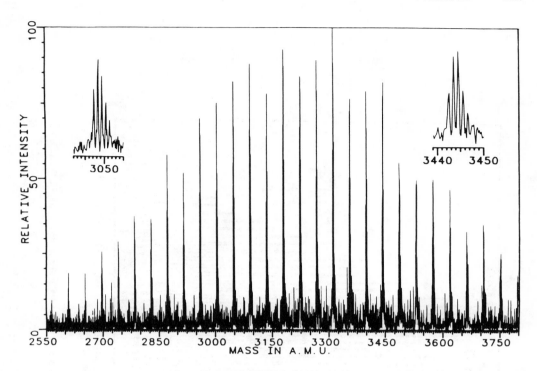

FIGURE 9.   LD-FTMS mass spectrum of poly(ethyleneglycol) of average molecular weight 3350 showing unit mass resolution. (From Brown, R. S. and Wilkins, C. L., *Fourier Transform Mass Spectrometry,* Buchanan, M. V., Ed., (ACS Symp. Ser. No. 359), American Chemical Society, Washington, D.C., 1987, 127. With permission.)

TOF. Some skewing to lower mass due to the fragmentation in the SIMS-TOF spectrum also occurs.

In addition to the PEG spectrum (Figure 9), results from a synthetic biopolymer poly(phenyl alanine) with an average molecular weight of 2000 were shown.[82] It was chosen as an inexpensive model for higher molecular weight peptides. Ions with masses approaching 4000 amu were produced. This is important in that peptides of masses greater than 2000 have failed to yield any high mass ions by LD-FTMS. Although this spectrum was encouraging, there were interpretation problems. The ions produced did not correspond to simple cation attachment species or to any reasonable fragmentation. It was suggested that different end groups than those which occur in natural peptides might be present as a result of introduction of a species to control the molecular weight of the sample during synthesis. The obtainable resolution is lower than would be expected from previous work with similar size polymers. LD-FTMS spectra of several peptides in the 1000- to 2000-amu range are also shown. Lower than expected resolution for large peptides[83] has also been observed with a tandem-quadrupole FTMS instrument. That is a hybrid instrument composed of a quadrupole mass spectrometer which acts as an ion source, coupled with an FTMS instrument, serving as the analyzer. It utilizes a liquid SIMS ionization source. The interested reader is referred to this reference for more information.

The first published application of LD-FTMS to polymers[84] appeared in 1985. It provides a description of the LD-FTMS instrumentation, which consisted of a pulsed carbon dioxide laser and a modified trapped ion cell with focusing lens (see Figure 13). Results were presented for a series of low molecular weight peptides and for two polymer samples. A positive ion cation attachment spectrum (Na^+ and K^+) was obtained for a sample of PEG 3350 and showed ions with a distribution to ~4000 amu. Fragmentation was minimal. A

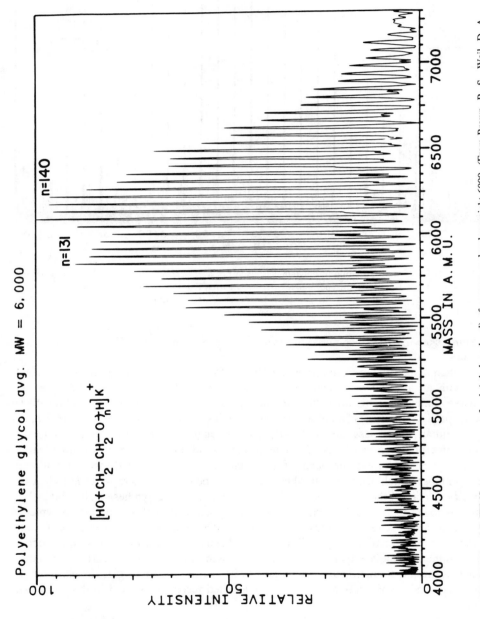

FIGURE 10. LD-FTMS mass spectrum of poly(ethyleneglycol) of average molecular weight 6000. (From Brown, R. S., Weil, D. A., and Wilkins, C. L., *Macromolecules*, 19, 1255, 1986. With permission.)

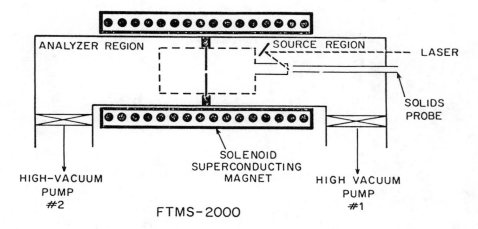

FIGURE 11.   Cross-sectional view of differentially pumped FTMS-2000. (From Wilkins, C. L. and Brown, R. S., *Mass Spectrometry in the Analysis of Large Molecules,* McNeal, C. J., Ed., John Wiley & Sons, New York, 1986, 191. With permission.)

higher molecular weight polyfluoroether, Krytox 16140, was demonstrated to produce abundant negative ions up to a mass of about 6500 amu (M^-). Excellent negative ion spectra are not uncommon for fluorinated compounds measured by LD-FTMS. A paper dealing solely with polymers was published in 1986,[79] parts of which have already been discussed with respect to conflicting LD-TOF work. In addition to the PEIs studied, the spectra for a series of PEGs with average molecular weights in the 1000- to 6000-amu range were shown. The PEG 6000 sample had cation attachment species extending beyond 7000 amu. This represents the highest mass organic ions yet published for LD-FTMS ionization. Again, few low mass fragment ions are observed. In more recent unpublished work from our laboratory, a 7-T FTMS instrument has produced laser-desorbed organic ions up to m/z 10,000 from a PEG sample.

Other polymers studied included PPG 4000 and PEG methyl ether of average molecular weight 5000 (PEGME 5000). Average molecular weight results are summarized for all of the polymers studied and compared with conventional end group titrations where applicable. The results for the PEI samples are compared with the results obtained with other mass spectrometers and different ionization methods. Two additional types of polymers were examined, poly(caprolactone diol) and polystyrene, both of average molecular weight 2000. Again, spectra for both samples were dominated by intense cation attachment species with little fragmentation observed.

Finally, several papers have appeared on the study of poly(*p*-phenylenes) (PPP) by LD-FTMS. The insolubility of the sample precluded the use of other ionization techniques such as FAB and FD. The insoluble nature of a material presents little problem to the laser ionization method since it requires only that the material be attached mechanically to the direct insertion probe of the mass spectrometer. MS was chosen in hope of distinguishing differences between various synthetic routes which produce products with very different properties. In the first of two papers[85] on this subject, LD-FTMS was applied to several samples of PPP, each prepared by a different synthetic method. Two different catalyst-oxidant combinations were used ($AlCl_3$-$CuCl_2$ and $FeCl_3$) to produce PPP samples from benzene. In addition to benzene, benzene-d_6 and *p*-dibromobenzene were also used as starting materials.

The mass spectra of the four samples of PPP (see Figure 14) are characterized by ions in the 400- to 1800-amu range which are reported to be M^+ species. This is somewhat unusual for LD spectra but perhaps not surprising considering the highly nonpolar nature of

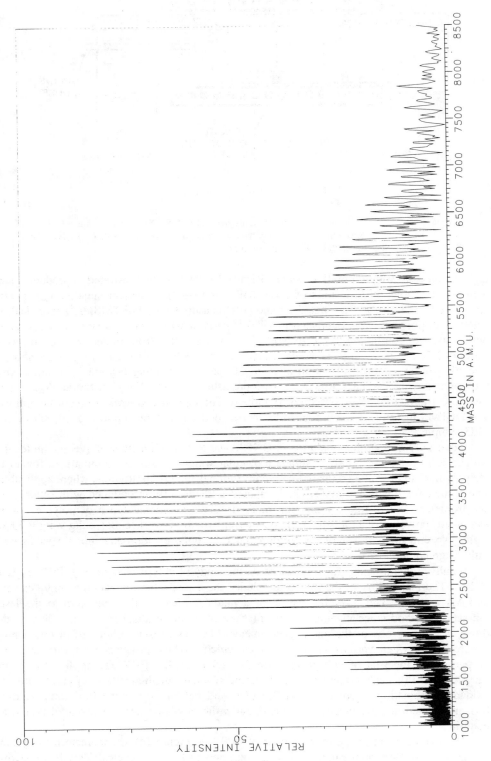

FIGURE 12.   LD-FTMS mass spectrum of poly(dimethylsiloxane).

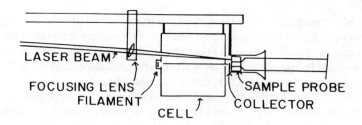

FIGURE 13. Cross-sectional view of LD-FTMS analyzer cell region. (From Wilkins, C. L., Weil, D. A., Yang, C. L. C., and Ijames, C. F., *Anal. Chem.*, 57, 520, 1985. With permission.)

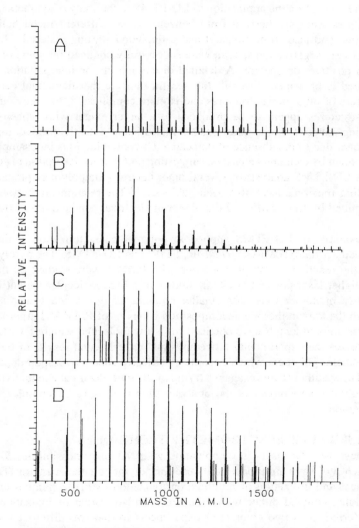

FIGURE 14. LD-FTMS spectra of poly(*p*-phenylene) synthesized by different routes. (From Brown, C. E., Kovacic, P., Wilkie, C. A., Cody, R. B., and Kinsinger, J. A., *J. Polym. Sci., Polym. Lett. Ed.*, 23, 453, 1985. With permission.)

this polymer. M^+ species have also been observed in the LD-FTMS spectra of porphyrins.[56] Major differences in the observed mass distributions of the different products are not apparent but there appears to be some significance to the observed differences in the amount of halogen incorporated by use of the various synthetic routes. The percentage of Cl incorporated in the final product is <1% when $AlCl_3$-$CuCl_2$ is employed but is at about 8% in the sample prepared with $FeCl_3$. The *p*-dibromobenzene starting material incorporates 6% Br. Average chain length results calculated from the LD data are in good agreement with those obtained by vapor phase osmometry, although perdeuterated PPP exhibited a lower molecular weight distribution, which may account for its somewhat higher solubility. The halogen content of the samples calculated from the LD-FTMS spectra was in good agreement with those obtained by elemental analysis. Halogens appear to be concentrated only in the end groups.

In a second report[86] on the application of LD-FTMS to the study of synthetically prepared PPP, the samples were synthesized from benzene by six different metal halide catalyst-oxidant systems, and also from biphenyl and *p*-terphenyl starting materials. In contrast to the previous paper, negative ion spectra were successfully obtained for some of the samples in addition to positive ion spectra. A detailed discussion on the interpretation of the mass spectra obtained is presented. Overall, the spectra indicate that the actual composition of PPP is a mixture of oligomeric structures and is more complicated than previously thought. Polynuclear structures appear to be produced to a larger extent with biphenyl than with benzene starting material. Some questions are raised about potential product discrimination by the technique, due to the absence of detectable Cl-containing ions in a sample which has a 13.3% Cl content by elemental analysis, suggesting preferential desorption of some species. The results of LD-FTMS analysis of several other heterocyclic polymers prepared by using different starting materials have also been published.[87] The relationship between structural data as determined by LD-FTMS and the observed electrical conductivity of the polymer is discussed.

A final example of LD-FTMS applied to polymer characterization is the study[88] of poly(methylmethacrylate) (PMMA) by NMR, FTIR, and LD-FTMS. The pyrolysis products produced by the reaction of red phosphorous with PMMA were studied by the preceding techniques. Higher laser power LD-FTMS studies of a high molecular weight PMMA produced primarily monomer ions with smaller amounts of dimer and trimer ions. This is consistent with the known thermal decomposition behavior of PMMA. Larger species of up to 22 monomer units in length were obtained for a low molecular weight PMMA. A review which summarizes the experimental procedures and experimental results of the analysis of intractable aromatic polymers which become electrically conductive upon doping has been published.[89] These authors caution against trying to attribute electrical conductivity properties to specific polymer structures, as the analysis shows them to be complex mixtures of dissimilar oligomers.

B. NEW LASER MASS SPECTROMETRY TECHNIQUES

Several new uses for lasers in combination with MS offer possibilities for significant advances which will have an impact on its use for polymer characterization. These include the development and utilization of multiphoton ionization (MPI) techniques and photodissociation of ions, coupled either with initial laser volatilization or ionization. These are sometimes referred to as two color laser experiments in that two different lasers, usually operating at different wavelengths, are employed. In the photodissociation experiment, the second laser is used to fragment ions which have already been formed by LD with the first laser (or by some other ionization technique). The absorption of photons (usually high energy UV radiation) causes the ion to absorb enough excess energy that it is dissipated by bond breaking. Most desorption ionization techniques produce mainly molecular ions, which, although important, do not provide structural information other than molecular weight.

Photodissociation offers a means to selectively enhance the fragmentation of an ion. Usually, high-powered pulsed lasers are employed, which makes the technique best suited for use with pulsed mass spectrometers. An added advantage to the use of an FTMS instrument is the possibility of isolating an ion of a particular m/z upon which photodissociation can be performed. This is done by utilizing a frequency sweep which bypasses the frequency of the ion of interest to excite the other ions and cause them to be ejected from the trap. The remaining ion species can then be subjected to photodissociation and a mass spectrum generated for fragment ions produced. This is a form of MS/MS experiment which is normally done in sector instruments or multiple quadrupole instruments via use of high kinetic energy collisions with inert gas molecules, leading to collisionally activated dissociation (CAD). It is thought that the use of photon excitation will be more efficient with higher mass species than CAD techniques. An added advantage of photodissociation methods with the FTMS instrument is the much high resolution which can be obtained on the fragment ions. Bowers and co-workers[90] have demonstrated this technique with small peptides and Hanson's group[91] has presented results from larger peptides ionized by liquid SIMS in a tandem quadrupole FTMS instrument. There are obvious advantages to being able to select the molecular ion of a single oligomer species in a polymer sample and probe its structure via photodissociation techniques coupled with MS. The only drawback is the need to have an appropriate chromophore in the molecule to enhance the absorption of photons.

The technique of multiphoton ionization has been utilized for some time in the study of small molecules by MS. Lubman[92] has published a recent review on the subject. By utilizing pulsed UV lasers which produce high-energy photons, organic ions can be directly ionized. Most organic species have ionization potentials in the 7- to 13-eV range. For example, a 193-nm photon produced from an ArF excimer laser has approximately 6.5 eV energy. Therefore, absorption of two photons at this wavelength is sufficient to cause ionization of most organics. As the wavelength of the photons used is increased, the energy per photon decreases, requiring the absorption of additional photons to effect ionization. Examples of this technique have been application to small peptides[93] and metal porphyrins.[94]

If the MPI technique is coupled with a laser which is used to volatilize neutral species into the gas phase, a potentially very powerful analytical method results. Recall that, in the laser desorption/ionization sources which have been described, for every ion produced it is estimated that 1000 to 10,000 neutrals are sputtered off the surface. In addition, the threshold laser power level for volatilization is much lower than that for ionization so that lower laser power can be employed to minimize sample decomposition. Grotemeyer and co-workers[95] have demonstrated this technique with porphyrins and peptides[96] with masses as high as 1300 amu. The same workers have shown[97] that the technique can be used to generate molecular ion species for insulin with a mass of 5927 amu. Small numbers of structurally useful fragment ions, which can be increased by utilizing higher laser power density, are also observed. These workers used a low power pulsed carbon dioxide laser to produce neutrals followed by the application of 270 nm radiation from an excimer-pumped dye laser which was frequency doubled. The ions produced were detected by a TOF mass spectrometer. An added benefit from the technique was much higher resolving power than is typical from a TOF instrument, sufficient to resolve individual masses. Similarly, Engelke and co-workers analyzed the 20 primary phenylthiohydantoin amino acids, using CO_2 laser desorption, followed by UV laser photodissociation (266 nm).[100] Excellent linear response and sensitivity were obtained. This work demonstrates the potential of two laser methods in MS.

Unlike the LD technique, which generally requires cationization or protonation reactions to induce ionization, potentially limiting its use primarily to polar species, the MPI technique produces spectra which are characterized by M^+ ions and therefore should not be limited to polar species. In fact, Sack and co-workers[98] have shown MPI results for polycyclic aromatics by GC-FTMS. This method offers the hope of successfully employing MS for the characterization of the large majority of commercial polymers, which are nonpolar in nature.

VI. REFERENCES

1. **Cottrell, J. S. and Greathead, R. J.**, Extending the mass range of a sector mass spectrometer, *Mass Spectrom. Rev.*, 5, 215, 1986.
2. **Campana, J. E.**, Time-of-flight mass spectrometry: a historical overview, *Anal. Instrum.*, 16, 1987.
3. **Laude, D. A., Johlman, C. L., Brown, R. S., Weil, D. A., and Wilkins, C. L.**, Fourier transform mass spectrometry: recent instrumental developments and applications, *Mass Spectrom. Rev.*, 5, 107, 1986.
4. **Schulten, H. R. and Lattimer, R. P.**, Applications of mass spectrometry to polymers, *Mass Spectrom. Rev.*, 3, 231, 1984.
5. **Lattimer, R. P., Harris, R. E., and Schulten, H. R.**, Applications of mass spectrometry to synthetic polymers, *Rubber Chem. Technol.*, 58, 577, 1985.
6. **Wiley, R. H.**, The mass spectral characterization of oligomers, *Macromol. Rev.*, 14, 379, 1979.
7. **Schuetzle, T. L., Riley, J. E., deVries, J. E., and Prater, T. J.**, Applications of high performance mass spectrometry to the surface analysis of materials, *Mass Spectrom. Rev.*, 3, 527, 1984.
8. **Beckey, H. D.**, *Principles of Field Ionization and Field Desorption Mass Spectrometry*, Pergamon Press, Oxford, 1977, 77.
9. **Wood, G. W.**, Field desorption mass spectrometry: applications, *Mass Spectrom. Rev.*, 1, 63, 1982.
10. **Linden, H. B., Knoll, H., Pezsa, I., and Wanczek, K. P.**, An external field desorption ion source for analytical ion cyclotron resonance spectrometry, presented at the 35th ASMS Conference on Mass Spectrometry and Allied Topics, Denver, May 1987, 21.
11. **Lattimer, R. P., Harmon, D. J., and Hansen, G. E.**, Determination of molecular weight distributions of polystyrene oligomers by field desorption mass spectrometry, *Anal. Chem.*, 52, 1808, 1980.
12. **Matsuo, T., Marsuda, H., and Katakuse, I.**, Use of field desorption mass spectra of polystyrene and polypropylene glycol as mass references up to mass 10000, *Anal. Chem.*, 51, 1329, 1979.
13. **Craig, A. G., Cullis, P. G., and Derrick, J.**, Field desorption of polymers: polybutadiene, *Int. J. Mass Spectrom. Ion Phys.*, 38, 297, 1981.
14. **Lattimer, R. P. and Schulten, H. R.**, Field desorption of hydrocarbon polymers, *Int. J. Mass Spectrom. Ion Phys.*, 52, 105, 1983.
15. **Lattimer, R. P. and Hansen, G. E.**, Determination of molecular weight distributions of polyglycol oligomers by field desorption mass spectrometry, *Macromolecules*, 14, 776, 1981.
16. **McNeal, C. J.**, Symposium on fast atom and ion induced mass spectrometry of nonvolatile organic solids, *Anal. Chem.*, 54, 43A, 1982.
17. **Barber, M., Bordoli, R. S., Elliot, G. J., Sedgwick, R. D., and Tyler, A. N.**, Fast atom bombardment mass spectrometry, *Anal. Chem.*, 54, 645A, 1982.
18. **Burlingame, A. L., Baillie, T. A., and Derrick, P. J.**, Mass spectrometry, *Anal. Chem.*, 58, 166R, 1986.
19. **Goad, L. J., Prescott, C. M., and Rose, M. E.**, Poly(ethylene glycol) as a calibrant and solvent for fast-atom-bombardment mass spectrometry: application to carbohydrates, *Org. Mass Spectrom.*, 19, 101, 1984.
20. **Gilliam, J. M., Lanis, P. W., and Occolowitz, J. L.**, On-line accurate mass measurement in fast atom bombardment mass spectrometry, *Anal. Chem.*, 56, 2285, 1984.
21. **Lattimer, R. P.**, Fast atom bombardment mass spectrometry of polyglycols, *Int. J. Mass Spectrom. ion Processes*, 55, 221, 1983/84.
22. **Lattimer, R. P. and Schulten, H. R.**, Field desorption and fast atom bombardment mass spectrometry of poly(ethylene imine), *Int. J. Mass Spectrom. Ion Processes*, 67, 277, 1985.
23. **Bletsos, I. V., Hercules, D. M., vanLeyen, D., and Benninghoven, A.**, Time-of-flight secondary ion mass spectrometry of polymers in the mass range 500—10000, *Macromolecules*, 20, 407, 1987.
24. **Gardella, J. A. and Hercules, D. M.**, Static secondary ion mass spectrometry of polymer systems, *Anal. Chem.*, 52, 226, 1980.
25. **Briggs, D., Brown, A., Van Den Berg, J. A., and Vickerman, J. C.**, A comparative study of organic polymers by SIMS and FABMS, *Springer Ser. Chem. Phys.*, 25, 162, 1983.
26. **Campana, J. E. and Rose, S. L.**, Secondary ion mass spectrometry of polymers, *Int. J. Mass Spec. Ion Phys.*, 46, 483, 1983.
27. **Hummel, D. O., Dussel, H. J., and Manshausen, P.**, Pyrolysis mass spectrometry of multicomponent polymeric systems, *Prog. Polym. Spectrosc.*, 9, 1986.

28. **Liebman, S. A. and Levy, S. J.,** Pyrolysis gas chromatography, mass spectrometry and Fourier transform IR spectroscopy, in *Polymer Characterization,* Craver, C. D., Ed., (Adv. Chem. Ser. No. 203), American Chemical Society, Washington, D.C., 1983, 617.
29. **Voorhees, K. J., Darfee, S. L., and Baldwin, R. M.,** Liquifaction reactivity correlations using pyrolysis/ mass spectrometry/pattern recognition procedures, in *Polymer Characterization,* Craver, C. D., Ed., (Adv. Chem. Ser. No. 203), American Chemical Society, Washington, D.C., 1983, 677.
30. **Reiner, E. and Moran, J. F.,** Pyrolysis/gas chromatography/mass spectrometry of biological macromolecules, in *Polymer Characterization,* Craver, C. D., Ed., (Adv. Chem. Ser. No. 203), American Chemical Society, Washington, D.C., 1983, 705.
31. **Adams, R. E.,** Positive and negative chemical ionization pyrolysis mass spectrometry of polymers, *Anal. Chem.,* 55, 414, 1983.
32. **Udseth, H. R. and Friedman, L.,** Analysis of styrene polymers by mass spectrometry with filament-heated evaporation, *Anal. Chem.,* 53, 29, 1981.
33. **Lee, A. K. and Sedwick, R. D.,** Applications of mass spectrometry to copolymers of ethylenes and propylene oxides. I. Structural analysis, *J. Polym. Sci., Polym. Chem. Ed.,* 16, 685, 1978.
34. **Jonsson, G. P., Hedin, A. B., Hakansson, P. L., Sundqvist, B. U. R., Save, B. G. S., Nielsen, P. F., Roepstorff, P., Johansson, K. E., Kamensky, I., and Lindberg, M. S. L.,** Plasma desorption mass spectrometry of peptides and proteins adsorbed on nitrocellulose, *Anal. Chem.,* 58, 1084, 1986.
35. **Chait, B. T., Shpungin, and Field, F. H.,** Fission fragment ionization mass spectrometry of polyethers, *Int. J. Mass Spectrom. Ion Processes,* 58, 121, 1984.
36. **Loo, J. A., Williams, E. R., Furlong, J. J. P., Wang, B. H., and McLafferty, F. W.,** Desorption and ionization methods for Fourier transform mass spectrometry, presented at the 35th ASMS Conference on Mass Spectrometry and Allied Topics, Denver, May 1987, 27.
37. **Lai, S. T. F., Chan, K. W., and Cook, K. D.,** Electrohydrodynamic ionization mass spectrometry of poly(ethylene glycols), *Macromolecules,* 953, 1980.
38. **Chan, K. W. S. and Cook, K. D.,** Extended mass range by multiple charge: sampling quadrupuly charged quasimolecular ions of poly(ethylene glycol) 4000, *Org. Mass Spectrom.,* 18, 423, 1983.
39. **Callahan, J. H. and Cook, K. D.,** The effect of polymer-solvent interactions on sampling efficiency in electrohydrodynamic ionization mass spectrometry, presented at the 32nd Annual Conference on Mass Spectrometry and Allied Topics, May 1984, San Antonio, TX, 552.
40. **Honig, R. E. and Woolston, J. R.,** Laser-induced emission of electrons, ions, and neutral atoms from solid surfaces, *Appl. Phys. Lett.,* 2, 138, 1963.
41. **Berkowitz, J. and Chupka, W. A.,** Mass spectrometric study of vapor ejected from graphite and other solids by focussed laser beams, *J. Chem. Phys.,* 40, 2735, 1964.
42. **Vastola, F. J., Mumma, R. O., and Pirone, A. J.,** Analysis of organic salts by laser ionization, *Org. Mass Spectrom.,* 3, 101, 1970.
43. **Posthumus, M. A., Kistemaker, P. G., Meuzelaar, H. L. C., and Ten Noever de Brauw, M. C.,** Laser desorption-mass spectrometry of polar nonvolatile bioorganic molecules, *Anal. Chem.,* 50, 985, 1978.
44. **Cotter, R. J. and Yergey, A. L.,** Thermally produced ions in desorption mass spectrometry, *Anal. Chem.,* 53, 1306, 1981.
45. **Hillenkamp, F., Unsold, E., Kaufmann, R., and Nitsche, R.,** Laser microprobe mass analysis of organic materials, *Nature,* 256, 119, 1975.
46. **Seydel, U. and Lindner, B.,** Qualitative and quantitative investigations on mycobacteria with LAMMA, *Z. Anal. Chem.,* 308, 253, 1981.
47. **Denoyer, E., Van Grieken, R., Adams, F., and Natusch, D. F. S.,** Laser microprobe mass spectrometry. I. Basic principles and performance characteristics, *Anal. Chem.,* 54, 26A, 1982.
48. **Hercules, D. M., Day, R. J., Balasanmugam, K., Dang, T. A., and Li, C. P.,** Laser microprobe mass spectrometry. II. Applications to structural analysis, *Anal. Chem.,* 54, 280A, 1982.
49. **McCreery, D. A., Ledford, E. G., and Gross, M. L.,** Laser desorption Fourier transform mass spectrometry, *Anal. Chem.,* 54, 1435, 1982.
50. **Conzemius, R. J. and Cappellen, J. M.,** A review of the applications to solids of the laser ion source in mass spectrometry, *Int. J. Mass Spectrom. Ion Phys.,* 34, 197, 1980.
51. **Cotter, R. J. and Tabet, J. C.,** Laser desorption MS for nonvolatile organic molecules, *Am. Lab.,* 16, 86, 1984.
52. **Cotter, R. J.,** Lasers and mass spectrometry, *Anal. Chem.,* 56, 485A, 1984.
53. **Hillenkamp, F.,** Laser desorption technique of nonvolatile organic substances, *Int. J. Mass Spectrom. Ion Phys.,* 45, 305, 1982.
54. **Chuang, T. J.,** Photodesorption and adsorbate-surface interactions stimulated by laser radiation, *J. Vac. Sci. Technol., B,* 3, 1408, 1985.
55. **Brown, R. S. and Wilkins, C. L.,** Laser desorption Fourier transform mass spectrometry of chlorophyll A and chlorophyll B, *J. Am. Chem. Soc.,* 108, 2447, 1986.

56. **Brown, R. S. and Wilkins, C. L.**, Laser-desorption Fourier transform mass spectrometry of synthetic porphyrins, *Anal. Chem.*, 58, 3196, 1986.
57. **Kistemaker, P. G., Boerboom, A. J. H., and Meuzelaar, H. L. C.**, Laser pyrolysis mass spectrometry: some aspects and applications to technical polymers, *Dyn. Mass Spectrom.*, 4, 139, 1976.
58. **Merritt, C., Scaher, R. E., and Petersen, B. A.**, Laser pyrolysis-gas chromatographic-mass spectrometric analysis of polymeric materials, *J. Chromatogr.*, 99, 301, 1974.
59. **Lum, R. M.**, Direct analysis of polymer pyrolysis using laser microprobe techniques, *Thermochim. Acta*, 18, 73, 1977.
60. **Lum, R. M.**, Microanalysis of trace contaminants by laser-probe pyrolysis, *Am. Lab.*, 10, 47, 1978.
61. **Lum, R. M.**, Thermal decomposition of poly(butylene terephthalate), *J. Polym. Sci., Polym. Chem. Ed.*, 17, 203, 1979.
62. **Fredin, L., Hansen, G. P., Sampson, M. P., Margrave, J. L., and Behrens, R. G.**, High-temperature quadrupole mass spectrometry for studying vaporization from materials heated by a CO_2 laser, Report LA-10744, Order No. DE87000806, National Technical Information Service, Baltimore, 1986.
63. **Estler, R. C. and Nogar, N. S.**, Mass spectrometric identification of wavelength dependent UV laser photoablation fragments from poly(methyl methacrylate), *Appl. Phys. Lett.*, 49, 1175, 1986.
64. **Nitsche, R., Kaufmann, R., Hillenkamp, F., Unsold, E., Vogt, H., and Wechsung, R.**, Mass spectrometric analysis of laser-induced microplasmas from organic samples, *Isr. J. Chem.*, 17, 181, 1978.
65. **Unsold, E., Hillenkamp, F., Renner, G., and Nitsche, R.**, Investigations on organic materials using a laser microprobe mass analyzer, *Adv. Mass Spectrom.*, 7B, 1425, 1978.
66. **Gardella, J. A., Hercules, D. M., and Heinen, H. J.**, Mass spectrometry of molecular solids: laser microprobe mass analysis (LAMMA) of selected polymers, *Spectros. Lett.*, 13, 347, 1980.
67. **Graham, S. W. and Hercules, D. M.**, Laser desorption mass spectra of biomedical polymers; Biomer and Avcothane, *Spectros. Lett.*, 15, 1, 1982.
68. **Gardella, J. A., Graham, S. W., and Hercules, D. M.**, in *Structural Analysis of Polymeric Materials by Laser Desorption Mass Spectrometry*, Craver, C. D., Ed., (Adv. Chem. Ser. No. 203), American Chemical Society, Washington, D.C., 1983, 635.
69. **Vanderborgh, N. E. and Jones, C. E. R.**, Laser microprobe mass analysis studies on coal and shale samples, *Anal. Chem.*, 55, 527, 1983.
70. **Verdun, F. R., Krier, G., and Muller, J. F.**, Increased sensitivity in laser microprobe mass analysis by using resonant two-photon ionization processes, *Anal. Chem.*, 59, 1383, 1987.
71. **Larciprete, R. and Stuke, M.**, Direct observation of excimer-laser photoablation products from polymers by picosecond-UV-laser mass spectroscopy, *Appl. Phys.*, B, 42, 181, 1987.
72. **Coates, M. L. and Wilkins, C. L.**, Laser-desorption Fourier transform mass spectra of polysaccharides, *Anal. Chem.*, 59, 197, 1987.
73. **Coates, M. L. and Wilkins, C. L.**, Laser desorption Fourier transform mass spectra of malto-oligosaccharides, *Biomed. Mass Spectrom.*, 12, 424, 1985.
74. **Mattern, D. E. and Hercules, D. M.**, Laser mass spectrometry of polyglycols: comparison with other mass spectral techniques, *Anal. Chem.*, 57, 2041, 1985.
75. **Holm, R., Karas, M., and Vogt, H.**, Polymer investigations with the laser microprobe, *Anal. Chem.*, 59, 371, 1987.
76. **van Breemen, R. B., Snow, M., and Cotter, R. J.**, Time-resolved laser desorption mass spectrometry. I. Desorption of preformed ions, *Int. J. Mass Spectrom. Ion Phys.*, 49, 35, 1983.
77. **Olthoff, J. K., Lys, I., Demirev, P., and Cotter, R. J.**, Modification of Wiley-Mclaren TOF analyzers for laser desorption, *Anal. Instr.*, 16, 93, 1987.
78. **Cotter, R. J., Honovich, J. P., Olthoff, J. K., and Lattimer, R. P.**, Laser desorption time-of-flight mass spectrometry of low-molecular-weight polymers, *Macromolecules*, 19, 2996, 1986.
79. **Brown, R. S., Weil, D. A., and Wilkins, C. L.**, Laser desorption-Fourier transform mass spectrometry for the characterization of polymers, *Macromolecules*, 19, 1255, 1986.
80. **Brown, R. S. and Wilkins, C. L.**, unpublished data, 1986.
81. **Wilkins, C. L. and Brown, R. S.**, Current status of high-mass analysis by Fourier transform mass spectrometry, in *Mass Spectrometry in the Analysis of Large Molecules*, McNeal, C. J., Ed., John Wiley & Sons, New York, 1986, 191.
82. **Brown, R. S. and Wilkins, C. L.**, Analytical applications of laser desorption-Fourier transform mass spectrometry for nonvolatile molecules, in *Fourier Transform Mass Spectrometry*, Buchanan, M. V., Ed., (ACS Sump. Ser. No. 359), American Chemical Society, Washington, D.C., 1987, 127.
83. **Hunt, D. F., Shabanowitz, J., Yates, J. R., Nian-Zhou Zhu, D., Russell, D. H., and Castro, M. E.**, Tandem quadrupole Fourier-transform mass spectrometry of oligopeptides and small proteins, *Proc. Natl. Acad. Sci. U.S.A.*, 84, 620, 1987.
84. **Wilkins, C. L., Weil, D. A., Yang, C. L. C., and Ijames, C. F.**, High mass analysis by laser desorption Fourier transform mass spectrometry, *Anal. Chem.*, 57, 520, 1985.

85. **Brown, C. E., Kovacic, P., Wilkie, C. A., Cody, R. B., and Kinsinger, J. A.,** Laser desorption/Fourier transform mass-spectral analysis of molecular weight distribution and end-group composition of poly(*p*-phenylene)s synthesized by various routes, *J. Polym. Sci. Polym. Lett. Ed.,* 23, 453, 1985.

86. **Brown, C. E., Kovacic, P., Wilkie, C. A., Kinsinger, J. A., Hein, R. E., Yaniger, S. I., and Cody, R. B.,** Polynuclear and halogenated structures in polyphenylenes synthesized from benzene, biphenyl, and *p*-terphenyl under various conditions: characterization by laser desorption/Fourier transform mass spectrometry, *J. Polym. Sci. Polym. Chem. Ed.,* 24, 255, 1986.

87. **Brown, C. E., Kovacic, P., Cody, R. B., Hein, R. E., and Kinsinger, J. A.,** Laser desorption/Fourier-transform mass spectral analysis of heterocyclic aromatic polymers, *J. Polym. Sci., Part C: Polym. Lett.,* 24, 519, 1986.

88. **Brown, C. E., Wilkie, C. A., Smukalla, J., Cody, R. A., and Kinsinger, J. A.,** Inhibition by red phosphorus of unimolecular thermal chain-scission in poly(methyl methacrylate): investigation by NMR, *J. Polym. Sci., Part A, Polym. Chem.,* 24, 1297, 1986.

89. **Brown, C. E., Kovacic, P., Wilkie, C. A., Cody, R. B., Hein, R. E., and Kinsinger, J. A.,** Laser desorption/Fourier transform mass spectral analysis of various conducting polymers, *Synth. Met.,* 15, 265, 1986.

90. **Bowers, W. D., Delbert, S. S., and McIver, R. T.,** Consecutive laser-induced photodissociation as a probe of ion structure, *Anal. Chem.,* 58, 969, 1986.

91. **Hanson, C. D., Castro, M. E., Russell, D. H., Hunt, D. F., and Shabanowitz, J.,** Fourier transform mass spectrometry of large (m/z >5,000) biomolecules, in *Fourier Transform Mass Spectrometry,* Buchanan, M. V., Ed., (ACS Symp. Ser. No. 359), American Chemical Society, Washington, D.C., 1987, 100.

92. **Lubman, D. M.,** Optically selective molecular mass spectrometry, *Anal. Chem.,* 59, 31A, 1987.

93. **Tembreull, R. and Lubman, D. M.,** Resonant two-photon ionization of small peptides using pulsed laser desorption in supersonic beam mass spectrometry, *Anal. Chem.,* 59, 1003, 1987.

94. **Morris, J. B. and Johnston, M. V.,** Multiphoton ionization of transition-metal tetraphenylporphines. Metal complexes which display molecular ionization, *Int. J. Mass Spectrom. Ion Proc.,* 73, 75, 1986.

95. **Grotemeyer, J., Boesl, U., Walter, K., and Schlag, E. W.,** A general soft ionization method for mass spectrometry: resonance-enhanced multiphoton ionization of biomolecules, *Org. Mass Spectrom.,* 21, 645, 1986.

96. **Grotemeyer, J., Boesl, U., Walter, K., and Schlag, E. W.,** *Org. Mass Spectrom.,* 21, 595, 1986.

97. **Grotemeyer, J., Boesl, U., Walter, K., and Schlag, E. W.,** Ionization and fragmentation of biomolecules investigated by laser mass spectrometry, presented at the 35th ASMS Conference on Mass Spectrometry and Allied Topics, Denver, May 1987, 191.

98. **Sack, T. M., McCrery, D. A., and Gross, M. L.,** Gas chromatography multiphoton ionization Fourier transform mass spectrometry, *Anal. Chem.,* 57, 1290, 1985.

99. Nicolet Analytical Instruments, Nicolet Analytical Instruments Guide, Madison, WI, 1987.

100. **Engelke, F., Hahn, J. H., Henke, W., and Zare, R. N.,** Determination of phenylthiohydantoin-amino acids by two-step laser desorption/multiphoton ionization, *Anal. Chem.,* 59, 909, 1987.

Chapter 2

LASER OPTICAL STUDIES OF POLYMER ORGANIZATIONS

Jean-Marc Haudin and Patrick Navard

TABLE OF CONTENTS

I. INTRODUCTION

The early years of the polymer science were mainly devoted to understanding the structure of a polymer chain and its intrinsic static properties, and to synthesis.[1-3] New experimental and theoretical tools recently appeared, like neutron scattering or photon-correlation spectroscopy on the experimental side and molecular dynamics, renormalization, scaling, and reptation theories.[4-7] Aside from these spectacular advances in the understanding of the nature and properties of polymer chains, their macroscopic organizations and their influence on properties also made very significant advances, leading to the design of new products and new processing techniques[8,9] such as reactive injection molding, thermotropic polymers, flexible polymer gel spinning, interpenetrating networks, or conducting polymers. The tools for studying the structure of these macroscopic organizations, like the semicrystalline spherulite, are mainly electron and optical microscopies and light, neutron, and X-ray scattering. This chapter will be devoted to the description of the uses of laser light scattering for understanding macroscopic polymer organizations. Excluded from this subject are thus the properties of a single chain, gel states, the structure of a polymer single crystal, polymer mixtures, and block copolymers. We will mainly focus on two kinds of organization, the semicrystalline one (spherulites, rods, sheaves, fibrils) and the mesomorphic one (nematic and cholesteric states).

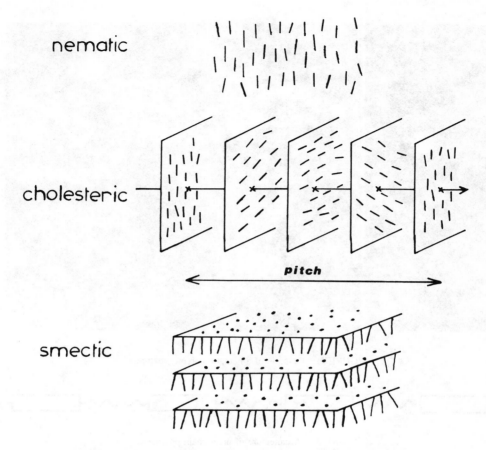

FIGURE 1. Schematic representation of nematic, cholesteric, and smectic phases.

II. THE DIFFERENT LEVELS OF POLYMER ORGANIZATION

Several levels of polymer organization can be found, and some can be studied by laser light scattering. Dilute and concentrated solutions, melt and glassy states, gel, and rubber are all characterized by various kinds of spatial correlation between neighboring chains. The correlation is mostly local, and does not extend to micron-size distances. Laser spectroscopies are very useful,[10,11] as well as birefringence measurements.[12,13] All these techniques give information on polymer conformation, size, chemical structure, aggregation, and response to a deformation. These subjects will not be treated in this chapter. Mesomorphic (liquid crystalline) and crystalline states are different since the correlation extends to a large number of chains, reaching or exceeding distances commensurable with the wavelength of light. The light scattered by these large entities brings important information on their structure and size, and this will be described here.

A. MESOMORPHIC STATES

It is now well known that aside from the two major thermodynamic states, liquid and crystal, there exists several other stable organizations, called liquid crystalline or mesomorphic. They are based on the decoupling between orientation and position. For this to occur, the molecules must have a large shape anisotropy, and this occurs when the polymer chain is rigid and long. Three types of mesomorphic states are described: the nematics, cholesterics, and smectics (Figure 1).

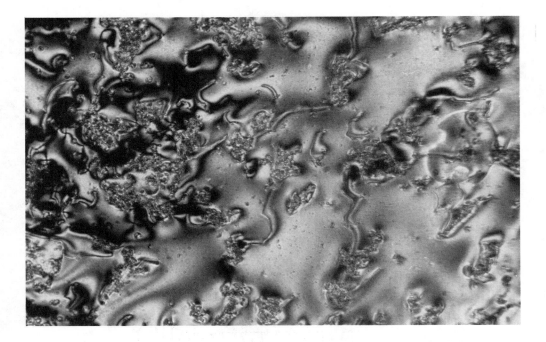

FIGURE 2.   Example of a nematic polymer texture.

FIGURE 3.   Architecture of main-chain polymers.

The nematic phase is very close to the liquid state, except that the long axis of the polymer has some degree of long range orientational order. The lower limit, no long range orientational order, is the liquid state. The other case is the nematic state. The molecules are free to move, mainly in the direction of the "mean direction" called director.

The cholesteric phase is a distortion of the nematic phase. The director follows a helical path, characterized by a pitch. This phase is thermodynamically similar to the nematic one, and occurs when an asymmetric (chiral) center is present.

The smectic phase is actually a family with several varieties. All of them are characterized by a positional order along the long axis and are thus closer to the crystalline state than to the nematic state. The molecules are packed into layers, the layers being nearly uncorrelated in position. The molecules are parallel to the director but the kind of packing and the orientation of the director about the layer plane specify the smectic type.

The nematic, cholesteric, and smectic phases present various types of defect texture. A texture obtained for a polymer nematic phase is shown in Figure 2. Two main polymer structures can give a mesomorphic state. When the chain itself is the oriented unit, one speaks of *main-chain mesomorphic polymers*[14-16] (Figure 3). Another possibility is to have a flexible backbone bearing rigid side chains which will orient. This is the case of the *side-chain mesomorphic polymers*[17,18] (Figure 4). Other cases might occur, as the disk-like polymers, called discotics.[19] Another distinction has to be made between thermotropic and lyotropic phases. A thermotropic polymer undergoes, upon heating, the following phase sequence: crystal → smectic or/and nematic (or cholesteric) → liquid melt. A lyotropic polymer needs a solvent in order to display a mesomorphic phase, above a certain critical concentration.

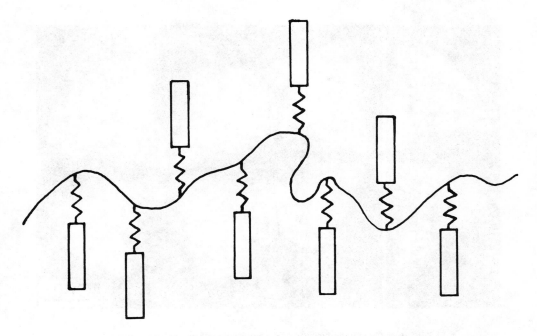

FIGURE 4.   Architecture of side-chain polymers.

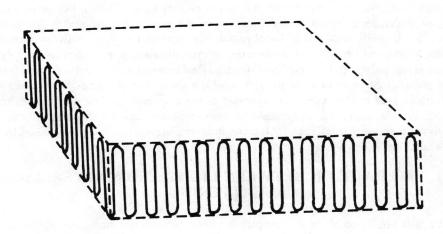

FIGURE 5.   A crystalline lamella showing the chain folding.

B. THE CRYSTALLINE STATE

The basic unit is the single crystal. Polymer single crystals[20] have been known for 30 years, and their structure is quite well understood. If the molecule is flexible, the crystal has the shape of a lamella, and chain molecules fold and reenter into the same crystal (Figure 5). Still a subject of controversy are the nature of the lamella surface and where a single chain reenters into the crystal. Other kinds of polymer crystals exist, either with rigid polymers or with flexible polymers crystallized under special conditions. The extended chain crystal is formed with chain molecules which do not fold. This is the case in high-modulus fibers.[21] The classical tools for studying all these crystals are neutron and X-ray scattering, electron microscopy, and Raman spectroscopy.

Another level is how these single crystals will gather, since most polymers are semi-

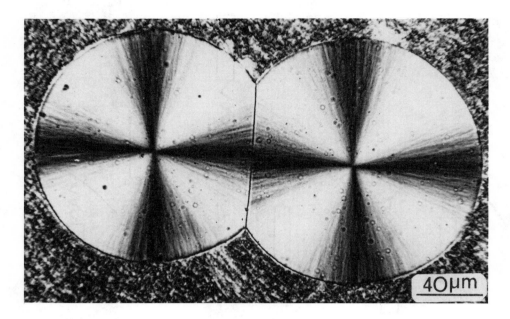

FIGURE 6.   Micrograph between crossed polarizers of two polymer spherulites.

crystalline, i.e., they are a mixture of amorphous and crystalline phases. For semicrystalline polymers which form crystalline lamellae, the most common organization is the spherulite (Figure 6). It consists of radial lamellae with an amorphous phase between them, thus forming a spherical entity.[22] Since this mode of crystallization is the one of major polymers (polyethylene, polypropylene, etc.) and since the final properties of the polymer are a function of the physical characteristics of the spherulites, a great deal of studies have been carried out to understand the formation and structure of these spherulites. One of the experimental methods is light scattering. Polymers may also crystallize under various structural forms such as rods, sheaves, fibrils, etc. All these structural organizations can be studied by laser light scattering.

III. INTERACTION BETWEEN LIGHT AND MATTER. LIGHT SCATTERING

We will briefly recall in this chapter the basic physics describing the light scattering phenomenon. The reader is referred to classical text books for further readings.[23-25] Polymers are dielectric materials, this property being at the origin of their optical behavior. Since all the experiments and theoretical treatments will involve polarized light, we will first recall some results concerning the polarization of dielectrics. We will then consider light scattering and more specifically small-angle light scattering, which is used to probe large entities (more than a few molecules). Two cases will be discussed in detail, small-angle light scattering by spheres and by rods.

A. POLARIZATION OF DIELECTRICS

The polymer specimen is assumed to be illuminated by a monochromatic light beam. As light is an electromagnetic wave, this beam can be characterized by two perpendicular vectors, varying sinusoidally in phase; the electric field vector E and the magnetic field vector H. We also assume that this incident wave is linearly polarized, so that E and H remain parallel to fixed directions. In a vacuum, before entering the dielectric medium, the variations of the electric field vector can be described by the following equation:

$$E = E_0 \exp i\omega\left(t - \frac{x}{c}\right) \tag{1}$$

where t is the time and x the coordinate along the propagation direction, c is the velocity of propagation of light in a vacuum, and ω is the circular frequency of the wave. All the points in a plane perpendicular to the propagation direction are subject to the same electric field; this is the definition of a plane wave in a vacuum.

Consider now what happens in the dielectric medium on the atomic scale. In an atom, the positively charged nucleus is moved into the direction of the applied electric field, while the center of gravity of the electrons is moved into the opposite direction. Therefore, the center of gravity of the positive charges does not coincide any longer with the one of the negative charges, and the atom can be regarded as an electric dipole of moment p. If $\Delta\tau$ is a volume element surrounding a point M and containing a great number of atoms, the polarization of the dielectric in M is defined by

$$P = \lim_{\Delta\tau \to 0} \frac{\sum_{\Delta\tau} p}{\Delta\tau} \tag{2}$$

where $\sum_{\Delta\tau} p$ is the vectorial sum of the moments of the atomic dipoles contained in $\Delta\tau$.

In the case of anisotropic materials, such as polymer crystals, the equation relating the polarization vector to the electric field has a tensorial form, i.e.:

$$P = \epsilon_0 \underline{\underline{\chi_e}} E \tag{3}$$

where ϵ_0 is the permittivity of free space, $\underline{\underline{\chi_e}}$ is a second-order tensor, which is called electric susceptibility tensor.

The electric induction vector D is defined by

$$D = \epsilon_0 E + P \tag{4}$$

From Equation 3 it follows that

$$D = \epsilon_0(\underline{\underline{1}} + \underline{\underline{\chi_e}})E = \epsilon_0 \underline{\underline{\epsilon}} E \tag{5}$$

where $\underline{\underline{1}}$ is the second order unit tensor. The tensor $\underline{\underline{\epsilon}} = \underline{\underline{1}} + \underline{\underline{\chi_e}}$ is the relative permittivity tensor.

In a vacuum, where the electric field is defined by Equation 1, the propagation of incident light is governed by the following wave equation:

$$\frac{\partial^2 E}{\partial x^2} - \frac{1}{c^2}\frac{\partial^2 E}{\partial t^2} = 0 \tag{6}$$

In the dielectric material, the propagation equations are derived from the Maxwell equations, which give the interrelations between the electric field E, the electric induction D, the magnetic field H, and the magnetic induction B. The tensor $\underline{\underline{\epsilon}}$ is supposed to be time independent. We also assume that the equation relating B to H has the same form as

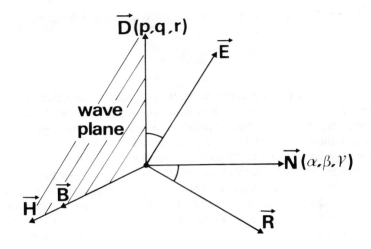

FIGURE 7.   Respective positions of vectors *E, D, H,* and *B*.

TABLE 1
Approximations Used in Light-Scattering Theories

Approximation	Conditions of validity
Rayleigh	$ka \ll 1$; $|m|\ ka \ll 1$
Rayleigh-Gans-Debye (RGD)	$|m - 1| \ll 1$; $2\ ka\ |m - 1| \ll 1$
Anomalous diffraction (AD)	$|m - 1| \ll 1$; $ka \gg 1$

in a vacuum, i.e., $B = \mu_0 H$, where μ_0 is the magnetic permeability of free space. With such assumptions, the Maxwell equations can be reduced to

$$\text{div } D = 0 \tag{7}$$

$$\nabla^2 E - \frac{\underline{\underline{\epsilon}}}{c^2} \frac{\partial^2 E}{\partial t^2} = grad \text{ div } E \tag{8}$$

The Maxwell equations make it possible to determine the respective positions of vectors *E, D, H,* and *B*. In Figure 7, *N* is the unit vector perpendicular to the *(D, H)* plane, which is called the wave plane. *R* is the unit vector normal to the *(E, H)* plane; it defines the direction of propagation of the wave. *D, E, N, R* belong to the same plane, perpendicular to *H*.

B. SCATTERING FUNCTIONS AND APPROXIMATIONS

As a first case, we will consider a single particle of arbitrary size and shape illuminated by a very distant light source. What is of interest is the intensity, phase, and polarization state of the scattered light at a large distance from the particle. Knowing the optical properties of the particle allows one to write the Maxwell equations. They can be solved only in a few cases, one case of interest here being the isotropic sphere in an isotropic medium of different refractive index. In all the other cases, approximations are necessary to solve the scattering functions.

Three approximations have been used in light scattering theories: Rayleigh approximation, Rayleigh-Gans-Debye (RGD) approximation (also called Rayleigh-Gans approximation), and the Anomalous Diffraction (AD) approximation. Table 1 gives the conditions under which each approximation is valid. In this table, a is the characteristic dimension of the scattering units (for spheres a will be the radius) and m is the refractive index of the scattering units, relative to the medium in which they are embedded. $k = 2\pi/\lambda$ where λ is the wavelength of light in the medium.

The Rayleigh approximation is valid for particles of size much less than the optical wavelength λ. The RGD approximation has been applied very often, sometimes beyond its range of validity. It has been widely used for the theoretical treatment of small-angle light scattering (SALS) and most of the calculations reported in this chapter are based upon it. The AD approximation is well known as a means of calculating the intensity of SALS for dielectric particles. It is appropriate for large particles, which scatter strongly close to the forward direction.

1. Rayleigh-Gans-Debye (RGD) Scattering

The basis of the theory of RGD scattering (also called Rayleigh-Gans scattering) is to suppose that, within a scattering particle, each volume element is described by Rayleigh scattering. The RGD approximation implies that the incident wave goes through the particle nearly as if there were no particle at all. Similarly, the wave scattered by a volume element in a certain direction can leave the particle without being modified or distorted by the presence of the other volume elements.

The physical basis of RGD scattering is therefore very simple: each volume element gives Rayleigh scattering and does so independently of the other volume elements. The waves scattered in a given direction by all these elements will interfere because of the different positions of the volume elements in space. In order to calculate the interference effects, it is necessary to refer the phases of all the scattered waves to a common origin and then to add the complex amplitudes.

In a first step, Rayleigh scattering has to be considered since the RGD approximation implies that each volume element within the scattering particle is described by Rayleigh scattering. Because of its small size, each volume element may be considered to be in a homogeneous applied electric field. As described previously, this applied field induces the polarization of the dielectric material. The moment of the dipole induced in a volume element dr is given by:

$$dp = Pdr = \epsilon_0 \underline{\underline{\chi_e}} \, Edr \tag{9}$$

where P is the polarization vector and $\underline{\underline{\chi_e}}$ the electric susceptibility tensor. This oscillating induced dipole radiates an electromagnetic wave with the same wavelength as the incident wave: the scattered wave.

If the volume element dr within the scattering particle is subjected to an incident field $E = E_0 \exp i\omega t$, the electric field of the scattered wave is described by[26]

$$dE_s = \frac{\omega^2}{4\pi\epsilon_0 c^2 d} [P_0 - (P_0 \cdot s')s'] \exp[i(\omega t - kd)]dr \tag{10}$$

where

$$P_0 = \epsilon_0 \underline{\underline{\chi_e}} \, E_0 \tag{11}$$

Equation 10 is valid at a large distance d from the dipole dp, and s' is the unit vector

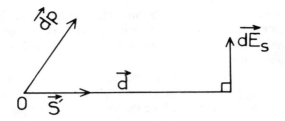

FIGURE 8.   Position of the elementary electric field dE_s at a great distance from the oscillating dipole dp.

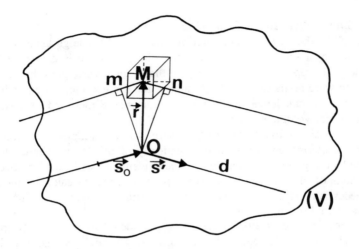

FIGURE 9.   Path difference between the scattered rays originating at two scattering centers O and M.

along the d direction (Figure 8). Equation 10 also shows that dE_s lies in the (P_0, d) plane and is perpendicular to d.

The electric field E_s scattered by the whole particle is obtained by summing up the elementary fields dE_s scattered by the different volume elements dr. To make such a calculation, it is necessary to take into account the phase differences resulting from the different positions of the volume elements in space, which requires referring all the phases to a common origin O (Figure 9). In O, the applied electric field is assumed to be $E = E_0 \exp i \omega t$. Consider now a volume element dr at a distance r from O. The path length for the ray going through M is greater by $\delta = mM + Mn$, m and n being the projections of O on the rays through M (Figure 9). We define the directions of the incident and scattered rays by the unit vectors s_0 and s'. Vectorially

$$\delta = s_0 \cdot OM - s' \cdot OM = r \cdot (s_0 - s') = r \cdot s \qquad (12)$$

It is assumed that the differences in path length are small compared with d, so that a common d can be used for all the volume elements. Thus, the electric field dE_s scattered by the element dr at a distance r can be written:

$$dE_s = \frac{\omega^2 \exp[i(\omega t - kd)]}{4\pi\epsilon_0 c^2 d} [P_0(r) - (P_0(r) \cdot s')s'] \exp[-ik(r \cdot s)]dr \qquad (13)$$

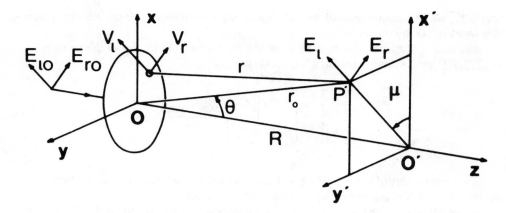

FIGURE 10. Definition of geometrical and electrical parameters describing the scattering process. (From Meeten, G. H., *Opt. Acta*, 29, 759, 1982. With permission.)

By integration over the whole volume of the scattering particle:

$$E_s = \frac{\omega^2 \exp[i(\omega t - kd)]}{4\pi\epsilon_0 c^2 d} \int_V [P_0(r) - (P_0(r) \cdot s')s'] \exp[-ik(r \cdot s)] dr$$

$$= E_{so} \exp[i(\omega t - kd)] \tag{14}$$

where

$$E_{so} = C_1 \int_V [P_0(r) - (P_0(r) \cdot s')s'] \exp[-ik(r \cdot s)] dr \tag{15}$$

and

$$C_1 = \frac{\omega^2}{4\pi\epsilon_0 c^2 d} = \frac{\pi}{\epsilon_0 \lambda_0^2 d} \tag{16}$$

λ_0 being the wavelength of the light in vacuum.

2. Anomalous Diffraction Theory

The scalar theory of anomalous diffraction[24] has been generalized by Meeten[27] to include polarized incident light illuminating an optically anisotropic object. Light of any polarization state incident on the object is described by orthogonal electric vectors $\{E_{\ell o}, E_{ro}\}$ linearly polarized parallel and perpendicular, respectively, to the scattering plane OP'O' in Figure 10. The object is supposed to be positioned close to and on the $-z$ side of the (x,y) plane which can be called the "shadow plane".[24]

Due to the presence of the object, electric vectors $\{V_\ell, V_r\}$ in the shadow plane are given by

$$\begin{bmatrix} V_\ell \\ V_r \end{bmatrix} = \begin{bmatrix} K_2 & K_3 \\ K_4 & K_1 \end{bmatrix} \begin{bmatrix} E_{\ell o} \\ E_{ro} \end{bmatrix} \tag{17}$$

where K_n are the components of the full Jones' matrix which relates the emergent light to that incident on the object.

The aim of scattering theory is to relate the electric vectors $\{E_\ell, E_r\}$ of the scattered light to $\{E_{\ell o}, E_{ro}\}$ by a scattering matrix S such that:

$$\begin{bmatrix} E_\ell \\ E_r \end{bmatrix} = \begin{bmatrix} S_2 & S_3 \\ S_4 & S_1 \end{bmatrix} \begin{bmatrix} E_{\ell o} \\ E_{ro} \end{bmatrix} \frac{\exp(-ikr_0)}{ikr_0} \tag{18}$$

The components of the scattering matrix above use the notation of Van de Hulst.[24] They completely describe the scattering properties of the object.

In the scalar case, the normalized scalar complex scattering amplitude $S(\theta)$ is given by[24]

$$S(\theta) = \frac{k^2}{4\pi}(1 + \cos\theta)\iint_A [1 - \exp(-i\Phi)]\exp(-ikr')dxdy \tag{19}$$

where $\Phi(x,y)$ is the phase lag of a ray passing through the object at (x,y) and parallel to the z-axis; θ is defined in Figure 10, $r' = r - r_0$, and A is the area of the object projected onto the shadow plane. By the same arguments used for deriving Equation 19, it is possible to express the components of the scattering matrix S:

$$S_1 = \frac{k^2}{2\pi}\iint_A (1 - K_1)\exp(-ikr')dxdy \tag{20}$$

$$S_2 = \frac{k^2}{2\pi}\cos\theta\iint_A (1 - K_2)\exp(-ikr')dxdy \tag{21}$$

$$S_3 = \frac{-k^2}{2\pi}\cos\theta\iint_A K_3\exp(-ikr')dxdy \tag{22}$$

$$S_4 = \frac{-k^2}{2\pi}\iint_A K_4\exp(-ikr')dxdy \tag{23}$$

The relations between the S components and the Jones' matrix allow to fully describe the scattering of an object with given optical properties, shape, and size. This will be used later for studying scattering by a spherulite.

C. GEOMETRY AND SCATTERED AMPLITUDE AND INTENSITY
1. Geometry

Scattering is independent of any specific laboratory coordinate system. Experiments are conducted in the laboratory coordinate system, and it is necessary to define correctly the scattering geometry. The plane defined by the initial and final wave vectors is the scattering plane. Four geometries can be described, depending both on the relative positions of the polarizers and on the position of the scattered wave vector. In real SALS experiments (see Section IV), the position of the analyzer is fixed and perpendicular to the direction of the initial wave vector. This defines two basic geometries: VV (vertical-vertical) when the two polarizers have parallel polarization directions, and VH (vertical-horizontal) when the two polarizers have perpendicular polarization directions. In most cases, with this arrangement,

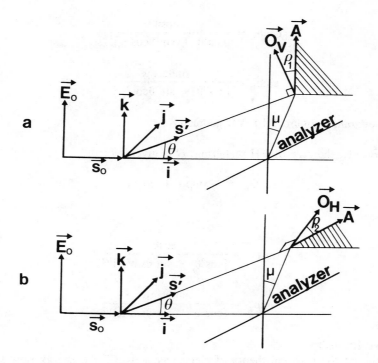

FIGURE 11.   Definition of vector O (a) A vertical (VV geometry) and (b) A horizontal (VH geometry).

$I_{VH} = I_{HV}$, except when some optical rotation is present, and $I_{HH} = I_{VV}$ when the system is random or has a symmetry axis around the initial wave vector.

2. Scattered Amplitude

As said, the scattered light is generally analyzed using a linear polarizer whose plane is perpendicular to the propagation direction s_0 of the incident wave. As a consequence, the amplitude of the scattered light is then obtained by

$$E'_{so} = E_{so} \cdot O \tag{24}$$

where E_{so} is given by the optical analysis, i.e., by Equation 15 if the RGD approximation is used. O is a unit vector which is perpendicular to the propagation direction s' of the scattered ray and lies in a plane containing the polarization direction A of the analyzer and perpendicular to the analyzer plane.

If the polarization direction of the polarizer is vertical, the two particular positions of the polarization direction A of the analyzer are

1. A is vertical (Figure 11a): VV geometry.

In the (i, j, k) coordinate system defined in Figure 11, O has the following expression:

$$O_v = -\sin\rho_1 i + \cos\rho_1 k \tag{25}$$

where the angle ρ_1 is defined by

$$\cos\rho_1 = \frac{\cos\theta}{[\cos^2\theta + \sin^2\theta\cos^2\mu]^{1/2}} \tag{26}$$

$$\sin\rho_1 = \frac{\sin\theta\cos\mu}{[\cos^2\theta + \sin^2\theta\cos^2\mu]^{1/2}} \tag{27}$$

2.   *A* is horizontal (Figure 11b): VH geometry

If we consider the case of VH scattering, *O* is given by

$$O_H = -\sin\rho_2 i + \cos\rho_2 j \tag{28}$$

where

$$\cos\rho_2 = \frac{\cos\theta}{[\cos^2\theta + \sin^2\theta\sin^2\mu]^{1/2}} \tag{29}$$

$$\sin\rho_2 = \frac{\sin\theta\sin\mu}{[\cos^2\theta + \sin^2\theta\sin^2\mu]^{1/2}} \tag{30}$$

3. Scattered Intensity

According to the classical theory of electromagnetism,[28] the energy flow scattered in the *s'* direction per unit area and per second is given by the modulus of the $E \times H$ vector, which is called Poynting's vector. The mean rate of energy flow per unit area, which is the intensity of the scattered light, is obtained by averaging over a long time. In an electromagnetic wave, *E* and *H* are at right angles to each other. Further, within the RGD approximation, both incident and scattered waves are supposed to travel through a homogeneous isotropic medium with an average refractive index n. In such a medium, the light velocity is v = c/n, and

$$|E| = \frac{c}{n}|B| = \frac{\mu_0 c}{n}|H| \tag{31}$$

Thus, the intensity scattered in the *s'* direction is given by

$$I_s = <|E_s \times H_s|> = \frac{1}{2}n\epsilon_0 c|E_{so}|^2 \tag{32}$$

The energy of the incident wave is defined by

$$I_0 = \frac{1}{2}n\epsilon_0 cE_0^2 \tag{33}$$

Hence,

$$I_s = \frac{I_0}{E_0^2}|E_{so}|^2 \tag{34}$$

If the scattered light is analyzed as described in the present section, $|E_{so}|^2$ has to be replaced by $|E'_{so}|^2$ given by Equation 24.

D. SMALL-ANGLE LIGHT SCATTERING (SALS) BY SPHERES

Studying spherical objects by SALS is very useful since such objects frequently occur (rain droplets, aerosols, powders, polymer lattices, polymer spherulites, etc.) and also because their theoretical treatment is not too difficult. The simplest case is an isotropic sphere embedded in an isotropic medium. It is possible to find the solution of Maxwell's equations which describe the field arising from a plane monochromatic wave incident upon a spherical surface, across which the properties change abruptly. Isotropic spheres are a limiting case of birefringent semicrystalline polymer spherulites and they are thus of direct interest. The approximate theories which are used for spherulites will have to agree with the exact case of an isotropic sphere when the spherulite birefringence will tend to zero.

1. Isotropic Sphere. Mie Theory

Using the electromagnetic theory, Mie[29] obtained in 1908 an exact solution for the scattering of a plane monochromatic wave by an isotropic sphere of any diameter embedded in an isotropic medium. A large amount of work has been published concerning theoretical developments as well as experimental checks and uses of this theory. Meeten[27] used the same theory to predict the intensity scattered at small angles for polarized VV scattering (Figure 11a) and depolarized VH scattering (Figure 11b). The relative intensities I_{VV} and I_{VH} are functions of the polar angle θ and of the azimuthal angle μ (Figures 10 and 11). μ is measured from the zero position defined by the direction of the electric vector of the incident light.

The intensity I is expressed by a dimensionless number G, called gain:

$$G = 4\pi I r_0^2 / A \tag{35}$$

where A is the cross section area of the scatterer projected onto the incident beam, and r_0 is the distance from the scatterer.

Any scatterer is fully described by the scattering matrix (Equation 18)

$$\begin{bmatrix} S_2 & S_3 \\ S_4 & S_1 \end{bmatrix} \tag{36}$$

using the notation of Van de Hulst.[24]

For SALS, it is known that[27]

$$I_{VV} = k^{-2}r_0^{-2}|S_1\sin^2\mu + S_2\cos^2\mu + (S_3 - S_4)\sin\mu\cos\mu|^2 \tag{37}$$

$$I_{VH} = k^{-2}r_0^{-2}|S_3\sin^2\mu - S_4\cos^2\mu + (S_2 - S_1)\sin\mu\cos\mu|^2 \tag{38}$$

where $k = 2\pi/\lambda$ and λ is the wavelength of the light in the (transparent) medium surrounding the scatterer. Here k is assumed to be constant everywhere outside the scatterer. Equations 37 and 38 are independent of the scattering theory used.

For the isotropic sphere,[24] $S_3 = S_4 = 0$. Thus the angular gains can be written

$$G_{VV} = 4x^{-2}|S_1\sin^2\mu + S_2\cos^2\mu|^2 \tag{39}$$

$$G_{VH} = x^{-2}|S_1 - S_2|^2\sin^2 2\mu \tag{40}$$

where x is the dimensionless scatterer size parameter ka for a sphere of radius a. Generally it is seen from Equation 40 that a pattern with fourfold symmetry arising from $\sin^2 2\mu$ is found for VH scattering provided that $S_1 \neq S_2$.

Mie's scattering theory provides S_1 and S_2 exactly[24] for an isotropic and homogeneous sphere. For a sphere of radius a and refractive index m_2, in a medium of refractive index m_1, the magnitudes of S_1 and S_2 are completely determined by x, m, and the polar scattering angle θ. Here x = ka and m is the relative refractive index of the sphere given by m_2/m_1.

Analytic expressions for S_1 and S_2 are in the form of infinite series, requiring numerical evaluation. Wickramasinghe[30] provides a useful summary of the equations and we largely use his notation in the following. He gives

$$S_1 = \sum_{n=1}^{\infty} f_n(a_n\pi_n + b_n\tau_n)$$

$$S_2 = \sum_{n=1}^{\infty} f_n(b_n\pi_n + a_n\tau_n) \tag{41}$$

where, with y = mx, and primes denoting differentiation with respect to the argument,

$$f_n = \frac{2n + 1}{n(n + 1)}$$

$$a_n = \frac{x\Psi_n'(y)\Psi_n(x) - y\Psi_n'(x)\Psi_n(y)}{x\Psi_n'(y)\xi_n(x) - y\xi_n'(x)\Psi_n(y)}$$

$$b_n = \frac{y\Psi_n'(y)\Psi_n(x) - x\Psi_n'(x)\Psi_n(y)}{y\Psi_n'(y)\xi_n(x) - x\xi_n'(x)\Psi_n(y)} \tag{42}$$

The functions Ψ_n and ξ_n are Riccati-Bessel, defined in terms of Bessel functions, J_n, i.e.,

$$\Psi_n(y) = \left[\frac{\pi y}{2}\right]^{1/2} J_{n+(1/2)}(y)$$

$$\xi_n(x) = \left[\frac{\pi x}{2}\right]^{1/2} [J_{n+(1/2)}(x) + i(-1)^n J_{-n-(1/2)}(x)] \tag{43}$$

where $i^2 = -1$.

The angular functions $\pi_n(\cos\theta)$ and $\tau_n(\cos\theta)$ are defined by

$$\pi_n = \frac{dP_n}{d\cos\theta}$$

$$\tau_n = \pi_n\cos\theta - \sin^2\theta \frac{d\pi_n}{d\cos\theta} \tag{44}$$

where P_n denotes a Legendre polynomial with $\cos\theta$ as its argument.

For VV scattering, separate values of S_1 and S_2 are required in Equation 39. For VH scattering, separate calculations of S_1 and S_2, and their subsequent subtraction in Equation 40, lead to large errors because of the closeness of S_1 to S_2 in magnitude; it should be noted that $S_1 = S_2 = S(0)$ when $\theta = 0$. The best method is to directly compute the difference as

$$S_1 - S_2 = \sum_{n=1}^{\infty} f_n(\pi_n - \tau_n)(a_n - b_n) \tag{45}$$

As was introduced a dimensionless parameter G for the intensity, we set a dimensionless parameter U relating the polar angle θ, the wavelength λ of the light in the surrounding medium and the radius a of the sphere

$$U = \frac{4\pi a}{\lambda} \sin(\theta/2) \tag{46}$$

Solving Equations 39 and 40 allows the calculation of the scattering gains G_{VV} and G_{VH} as a function of U. A full description of G_{VV} and G_{VH} has been given by Meeten and Navard.[31-33] We will just recall here the relevant results which will be used later when comparing the exact Mie theory to approximate ones.

The sphere is placed in O in Figure 10. The gains G_{VH} and G_{VV} are calculated in the x'y' plane. This corresponds to the classical experimental setup. In the VV geometry, the gain G_{VV} recorded in the x'y' plane is a succession of dark and bright rings with nearly no azimuthal dependence. The maximum intensity is at $\theta = 0$ (center of the pattern). In the VH mode, the pattern is weak, compared to the VV one, and is also a succession of dark and bright rings, but with an extinction at $\theta = 0$ (center) and at $\mu = n (\pi/2)$, n being an integer. The first order is the so-called four-leaf clover pattern.

Let us first consider the first order intensity maximum at $\mu = 45°$ in the VH mode and the first order intensity minimum in the VV mode (μ is in this case of no importance). Two parameters can be examined, the θ angular position converted into U units (U_{VH} and U_{VV} for the VH and VV modes, respectively) and the intensity converted into gain units (G_{VH} and G_{VV}). There is a certain critical sphere diameter which depends on the ratio of the refractive indices of the sphere and the medium, and above which U_{VH}, U_{VV}, G_{VH}, and G_{VV} strongly oscillate with the sphere diameter. For example, if $m_1 = 1.587$ and $m_2 = 1.332$, which corresponds to a polystyrene sphere in water, there is no oscillation for U or G below $2a = 2 \mu m$. An example of such oscillations in the VH mode for a sphere diameter around $9.9 \mu m$ is given in Figure 12. A similar behavior exists for the VV mode. These oscillations consisting of very narrow peaks and of a smooth periodic background are due to the resonant structure[34,35] of the Mie scattering coefficients a_n and b_n (Equation 42). The period of the oscillations is very small, and beyond any possible experimental check by varying sphere diameters. It is thus useful to average G over one period for comparisons with other theories or with experiments. Figure 13 gives the average G_{VH} as a function of U for a 6 μm-diameter sphere, with $m_1 = 1.587$ and $m_2 = 1.332$.

The main results of these studies are

1.   There is a complete agreement between the theory and experiments with well-characterized monodisperse polystyrene spheres.
2.   The theory is based on the scattering by a single sphere. This remains valid for many spheres in diluted dispersion, where the distance between two spheres is much greater than their diameter. As soon as they are touching, the theory has to be modified because of interferences, multiple scattering, or dependent scattering.[36]
3.   In the VH geometry, the intensity is maximum at $\mu = \pm 45°$, giving the so-called cloverleaf pattern. This is in contrast with many reports, stating that G_{VH} should be zero for isotropic spheres.[37]
4.   In the VH geometry, the value of U_{VH} at $\mu = 45°$, where the gain is maximum, is in the range $U_{VH} = 2.8 - 3.0$.

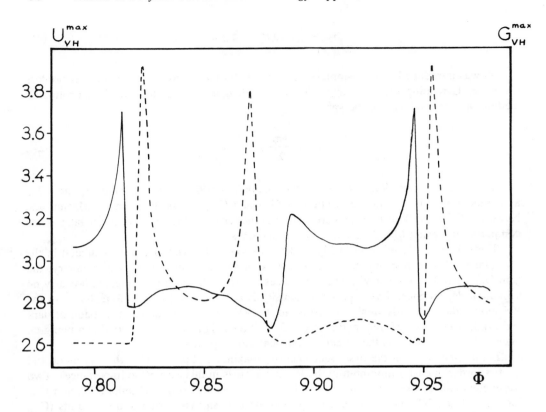

FIGURE 12.   Isotropic sphere. Mie theory. Gain G_{VH}^{max} (---) and U_{VH}^{max} (——) in a VH geometry. Φ is the sphere diameter.

2. Isotropic and Anisotropic Spheres. Anomalous Diffraction Theory

For an anisotropic sphere as well as for an isotropic one, it can be shown that $S_3 = S_4 = 0$. From Equations 20 to 23, Meeten[27] has derived I_{VV} and I_{VH} for a single sphere:

$$I_{VV} = \frac{1}{k^2 r_0^2} \{|S_1|^2 + |S_2 - S_1|^2 \cos^4\mu + 2Re[S_1^*(S_1 - S_2)]\cos^2\mu\} \tag{47}$$

$$I_{VH} = \frac{1}{k^2 r_0^2} \left| S_{is}\sin^2\frac{\theta}{2} + S_{an}\cos^2\frac{\theta}{2} \right|^2 \sin^2 2\mu \tag{48}$$

where Re denotes the real part of a complex number and S_1^* is the conjugate of S_1 and with:

$$S_1 = S_{is} + S_{an}$$

$$S_2 = (S_{is} - S_{an})\cos\theta$$

$$S_{is} = k^2 \int_0^a \left\{ 1 - \frac{1}{2}[\exp(-i\Phi_e) + \exp(-i\Phi_o)] \right\} J_0(q\rho)\rho d\rho \tag{49}$$

$$S_{an} = k^2 \int_0^a \frac{1}{2}[\exp(-i\Phi_o) - \exp(-i\Phi_e)]J_2(q\rho)\rho d\rho \tag{50}$$

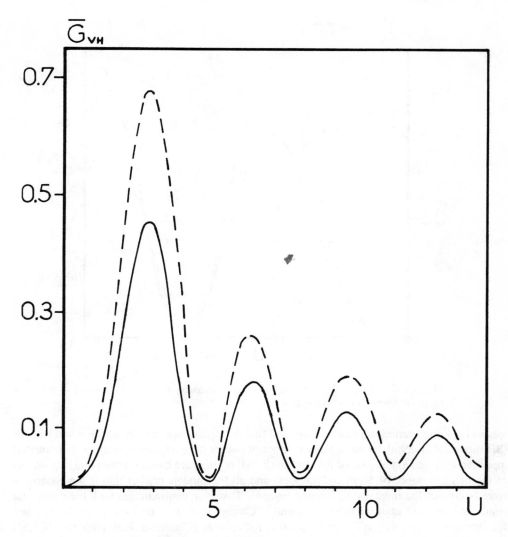

FIGURE 13.   Isotropic sphere. Mie theory. Average gain G_{VH} as a function of the parameter U. Sphere diameter $\Phi = 6 \pm 0.075$ μm (—) and $\Phi = 6 \pm 0.3$ μm (---).

J_0 and J_2 are integer-order Bessel functions, $q = 2 \, k \, \sin(\theta/2)$. ρ is the radius for the considered light ray, and Φ_e and Φ_o are the phase differences for the extraordinary ray and the ordinary ray, respectively.

Equations 47 and 48 are valid for anisotropic and isotropic spheres ($S_{an} = 0$ for an isotropic sphere). A comparison between the AD approximation and the exact Mie theory is given in Figure 14. It has to be recalled[27] that this approximation is inaccurate if there is too large a refractive index mismatch between the scatterer and its surroundings, at large scattering angles and for particles of shape and size such that ray propagation through the particle and diffraction by it are not valid assumptions.

The AD approximation can be very useful for modeling the scattering by anisotropic polymer particles. To our knowledge, it has not been used, except for spheres as seen above.

3. Isotropic and Anisotropic Spheres: the Rayleigh-Gans-Debye Approximation

This approximation has been widely used for studying supermolecular polymer organizations like semicrystalline spherulites, rods, or mesomorphic textures. The case of a single

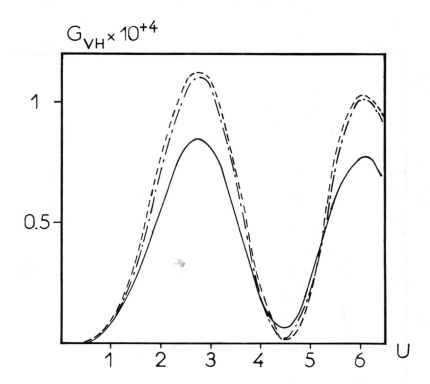

FIGURE 14.   Comparison of Mie theory (——), RGD theory (---), and AD theory (·-·) for an isotropic sphere of diameter 4 μm.

sphere is of importance and has been applied first to polymer spherulites by Stein and Rhodes (SR model),[38] in 1960. Then, a large number of papers have appeared, dealing with various modifications and extensions of this approach. All of them are based on the same hypotheses as the original paper of Stein and Rhodes, and all the models predict that if the spherulite were isotropic, the intensity I_{VH} should be zero. This is in contradiction with the exact Mie result and the AD approximation. Recently, Champion et al.[39] derived again the I_{VH} and I_{VV} functions using the RGD approximation for spheres [Champion-Killey-Meeten (CKM) model]. Their result is in agreement with the Mie theory when the birefringence of the sphere tends to zero and is in agreement with the SR model when the birefringence is large. There is no doubt that the CKM model will change some of the previously drawn conclusions about the light scattered by polymer spherulites. Nevertheless, no work has yet appeared using this model. Therefore, we will only describe the CKM model and compare it to the SR one. We will then describe the SR one. All the comparisons to experiments up to date having been made with the SR model, we will only use it in Sections V and VI.

a. Champion-Killey-Meeten (CKM) Model[39]

The main improvement brought by the CKM model is the derivation of the matrix S (Equation 18) for a sphere in the RGD approximation. It gives a general description of the scattering, independent of the state of polarization of the light. This prevents from confusing the properties of the analyzer with the ones of the scatterer, as seems to have been done in all other papers dealing with the RGD theory.

As seen previously, only S_1 and S_2 are not equal to zero for a sphere. The CKM model gives[39]

$$S_1 = (2ik^3a^3/3U^3)\{3(\overline{\mu} - 1)(\sin U - U\cos U) + \Delta\mu[U\cos U - 4\sin U + 3SiU]\} \quad (51)$$

$$S_2 = (2ik^3a^3/3U^3)\{3(\overline{\mu} - 1)\cos\theta(\sin U - U\cos U)$$

$$- \Delta\mu[1 + \cos^2(\theta/2)][U\cos U - 4\sin U + 3SiU]\} \qquad (52)$$

where U retains its previous meaning (Equation 46), and $\Delta\mu = (m_r - m_t)/m$, $\overline{\mu} = (m_r + 2 m_t)/3$ m, m_r, m_t being the radial and tangential refractive indices of the sphere, and m the refractive index of the surrounding medium. SiU is the sine integral defined by

$$SiU = \int_0^U \frac{\sin x}{x} \, dx \qquad (53)$$

Using Equations 37 and 38, I_{VV} and I_{VH} are

$$I_{VV} = C\left[(\overline{\mu} - 1)f_1 + \frac{1}{3}\Delta\mu f_2\right.$$

$$\left. - \left\{2(\overline{\mu} - 1)f_1\sin^2(\theta/2) + \frac{1}{3}\Delta\mu f_2[2 + \cos^2(\theta/2)]\right\}\cos^2\mu\right]^2 \qquad (54)$$

$$I_{VH} = C\left\{2(\overline{\mu} - 1)f_1\sin^2(\theta/2) + \frac{1}{3}\Delta\mu f_2[2 + \cos^2(\theta/2)]\right\}^2\sin^2\mu\cos^2\mu \qquad (55)$$

where $C = (2k^2a^3/r_0)^2$, $f_1 = (\sin U - U \cos U)/U^3$, and $f_2 = (U \cos U - 4 \sin U + 3 SiU)/U^3$.

A comparison between the predictions of the CKM model for an isotropic sphere ($m_t = m_r$) and the Mie and AD theories is given in Figure 14. The agreement is very good and gives a good confidence in these three theories.

b. Stein-Rhodes (SR) Model[38]

The SR model for spheres is a good illustration of what is generally called in the literature the model approach. It involves calculating the scattering by summing the scattered amplitudes arising from all the volume elements constituting the scattering particle. This particle has a definite shape (here a sphere) and the arrangement of the scattering elements within the particle is geometrically defined, which provides an expression of the electric susceptibility tensor $\underline{\underline{\chi_e}}$. The polarization P_0 is calculated using Equation 11. The scattered amplitude is then obtained from Equations 15 and 24. It must be noted that the notations in the previously published papers are generally different from the ones used in Section B.1. For instance, Equations 11 and the combination of Equations 15 and 24 are frequently written:

$$M = \underline{\underline{\alpha}} E_0 \qquad (56)$$

$$E = C\int(M \cdot O)\exp[-ik(r \cdot s)]dr \qquad (57)$$

$\underline{\underline{\alpha}}$ is called the polarizability tensor and M is improperly named "the induced dipole moment in the volume element". Although the notations used herein are the correct ones for describing dipole scattering, the notations of the original papers will, however, be used to present the theoretical results reported in these papers.

Anisotropic spheres (three-dimensional spherulites) were first considered by Stein and

Rhodes[38] who calculated the SALS patterns to be expected from a homogeneous anisotropic sphere embedded in an isotropic medium. They assumed in their derivation that any volume element of the sphere was uniaxially birefringent with its optic axis in the sphere radius direction. Their original equations were slightly modified to correct an error in sign in the VV SALS equation,[40,41] and to take into account the correct expression of vector O^{42}, given in Section C.

Later, Van Aartsen[43] rederived their equations and extended them to include the case where the optic axis c makes an angle β with the spherulite radius. A similar calculation was performed by Keijzers.[44] Van Aartsen considered uniaxial crystals in which α_1 and α_2 are the polarizabilities in the direction of the optic axis and perpendicular to it, respectively. When the optic axis is parallel to the spherulite radius, $\beta = 0$ and the equations given by Van Aartsen are similar to those of Stein and Rhodes in their corrected form. For $\beta = 90°$, Samuels[40] has shown that the equations are the same as for $\beta = 0$, when examined in terms of radial α_r and tangential α_t polarizabilities. The correct form for these equations is[45,46]

$$I_{VV} = AV_0^2\cos^2\rho_1(3/U^3)^2\{(\alpha_r - \alpha_s)(SiU - \sin U) + (\alpha_t - \alpha_s)(2\sin U - U\cos U - SiU)$$

$$+ (\alpha_r - \alpha_t)[\cos^2(\theta/2)/\cos\theta]\cos^2\mu(4\sin U - U\cos U - 3SiU)\}^2 \tag{58}$$

$$I_{VH} = AV_0^2\cos^2\rho_2(3/U^3)^2\{(\alpha_r - \alpha_t)[\cos^2(\theta/2)/\cos\theta]\sin\mu\cos\mu$$

$$(4\sin U - U\cos U - 3SiU)\}^2 \tag{59}$$

where I_{VV} and I_{VH} denote scattered intensities for VV and VH scattering, respectively. A is a proportionality factor; V_0 is the volume of the anisotropic sphere; α_r and α_t are the radial and tangential polarizabilities of the sphere, respectively; α_s is the polarizability of the surroundings; θ and μ are the radial and azimuthal scattering angles defined in Figure 10; $\cos \rho_1$ and $\cos \rho_2$ are given by Equations 26 and 29, respectively; U has the same definition as in Equation 46; and SiU is the sine integral defined by Equation 53.

For an isotropic sphere ($\alpha_r = \alpha_t$), the scattered intensity for the VH geometry is zero, which is in contradiction with the exact Mie theory and with the AD approximation. This problem is avoided in the recent CKM model where I_{VH} is now different from zero for an isotropic sphere (see Equation 55).

The comparison of CKM and RS models (Equations 54 and 55, and 58 and 59) shows a good agreement when the spherulite size and the birefringence are large.[32] In the other cases, the disagreement can be large. The accuracy of the RS model is a function of the birefringence, the size, and the optical properties of the scatterer. This shows that it is difficult to draw unambiguous conclusions on the size of the scatterer only from SALS.

E. SMALL-ANGLE LIGHT SCATTERING BY RODS

Rodlike scattering is also a type of scattering which is frequently found.[47-54] Within the framework of the RGD approximation, Rhodes and Stein[47] have derived a two-dimensional model for the scattering from a distribution of anisotropic rods of a given length L and infinitesimal thickness lying in the yz plane of the film (Figure 15). The scattering equations take the form:

$$I_{VV} = \rho_0^2 L^2 \int_0^\pi N(\alpha)(\delta\cos^2\alpha' + b_t)^2\left[\frac{\sin(kaL/2)}{(kaL/2)}\right]^2 d\alpha \tag{60}$$

$$I_{VH} = \rho_0^2 L^2 \int_0^\pi N(\alpha)\delta^2\sin^2\alpha'\cos^2\alpha'\left[\frac{\sin(kaL/2)}{(kaL/2)}\right]^2 d\alpha \tag{61}$$

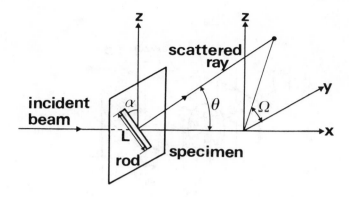

FIGURE 15. The coordinate system for rod scattering.

Here, ρ_0 is the scattering power of a rod per unit length per unit incident field strength when the rod is oriented at $\alpha = 0°$, k is $2\pi/\lambda$ where λ is the wavelength of light in the medium and α is the tilt angle of the rod as measured from the z axis. The azimuthal angle Ω is measured in the yz plane from the y axis and a $= -\sin(\alpha + \Omega)\sin\theta$. The optical anisotropy of the rod is $\delta = b_\ell - b_t$ where $b_\ell = b'_\ell - b_s$ and $b_t = b'_t - b_s$. Here, b'_ℓ is the longitudinal polarizability of the rod, b'_t is the tangential polarizability, and b_s is the polarizability of the surroundings. The quantity α' is equal to $\alpha + \omega_0$, where ω_0 is the angle the direction of maximum polarizability makes with the long axis of the rod. The function $N(\alpha)$ describes the rod orientation distribution.

Kawai and co-workers[55] have extended Stein's formulation for two dimensions to three dimensions, assuming the rod to be both finite in length but infinitesimally thin in radius, and the principal optical axes of the scattering elements to be oriented within the rod with given polar and azimuthal angles to the rod axis. This three-dimensional approach was generalized by Van Aartsen[56] and Hayashi and Kawai,[57] who considered rods finite in length as well as in radius. In Van Aartsen's model[56] the optic axes of the scattering elements are parallel to the rod axis whereas, in Hayashi and Kawai's formulation,[57] they are assumed to be oriented at given polar and azimuthal angles to the rod axis. Other types of rod shape were also investigated: rectangular parallelepipeds[58] and rods with a cross section of lozenge shape.[56]

F. EFFECTS OF INTERPARTICLE INTERFERENCE

In the initial theories, only the scattering from individual particles (spheres, disks, sheaves, rods) which scatter independently of each other has been considered. The total intensity is found by adding scattered intensities from the individual particles. A modification of these theories consists in considering interparticle interference. In such a case the phase differences between the waves scattered by the different particles are taken into account.

A model for interspherulitic interference has been derived by Stein and Picot[59] and Prud'homme and Stein.[60] It considers an assembly of N spherulites, each spherulite being located at a distance R_i from the center of the first one. The total scattered amplitude is written as

$$E = C\left\{\sum_{i=1}^{N}\int_{r_i}[M(r_i)\cdot O]\exp[-ik(R_i + r_i)\cdot s]dr_i + \int_{r_s}(M_s\cdot O)\exp[-ik(r_s\cdot s)]dr_s\right\} \quad (62)$$

The first term of Equation 62 gives the scattering amplitude of the N spherulites, the second term gives the scattering amplitude of the surroundings, $M(r_i)$ represents the induced

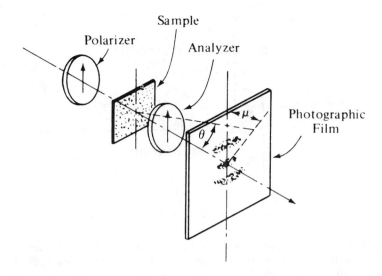

FIGURE 16.   The experimental arrangement for photographic light scattering. (From Stein, R. S. and Rhodes, M. B., *J. Appl. Phys.*, 31, 1873, 1960.
With permission.)

dipole moment of the particle i at position r_i, and M_s is the induced dipole moment of the
surroundings. The scattering from the surroundings may be obtained by subtracting from
the integral over the entire scattering area, the sum of the integrals over the space occupied
by the spherulites as though it was filled with matter having the polarizability of the surroundings. From Equation 62, one obtains

$$E = \sum_{i=1}^{N} f_i \exp[-ik(R_i \cdot s)] \tag{63}$$

where

$$f_i = C\int\{[M(r_i) - M_s] \cdot O\}\exp[-ik(r_i \cdot s)]dr_i \tag{64}$$

The intensity of scattering is given by

$$I \sim \sum_{i=1}^{N} |f_i|^2 + \sum\sum_{i\neq j} f_i f_j^* \exp[-ik(R_i - R_j) \cdot s] \tag{65}$$

Equation 65 was treated by Kawai and co-workers in terms of a paracrystal model of
the Hoseman type.[61,62] This approach makes it possible to introduce into the model fluctuations of the particle orientation, and fluctuations of the distance between adjacent particles.
It was applied to two types of scattering particles: rods[61] and sheaves.[62]

IV. EXPERIMENTAL SETUPS

A. THE PHOTOGRAPHIC TECHNIQUE
A simple experimental method involves recording the whole SALS pattern on a photographic film.[38,63-66] The apparatus is ordinarily very simple, as shown schematically in
Figure 16. Monochromatic light is passed directly through a polarizer, the specimen, and
an analyzer. The resulting primary and scattered intensity is recorded on the photographic

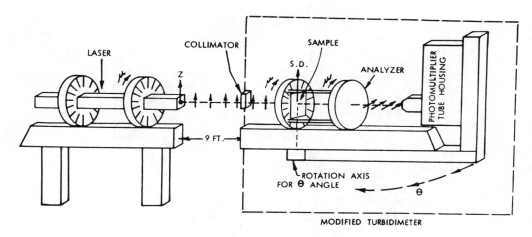

FIGURE 17. Diagram of a photometric SALS experiment. (From Samuels, R. J., *J. Polym. Sci., Part C*, 13, 37, 1966. With permission.)

film. The photographic technique was developed using conventional mercury arc light sources with pinhole-and-lens optics to render the light parallel and with filters to make it monochromatic.[63] Modern equipment utilizes laser sources, which produce a parallel, plane-polarized monochromatic light beam without any extra optics.[64-66] For instance, He-Ne lasers with a red radiation at $\lambda = 6328$ Å are currently used in laboratories. Their power is usually low, e.g., 5 mW.

The experiments are usually performed in the two geometries, described in Section III.C.
1. The polarizer and analyzer both have their polarization directions vertical (the VV geometry).
2. The polarizer and analyzer axes are crossed at 90°, with one polarization direction vertical and the other horizontal (the VH geometry).

In order to obtain good photographic patterns, it is desirable to eliminate surface scattering either by working with films having a smooth surface, or by immersing the sample between glass slides using a fluid of matching refractive index. When studying the scattering by oriented samples, the birefringence of the sample may affect the scattering.[67,68] For uniaxially oriented specimens, the measurement conditions should be such that the orientation direction lies along the polarizer or the analyzer polarization direction.

B. THE PHOTOMETRIC TECHNIQUE

Quantitative studies of light scattering can be made by the photometric technique[65,66,69,70] in which the variation of scattered intensity with angle is measured directly using a detector, usually a photomultiplier tube. For instance, in the apparatus sketched in Figure 17 the photomultiplier measures the scattered intensity as a function of the radial (polar) angle θ at each azimuthal angle μ (angles θ and μ are defined in Figures 10 and 16). To make such a measurement, the laser, sample, and analyzer are tilted together at an angle μ. The measured intensities have to be corrected in several ways, which are inherent to the geometry of the light scattering apparatus: reflection, refraction, secondary scattering, etc.[44,69]

The early light-scattering devices were designed mainly for making static measurements. The later designs incorporated fast detection systems which enable the recording of scattered intensity during crystallization or deformation. For instance, Van Antwerpen and Van Krevelen[71] have described a device incorporating a moving photocell which linearly scans the intensity and displays the variation of intensity with θ-angle on an oscilloscope. High

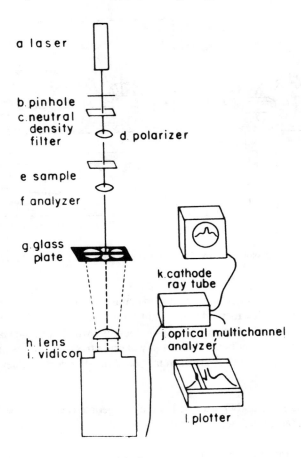

FIGURE 18.   Schematic diagram of the optical multichannel analyzer.
(From Russel, T. P., Koberstein, J., Prud'homme, R., Misra, A., Stein,
R. S., Parsons, J. W., and Rowell, R. L., *J. Polym. Sci., Polym. Phys.
Ed.*, 16, 1879, 1978. With permission.)

recording speed has been achieved by Pakula and Soukup[72] using a rotating system of slits
and a stationary detector onto which the radiation passing through the slit is focused. This
latter device permits the oscilloscope presentation of a two-dimensional scan of the scattering
pattern and has the advantage of not being confined to a linear scan. However, both of these
devices are limited in their precision. In view of these limitations, an automated light-
scattering apparatus was designed by Chu and Horne.[73] This system is capable of making
rapid radial and azimuthal scans with a good accuracy. The operation, data collection, and
reduction are fully automatic.

C. RECENT DEVELOPMENTS

In view of modern technology, electronic scanning is a fruitful approach.[74-76] Scattered
intensities from films of crystalline polymers are great enough to be conveniently detected
by vidicons (television camera tubes), recorded on video tape, and/or digitized and processed
in a variety of ways. Electronic scanning may be accomplished by using an optical multichannel analyzer (OMA). A schematic diagram of the OMA is shown in Figure 18.[75] The
light scattered by the sample passes through the analyzer and is imaged on a frosted glass
plate. A slit on the glass plate defined the azimuthal angle of scattering under consideration.
This real image is focused on the vidicon by a lens. Electronically, the vidicon defines a

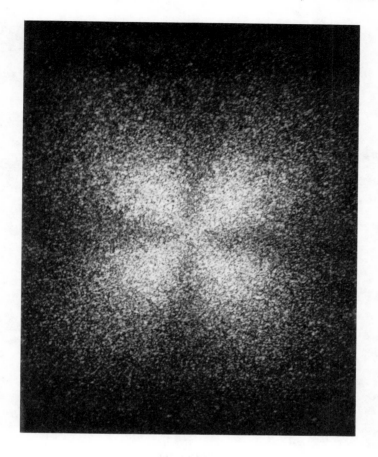

A

FIGURE 19.   Typical SALS patterns for spherulitic scattering. The polarization direction is vertical; (A) VH and (B) VV.

one-dimensional array of 500 channels corresponding to spatial increments of the image. In this way, the image intensity along any given azimuthal angle of the scattering pattern is recorded in the memory of the optical multichannel analyzer. The image intensity may then be displayed either on a cathode ray tube or on an X-Y plotter as a function of its spatial position or scattering angle.

Stein and co-workers[76] have then developed a two-dimensional sensitive detector. As for the one-dimensional position-sensitive detector described above, the device utilizes a vidicon detector. The vidicon detector and its controller are capable of scanning the scattered pattern in two dimensions. The data collected by the detector are transmitted to a microprocessor.

V. SPHERULITES AND RODS

Light scattering is widely used to characterize crystalline superstructures in polymeric solids. Two types of scattering patterns are frequently found in crystalline films: spherulitic scattering (Figure 19) and rodlike scattering (Figure 20). Both have been extensively studied in order to elucidate the crystalline superstructures in terms of their size, shape, and orientation of molecular axis within such microstructures.

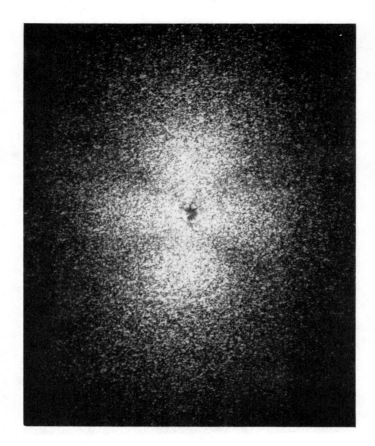

FIGURE 19B.

In the case of spherulitic scattering, the first theoretical approaches used to interpret the experimental patterns considered both two- and three-dimensional perfect spherulites. These calculations were then modified to take into account particular morphological features such as truncations, incomplete growth, internal disorder, and deformation.

Rodlike scattering has been found in many polymeric materials. Films of polytetrafluoroethylene, polychlorotrifluoroethylene, polyethylene terephthalate, polyethylene, and some polymer single crystals exhibit rodlike scattering. In these polymers, the fundamental structural unit is the folded-chain lamella and these materials can also crystallize with a spherulitic morphology. Cellulose, cellulose derivatives such as hydroxypropylcellulose, polypeptides, and reconstituted collagen films are another group showing rodlike scattering. A fundamental structural unit of these materials is the microfibril or fibril. These materials do not easily crystallize in spherulitic morphology.

The proposed models initially considered scattering by individual scattering particles (spherulite, rod). Then, interference between the scattering particles was introduced into the models.

A. LIGHT SCATTERING BY SPHERULITES (SR MODEL)

As shown in Figure 19, the typical scattering pattern given by a spherulitic film consists of four lobes, and looks like a four-leaf clover. Such patterns have been first interpreted using the SR model for an individual anisotropic sphere (see Section III.D.3). The theoretical Equations 58 and 59 have been programmed for a computer to produce the theoretical VV

A

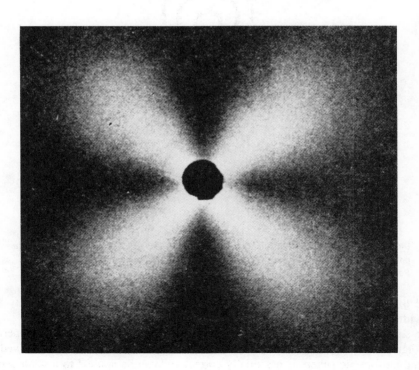

B

FIGURE 20. Typical SALS patterns for rodlike scattering; (A) VH and (B) VV. (From Samuels, R. J., *J. Polym. Sci., Part A2*, 7, 1197, 1969. With permission.)

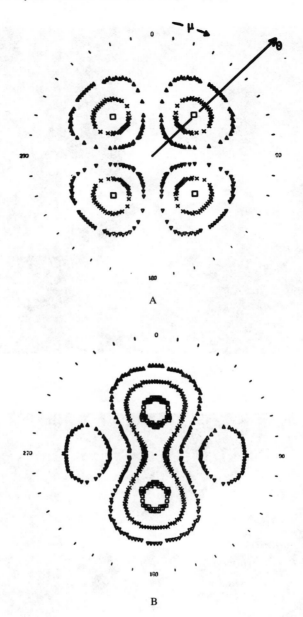

FIGURE 21.   Computed SALS patterns for unoriented isotactic polypropylene films. The polarization direction is vertical; (A) VH and (B) VV. (From Samuels, R. J., *J. Polym. Sci., Part A2*, 9, 2165, 1971. With permission.)

and VH SALS patterns displayed in Figure 21.[40,45,66] These are constant-intensity contour plots, with the symbols representing different intensity levels.

The similarity between the theoretical VH pattern in Figure 21A and the experimental one in Figure 19A suggests first that the theoretical VH SALS Equation 59 derived for an individual anisotropic sphere is essentially correct and can be applied to a nonindependent spherulite scattering system. The most obvious correspondence between the two patterns is that both have the same characteristic four-leaf clover shape. Further examination of the theoretical VH pattern shows that the maximum intensity in the four lobes occurs at azimuthal angles μ of 45, 135, 225, and 315°, respectively. This can be observed in the experimental pattern (Figure 19) and has been photometrically demonstrated as well.[65] Again in the theoretical VH SALS pattern the highest intensity (squares) in any lobe of the four-lobed pattern always concentrates, for a given spherulite diameter, at a particular radial angle θ_{max}. θ_{max} is represented by the distance of the intensity maximum of one of the lobes to the center of the whole pattern. Examination of Equation 59 shows that the symmetry of the VH SALS pattern and the extinctions along the polarization axes ($\mu = 0$ and 90°) result from the term $\sin \mu \cos \mu$ in Equation 59. This term is a maximum at $\mu = 45°$. The variation of the scattered intensity as a function of the radial scattering angle θ is essentially represented by the term $(3/U^3)$ (4 sin U − U cos U − 3SiU). The intensity is zero at zero angle (U = 0) and goes through a maximum with increasing θ angle. This maximum is observed at U = 4.09. The existence of an intensity maximum at $\mu = 45°$ and U = 4.09 makes possible to obtain the average spherulite radius in a polymer film from the VH SALS pattern. The determination of the spherulite size is one of the most important applications of the VH SALS measurement. This determination is performed in the following way: the distance from the center of the VH pattern to the intensity maximum in one lobe, in conjunction with the known sample-to-film distance, is a measure of the polar angle θ_{max}. This angle has to be corrected for refraction at the film surface, in order to obtain the polar angle within the polymer film.[44,77] The value of the spherulite radius R_0 is then deduced from:

$$R_0 = \frac{1.025}{\pi} \frac{\lambda}{\sin(\theta_{max}/2)} \qquad (66)$$

where λ is the wavelength of light in the medium. Since R_0 is inversely proportional to $\sin(\theta_{max}/2)$, small spherulites will have a cloverleaf pattern with maxima at large radial angles while large spherulites will have maxima at small radial angles. This is illustrated in Figure 22, where the effect of polyethylene spherulite size on the radial position of the VH SALS pattern is shown. These VH SALS patterns were obtained by keeping the sample-to-film distance constant during the experiment. The major advantage of Equation 59 is that I_{VH} does not depend on the polarizability α_s of the surroundings but only on the anisotropy term $(\alpha_r - \alpha_t)$. Furthermore the magnitude of A and $(\alpha_r - \alpha_t)$ affects only the absolute magnitude and not the position of θ_{max}. Thus, these parameters do not influence the calculation of the spherulite size from the VH SALS pattern.

It has to be recalled that Equation 59 is not correct, and must be replaced by Equation 55, in which the optical properties of the medium and the sphere and even the spherulite radius affect the position of θ_{max}. The exact extent of the approximations made in the SR model, and leading to Equation 59, has not been yet studied. It is thought that Equation 59 can be considered as a very good approximation when $(\alpha_r - \alpha_t)$ and the spherulite diameter are large.[32,39] All the following studies fall in this category. One exception could be the study of the beginning of the spherulite growth, when the characteristic dimension of the spherulite is small. However, as will be shown, its shape is not spherical, and Equations 55 or 59 cannot be applied.

Equation 58 shows that the VV SALS pattern is affected by the polarizabilities of both the spherulite and the environment. This equation contains three mathematical functions. The first function, (SiU − sin U), has spherical symmetry, the second function, (2 sin U − U cos U − SiU), has spherical symmetry in the low-angle regions, and the third function, $[\cos^2(\theta/2)/\cos\theta]\cos^2\mu$(4 sin U − U cos U − 3SiU), is the major contributor to the anisotropy of the VV SALS pattern. Figure 23 shows the computed intensity distribution for each of these three functions. The final VV pattern will be a combination of these three basic forms and will depend strongly on the magnitude and the sign of the polarizability weighting factors $(\alpha_r - \alpha_s)$, $(\alpha_t - \alpha_s)$, and $(\alpha_r - \alpha_t)$, respectively. The sensitivity of the VV SALS pattern

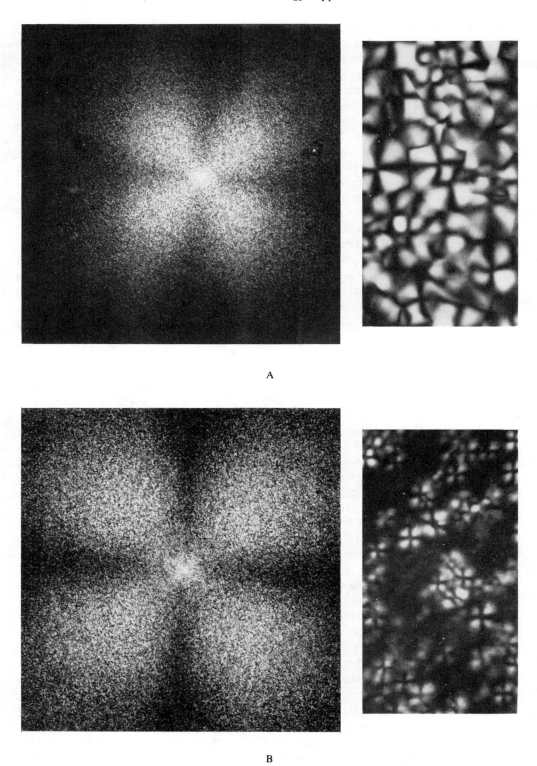

FIGURE 22.   Effect of spherulite size on the position of the VH SALS pattern. Average spherulite diameter (A) 30 μm and (B) 8 μm.

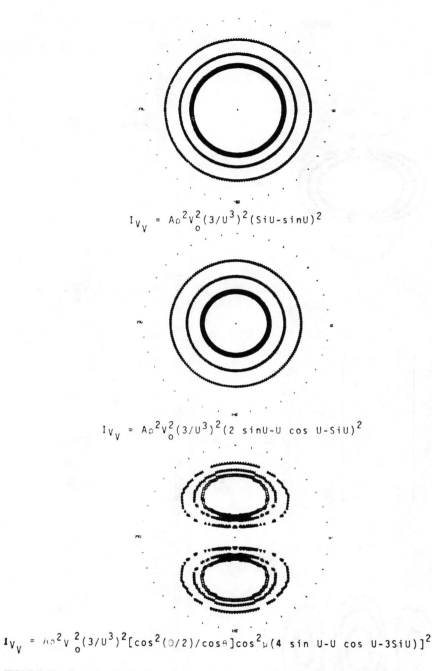

$$I_{V_V} = A_0^2 v_0^2 (3/U^3)^2 (SiU - sinU)^2$$

$$I_{V_V} = A_0^2 v_0^2 (3/U^3)^2 (2\ sinU - U\ cos\ U - SiU)^2$$

$$I_{V_V} = A_0^2 v_0^2 (3/U^3)^2 [cos^2(\theta/2)/cos\theta] cos^2\mu (4\ sin\ U - U\ cos\ U - 3SiU)]^2$$

FIGURE 23. Intensity distribution for each of the three functions in the VV SALS equation for an anisotropic sphere; (a) (top) $I_{VV} = AV_0^2 \cos^2\rho_1 (3/U^3)^2 (SiU - \sin U)^2$, (b) (middle) $I_{VV} = AV_0^2 \cos^2\rho_1 (3/U^3)^2 (2 \sin U - U \cos U - SiU)^2$, and (c) (bottom) $I_{VV} = AV_0^2 \cos^2\rho_1 (3/U^3)^2 \{[\cos^2(\theta/2)/\cos \theta] \cos^2\mu (4 \sin U - U \cos U - 3 siU)\}^2$. (From Samuels, R. J., *J. Polym. Sci., Part A2*, 9, 2165, 1971. With permission.)

to the magnitude and the sign of the polarizability of both the spherulite and the surrounding medium is shown in Figure 24. These theoretical patterns were calculated for a negative spherulite ($\alpha_t > \alpha_r$) with constant values of α_r and α_t, and varying values of α_s. The VV pattern (A) is obtained for $\alpha_s = \alpha_t$ and VV pattern (B) corresponds to $\alpha_s = \alpha_r$. These patterns represent only two of a large range of patterns that would be obtained if the

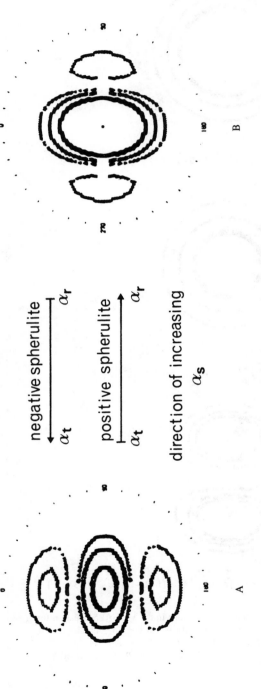

FIGURE 24. Effect of the background polarizability α_s and of the spherulite optical sign on the theoretical VV pattern for an anisotropic sphere. Pattern A corresponds to $\alpha_s = \alpha_t$ and pattern B to $\alpha_s = \alpha_r$. (From Samuels, R. J., *J. Polym. Sci., Part A2*, 9, 2165, 1971. With permission.)

birefringence of the spherulite is fixed and the refractive index of the background is changed. They show that a decrease in the background polarizability is likely to induce a change in a VV SALS pattern from an (A) type to the (B) type pattern. Conversely, for a positive spherulite, pattern (B) is expected to occur at a higher background refractive index than pattern (A) (Figure 24). Thus, although theoretically the same VV patterns can be obtained from either a positive or a negative spherulite, depending on the background refractive index, the direction of change in the VV pattern with changing the background refractive index will be determined by the sign of the birefringence of the spherulite. This theoretical prediction was confirmed experimentally by Samuels.[40] An important application of this theoretical result is the determination of the optical sign of spherulites when they are too small to be identified by optical microscopy. Samuels has shown that this determination is possible provided that a specially outlined procedure is followed.[40]

B. LIGHT SCATTERING BY ANISOTROPIC DISKS: TWO-DIMENSIONAL SPHERULITES

In films in which the thickness is less than the spherulite diameter, the spherulite is necessarily truncated and can be considered as two-dimensional. Such spherulites are often obtained in laboratory experiments by crystallization between two glass slides on a hot stage.

In order to interpret scattering by two-dimensional spherulites, Stein and Wilson[78] have derived the SALS equations for an optically anisotropic disk, and Clough et al.[42,79] extended them to include such variables as optic axis alignment. The two-dimensional spherulite is approximated by a disk with its plane parallel to the surface of the film. The model considers uniaxial crystals in which α_1 and α_2 are the polarizabilities in the direction of the optic axis and perpendicular to it, respectively. A scattering element in the disk is located by the polar coordinates (r, α). The optic axis is at an arbitrary angle β to the radius and may rotate about the radius through the angle ω ($\omega = 0$ when the optic axis lies in the plane of the disk).

The scattered amplitudes for the VV and VH geometries are calculated using Equations 56 and 57:

$$E_{VV} = -\pi CE_0 \cos\rho_1 \int_0^{R_0} \{[(\alpha_1 - \alpha_2)(\cos^2\beta + \sin^2\beta\cos^2\omega) + 2\alpha_2]J_0(x)$$

$$- (\alpha_1 - \alpha_2)[(\cos^2\beta - \sin^2\beta\cos^2\omega)\cos2\mu - (\sin2\beta\cos\omega)\sin2\mu]J_2(x)\}rdr \quad (67)$$

$$E_{VH} = \pi CE_0 \cos\rho_2(\alpha_1 - \alpha_2) \int_0^{R_0} \{[(\cos^2\beta - \sin^2\beta\cos^2\omega]\sin2\mu$$

$$+ [\sin2\beta\cos\omega]\cos2\mu\}J_2(x)rdr \quad (68)$$

where R_0 is the disk radius; $\cos\rho_1$ and $\cos\rho_2$ are given by Equations 26 and 29, respectively. $J_0(x)$ and $J_2(x)$ are Bessel functions of the variable x defined as

$$x = \frac{2\pi r}{\lambda}\sin\theta \quad (69)$$

For spherulites embedded in a matrix of polarizability α_s, α_1 and α_2 should be replaced by $(\alpha_1 - \alpha_s)$ and $(\alpha_2 - \alpha_s)$.

Various special cases of Equations 67 and 68 have been considered.[42]

1.   Disks with optic axes in the plane of the disk ($\omega = 0$)

The scattered intensities are given by

$$I_{VV} = AA_0^2\cos^2\rho_1\left[\frac{2}{W^2}\right]^2\{\alpha_1[1 - J_0(W)] + \alpha_2[WJ_1(W) - (1 - J_0(W))]$$

$$- (\alpha_1 - \alpha_2)\cos^2\xi[2(1 - J_0(W)) - WJ_1(W)]\}^2 \qquad (70)$$

$$I_{VH} = AA_0^2\cos^2\rho_2\left[\frac{2}{W^2}\right]^2\{(\alpha_1 - \alpha_2)\sin\xi\cos\xi[2(1 - J_0(W)) - WJ_1(W)]\}^2 \qquad (71)$$

where A is a proportionality factor, A_0 is the area of the disk, $\xi = \mu + \beta$ and W is defined as

$$W = \frac{2\pi R_0}{\lambda}\sin\theta \qquad (72)$$

These equations are similar to those previously given by Stein and Wilson[78] except that μ is replaced by $\xi = \mu + \beta$. This substitution corresponds to a rotation of both VV and VH SALS patterns through the angle β. In particular, when $\beta = 45°$, the intensity maxima in the VH pattern are located on the polarization axes ($\mu = 0$ and $90°$).

2.   Disks with random rotation of optic axis about the radius. If the optic axes twist randomly, there is no correlation between ω and r, and the terms $\cos \omega$ and $\cos^2\omega$ in Equation 68 may be averaged separately, resulting in:

$$I_{VH} = AA_0^2\cos^2\rho_2\left[\frac{3\cos^2\beta - 1}{2}\right]^2$$

$$\left[\frac{2}{W^2}\right]^2\{(\alpha_1 - \alpha_2)\sin\mu\cos\mu[2(1 - J_0(W)) - WJ_1(W)]\}^2 \qquad (73)$$

Thus the pattern shape and position are the same as those obtained from Equation 71 for $\beta = 0$, but the intensity is decreased by the factor of $[\frac{3\cos^2\beta - 1}{2}]^2$.

3.   Disks with helicoidal rotation of optic axis about the radius. If ω varies linearly with r, helicoidal orientation of the optic axes results. Such an orientation is usually encountered in ringed spherulites. In such a case, the integrals in Equations 67 and 68 must be evaluated numerically. Helicoidal periodicity gives scattering maxima at an angle related to the period of helical twist. For instance, in Figure 25, VH intensity contours are plotted for the case of $\beta = 90°$, illustrating the low-angle maxima resulting from the spherulite size and the higher angle maxima arising from the helicoidal orientation.[79] These results are consistent with experimental observations.

Computer calculations show that the theoretical SALS patterns corresponding to the three two-dimensional models described above have generally the same appearance as those found for three-dimensional spherulites.[42,78,79] For instance, a classical four-leaf clover pattern is usually obtained for the VH geometry. Useful information can be obtained from a

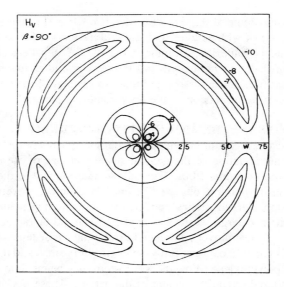

FIGURE 25. Intensity contour diagram for VH scattering of two-dimensional ringed spherulite. The number of complete rotations of the optic axes is n = 5. (From Stein, R. S., Erhardt, P., Van Aartsen, J. J., Clough, S., and Rhodes, M., *J. Polym. Sci., Part C*, 13, 1, 1966. With permission.)

TABLE 2
Correspondence of the Disk and Sphere SALS Equations

Disk

$$I_{VV} = AA_0^2\cos^2\rho_1\left[\frac{2}{W^2}\right]^2\{(\alpha_r - \alpha_s)J'$$
$$+ (\alpha_t - \alpha_s)(J'' - J')$$
$$+ (\alpha_r - \alpha_t)\cos^2\mu(J'' - 2J')\}^2$$

$$I_{VH} = AA_0^2\cos^2\rho_2\left[\frac{2}{W^2}\right]^2$$
$$\{(\alpha_r - \alpha_t)\sin\mu\cos\mu(2J' - J'')\}^2$$

$$J' = 1 - J_0(W); \quad J'' = WJ_1(W)$$

$$W = \frac{2\pi R_0}{\lambda}\sin\theta$$

A_0: area of the disk

$$R_0 = \frac{1.96}{\pi}\frac{\lambda}{\sin\theta_{max}}$$

Sphere

$$I_{VV} = AV_0^2\cos^2\rho_1\left[\frac{3}{U^3}\right]^2\{(\alpha_r - \alpha_s)S'$$
$$+ (\alpha_t - \alpha_s)(S'' - S')$$
$$+ (\alpha_r - \alpha_t)[\cos^2(\theta/2)/\cos\theta]\cos^2\mu(S'' - 3S')\}^2$$

$$I_{VH} = AV_0^2\cos^2\rho_2\left[\frac{3}{U^3}\right]^2$$
$$\{(\alpha_r - \alpha_t)[\cos^2(\theta/2)/\cos\theta]\sin\mu\cos\mu(3S' - S'')\}^2$$

$$S' = SiU - \sin U; \quad S'' = \sin U - U\cos U$$

$$U = \frac{4\pi R_0}{\lambda}\sin\frac{\theta}{2}$$

V_0: volume of the sphere

$$R_0 = \frac{1.025}{\pi}\frac{\lambda}{\sin(\theta_{max}/2)}$$

direct comparison between the scattering equations. In the literature, this comparison has been made essentially in the case where the angle β between the optic axis and the spherulite radius equals zero.[40,41,78] In this case, the scattering equations for individual disks and spheres are compared in Table 2. In these equations, α_r and α_t denote the radial and tangential polarizabilities of the scattering particle and α_s is the polarizability of the surroundings. It can be seen that for the small angles θ $(\cos^2(\theta/2)/\cos\theta \simeq 1)$, the equations for two- and three-dimensional spherulites can be cast into similar forms. Table 2 also shows that the disk radius R_0 can be determined from the angle θ_{max} at which maximum intensity appears. The constant 1.96 is almost twice that of 1.025 for spheres. However, the denominator sin θ_{max} is twice that of $\sin(\theta_{max}/2)$, so that the two equations for R_0 represent almost identical relations. For example, a three-dimensional spherulite with a radius of 1 μm will have a

VH SALS intensity maximum at $\theta_{max} = 23.8°$ if λ is 6320 Å. For the same wavelength of light in the medium, a disk with a radius of 1 μm will have a VH SALS intensity maximum at $\theta_{max} = 23.3°$. Thus the intensity maximum in the VH SALS pattern obtained from a disk and from a sphere of the same radius will fall in the same radial angular region.[40] Consequently, it is not possible, under these conditions, to distinguish between two- and three-dimensional spherulites by inspection of the light-scattering pattern.

C. MODIFICATIONS OF THE MODELS FOR SPHERULITIC SCATTERING

The models of scattering from individual anisotropic spheres or disks, described above, give a satisfactory interpretation of the general shape and size of the experimental SALS patterns from spherulitic polymer films. They allow measurement of the average spherulite radius from the intensity maxima observed in the VH patterns. Nevertheless, it has been found in many cases that the theoretical models for VH scattering predict intensities which are too low at small, and large, scattering angles.[80] The photographic VH SALS pattern may differ from the four-leaf clover pattern and exhibit the "four-tennis-racket" appearance described by Motegi et al.[81] In polymer films containing very small and imperfect spherulites, such as low-density polyethylene films, the VH pattern may be much more degenerate, with a large intensity scattered around the center of the pattern. These deviations from the ideal patterns for an isolated spherulite have been attributed to different morphological features of real spherulites: distribution of spherulite sizes, interspherulitic interference, impingement, incomplete growth, and internal disorder. The models of scattering from individual spherulites have been modified to take into account such effects.

1. Polydispersity of Spherulite Size and Interspherulitic Interference

If no correlation between the spherulites is assumed, the scattered light intensity is the sum of the intensities scattered by individual spherulites.[81] Thus, for the VH geometry

$$I_{VH} = \int_0^\infty N(R)I_{VH}(R)dR \tag{74}$$

where $N(R)$ defines the distribution of the spherulite radii. Motegi et al.[81] have considered different types of distribution (monodisperse, box, or gaussian). There is essentially no change in the intensity distribution, except for the shift of θ_{max} and some skew of the intensity distribution to lower scattering angles.

When interspherulitic interference is now considered, a modulation of the light scattering curves is produced if the number of spherulites illuminated by the incident radiation is small.[60] As this number increases, the modulation will become smaller and eventually will disappear for a sufficiently large number of spherulites. The position of the VH scattering maximum is not affected by this interference. It was postulated that this same interference effect may explain the speckled appearance of the experimental spherulitic patterns.

2. Impingement and Incomplete Growth

In a spherulitic sample, spherulites are not perfect spheres but are truncated by impingement with other spherulites. The problem of scattering by random assemblies of two-dimensional spherulites was studied by Stein and Picot.[82] They considered principally single or double truncation and examined only a few cases of multiple truncation. For a single truncation, Figure 26 shows the variation of VH patterns as the truncation increases. To get a valid comparison of the intensity levels the scattered intensities have been always calculated for a particle of area equivalent to the complete disk. The perturbations in the VH pattern are essentially an appearance of a link between the lobes which in the case of the complete disk are separated by zero intensity lines along the polarization axes. When the truncated

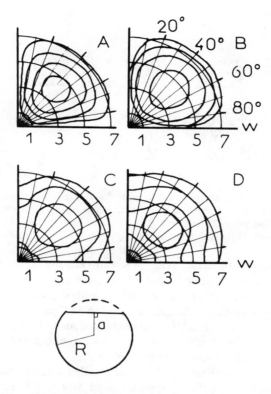

FIGURE 26. Theoretical VH patterns for perfect and truncated spherulites. The parameter G = a/R characterizes the degree of truncation; W
is the reduced variable defined in Table 2. (A) G = 1, (B) G = 0.7, (C)
G = 0.3, and (D) G = 0. (From Stein, R. S. and Picot, C., *J. Polym.
Sci.*, Part A2, 8, 2127, 1970. With permission.)

spherulite tends to a half-disk shape, the links become wider leading to a pattern which is
more circular (Figure 26D).

The case of multiple truncation has been described statistically by Prud'homme and
Stein.[83] In their model, the average size of a truncated spherulite is characterized by an
average distance a given by

$$a = \sum_{i=1}^{N} a_i/N \qquad (75)$$

where a_i is the distance between the center of the spherulite and any point of its perimeter.
The SD σ can be written as

$$\sigma = \left(\frac{\sum_{i=1}^{N} (a_i - a)^2}{N} \right)^{1/2} \qquad (76)$$

The severity of the truncation is then expressed by the statistical parameter σ^2/a^2. Figure
27 shows the VH light-scattering patterns calculated for different values of this truncation
parameter. For truncation parameters of 0.10 and 0.20, the tennis-racket type of pattern is
obtained. For higher values of the truncation parameter, the patterns become very disordered.
Another important conclusion of these calculations is that for a given average size a, trun-

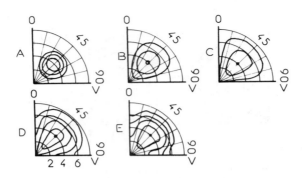

FIGURE 27.   Scattering contour plots for truncation parameters σ^2/a^2 equal to: (A) 0.0019, (B) 0.10; (C) 0.20, (D) 0.30, and (E) 0.40. (From Prud'homme, R. E. and Stein, R. S., *J. Polym. Sci., Polym. Phys. Ed.*, 11, 1683, 1973. With permission.)

cation shifts the maximum scattering position towards smaller angles than that at which the maximum would occur if the spherulites were not truncated and had a radius equal to a.

It must be noticed that for the two models described above, the truncations must be fairly large to produce significant changes in the VH SALS patterns: for the former, G must be lower than 0.7 and for the latter σ^2/a^2 must be higher than 0.10. High amounts of truncation are necessary to obtain disordered patterns. So, truncation can only account for a part of the departure from the perfect spherulite pattern.

Tabar et al.[84] have generalized the theory of Prud'homme and Stein to account for the effect of the impingement of growing spherulites on their VH SALS patterns. During the crystallization process, various stages of spherulite growth occur. In the early stages, probably no spherulites impinge. As the spherulites continue to grow, more and more of them will impinge on each other, and the extent of truncation for each individual spherulite increases. The theory is developed on the basis of results of computer-simulated two-dimensional spherulite growth and calculated scattered intensities. The impingement produces a lowering of the intensity of the scattering maximum and the diminishing of the overall sharpness of the scattering peak. This study concerned only simultaneously nucleated spherulites. It was extended to sporadically nucleated two-dimensional spherulites and to simultaneously nucleated three-dimensional spherulites.[85]

Another cause for tennis-racket type patterns is the incompleteness of development. Spherulites nucleate from bundle-like crystals which evolve into sheaves and eventually into complete spherulites. Such an evolution has been idealized by the fan model which consists in two opposite sectors of a disk[81,86] (Figure 28). The orientation of the fan is defined by the angle γ between its axis and the Z direction and the angle 2β characterizes the aperture of the fan. The scattered intensities are obtained by averaging the square of the amplitudes for all the orientations of the particle. Figure 29 gives the variation of the logarithm of the I_{VH} intensity along the 45° azimuthal direction when β varies from 90° to 1°. As seen in the figure, the scattered intensity at low scattering angles is remarkably increased by decreasing the sector angle. Simultaneously, the intensity is also increased at high values of the scattering angle. It was also shown[86] that for very small values of the sector angle, the calculated VH pattern becomes similar to that of a rod, with the exception that a scattering maximum is still seen at an angle close to that at which the spherulite scattering maximum occurs.

As a consequence, it was suggested by Motegi et al.[81] that the more sheaflike the crystalline texture, the more the VH pattern approaches the four-tennis-racket pattern, even showing the strongest intensity at low scattering angles. Such phenomena are often found in low density polyethylene films.[81] Moreover, the evolution of patterns occurring during the early stages of growth of polyethylene terephthalate spherulites has been described by

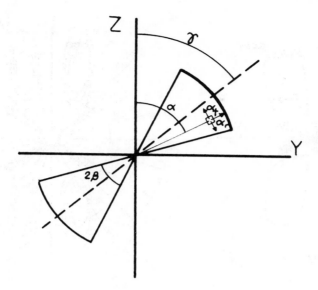

FIGURE 28. The fan model. Geometrical description. (From Picot, C., Stein, R. S., Motegi, M., and Kawai, H., *J. Polym. Sci., Part A2*, 8, 2115, 1970. With permission.)

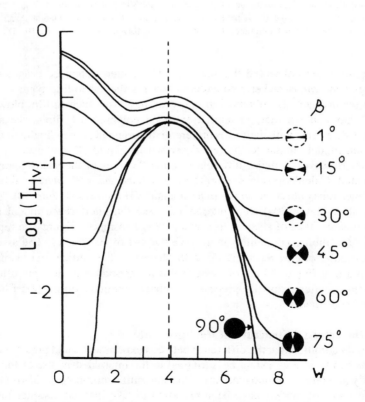

FIGURE 29. The fan model. Plot of variation of Log (I_{VH}) along the 45° azimuthal direction as a function of W. The intensity curves correspond to particules of identical area. (From Picot, C., Stein, R. S., Motegi, M., and Kawai, H., *J. Polym. Sci., Part A2*, 8, 2115, 1970. With permission.)

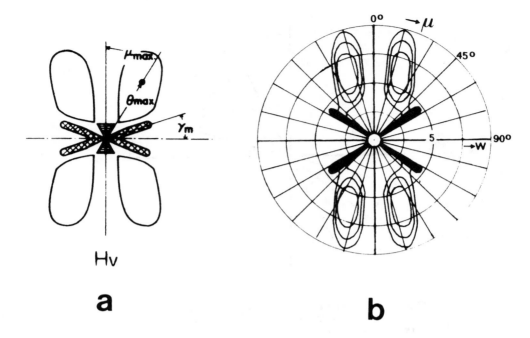

a   **b**

FIGURE 30. (a) Schematic representation of the experimental VH pattern obtained from tubular-extruded polybutene-1 films, and (b) calculated VH pattern for assemblies of 10 sheaves oriented at ± 35° with respect to the extrusion direction. (From Hashimoto, T., Todo, A., and Kawai, H., *Polym. J.*, 10, 521, 1978. With permission.)

Misra and Stein.[87] It was observed that the VH SALS patterns changed progressively from rodlike through four-tennis-racket type patterns towards the spherulitic type.

Sheaflike textures are also obtained in industrial processes such as film blowing. Hashimoto et al.[88] have shown that, in tubular-extruded polybutene-1 films, the sheaves are aligned nearly side by side with their axes preferentially oriented perpendicular to the machine direction. A fan model similar to the one described in Figure 28 was used to predict the SALS patterns. However, a major difference is that the axes of the sheaves are now preferentially oriented. Calculations of scattering from isolated sheaves[88] were unable to account for the four inner lobes observed in the experimental VH patterns (Figure 30). These lobes (double-hatched in Figure 30a) were interpreted using the interference model in terms of scattering from assemblies of sheaves as a whole.[62] For instance, Figure 30b represents the VH pattern for assemblies containing an average number of 10 sheaves. The assemblies are oriented at particular angles $\alpha_0 = \pm 35°$ with respect to the vertical extrusion direction. Patterns like those of Figure 30 are of great practical importance as they are often found in polymer specimens after industrial processing: tubular extruded films, cast films, and injection-molded specimens.

3. Effect of the Internal Structure of the Spherulite

It is generally admitted that the difference between experimental and predicted spherulitic SALS patterns can to a large extent be attributed to the internal structure of the spherulite, and particularly to internal inhomogeneities of the spherulitic morphology, also called internal disorder. Two types of models have been proposed to take internal disorder into account. The first ones concern modifications of the geometrical description of a spherulite as a homogeneous anisotropic sphere or disk. The second ones combine the model approach with a statistical approach.

a. *Geometrical Approach*

An interesting attempt to model the actual internal structure of a spherulite was made by Prud'homme et al.[89] As a polymer spherulite consists of thin crystalline lamellae separated by amorphous interlamellar layers, a two-dimensional spherulite is described as a planar spherulitic array of anisotropic rodlike lamellae, surrounded by isotropic amorphous material. The total scattered amplitude of the spherulite is calculated by summing up the amplitude scattered by each rod, the phases of the scattered waves being referred to a common origin. Theoretical patterns were computed for different lamellar spacings S. When S is 200 Å, the I_{VH} scattered intensity is exactly the same as the one obtained from the homogeneous anisotropic disk (Equation 71). When S is 1000 Å, there is little difference between the two models. When the interlamellar spacing is 3000 Å, the scattered intensity in the large-angle region increases. In polymer spherulites, the interlamellar spacing is usually about a few hundred angstroms, depending upon the thermal history. Therefore, the presence of lamellae in the spherulite has little effect on light scattering. This is due to the fact that the dimensions of the internal structure of the spherulite are much smaller than that of the wavelength of light.

b. *Statistical Approach*

Keijzers[44] has proposed a hybrid type scattering theory in which the scattered intensity is the sum of two terms: the first one is a perfect-spherulite term of the type previously described herein and the second one is a random orientation fluctuation term. This model was essentially applied to polymer specimens in which the spherulitic order is low: isotactic polystyrene samples with complete or incomplete spherulitic growth, and quenched isotactic polypropylene samples. The theory was successful in describing the scattering from such specimens and allowed a separation of the experimental intensities into spherulitic scattering and random orientation scattering.

An important limitation of the Keijzers approach is the treatment of the random contribution as a separate phase, scattering independently. It is likely that this contribution can be better treated as a perturbation from the ideal spherulite. Thus, a model like the one presented by Stein and Chu[90] seems to be more realistic. These authors assumed that there is orientational disorder within the spherulites and calculated the consequences of such a disorder on the scattering. The orientational disorder is specified in terms of fluctuations of the optic axes of the crystals. In a perfect spherulite, it is assumed that the optic axes make a definite angle β with respect to the spherulite radii. In a real spherulite, β can be written as

$$\beta(r,\alpha) = \beta_0 + \Delta(r,\alpha) \tag{77}$$

where (r, α) are the polar coordinates inside the spherulite and $\Delta(r, \alpha)$ is the fluctuation which depends upon location within the spherulite. To keep computation times reasonable, the calculation was limited to two dimensions and only two limiting cases were considered: (1) the case of radial disorder in which the optic axis orientation fluctuates only with the radial distance r, i.e., $\Delta = \Delta(r)$ and (2) the case of angular disorder in which β varies with α but not with r.

It was concluded that increasing radial disorder leads to increase the relative I_{VH} scattered intensity at angles larger than that of the scattering maximum, whereas angular disorder enhances relative intensity at angles less than the maximum. In both cases, the angle θ_{max} is not modified. Experimentally, enhanced intensities are observed for both small and large scattering angles, which indicates a combination of radial and angular disorder.

The approach of Stein and Chu[90] was generalized by Yoon and Stein,[91] who developed

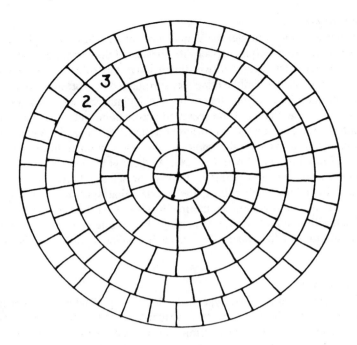

FIGURE 31.   Circular lattice representation for a two-dimensional spherulite. (From Yoon, D. Y. and Stein, R. S., *J. Polym. Sci., Polym. Phys. Ed.*, 12, 763, 1974. With permission.)

a lattice theory of orientational disorder in two dimensional spherulites. The spherulite was represented as a circular lattice of cells of equal area as shown in Figure 31. In the ith cell, the optic axis is given by β_i which deviates from the average value β_0 by $\Delta\beta_i$:

$$\beta_i = \beta_0 + \Delta\beta_i \qquad (78)$$

It is assumed that the angular differences $\Delta\beta_i$ are quantized in units of the increment $\pm \delta$. To avoid long-range disorder deviating from the mean value of β_0, it is also assumed that, the larger $\Delta\beta_i$, the greater is the tendency to return to β_0. The circular lattice is filled progressively, starting from the innermost cell. The assignment of the orientation-angle deviation $\Delta\beta_i$ for a lattice cell such as cell 3 in Figure 31 is done in consideration of its correlation with the two nearest-neighbor cells 1 and 2, which have already been filled. The variations of the relative I_{VH} scattered intensity calculated for the particular cases where only radial or angular disorder is considered are shown in Figure 32 and compared with curves for a perfect spherulite. These results are in good agreement with the predictions of Stein and Chu's theory in that excess intensity at large angles results from radial disorder, whereas excess intensity at small angles results from angular disorder. The curves for the disordered spherulites are multiplied by normalizing factors to make the intensities approximately agree with those for the perfect spherulite at the position of the first-order maximum. Thus, the intensity at this position is appreciably reduced by disorder.

D. SCATTERING FROM DEFORMED SPHERULITES

The change in scattering patterns upon stretching was first observed by Stein and Rhodes.[38] Various models[42,43,65] were introduced for describing the ways in which the deformation of the spherulites induces modifications of the SALS patterns. These models are generally based on the affine transformation of the spherical or circular spherulites into ellipsoids. In

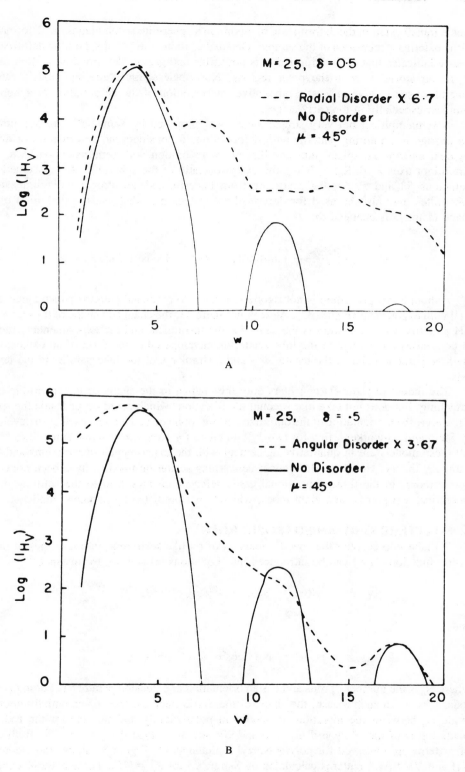

FIGURE 32. Variation of the I_{VH} scattered intensity at $\mu = 45°$ for a disordered two-dimensional spherulite. M is the number of lattice cells along the spherulite radius. $\delta = 0.5$ rad. (A) Radial disorder and (B) angular disorder. (From Yoon, D. Y. and Stein, R. S., *J. Polym. Sci., Polym. Phys. Ed.*, 12, 763, 1974. With permission.)

such a transformation the deformation of a point in a spherulite accompanies the deformation of the external dimensions of the sample. Detailed examination[92,93] of spherulite deformation clearly indicates that such deformation is not affine and greater deformation is often found in the equatorial than in the polar regions. Nevertheless, the scattering models can be generally used to give at least a qualitative interpretation of the deformation of spherulite samples, especially in the early stages.

A semiempirical but very simple theory was proposed by Samuels[65,66] to account for the change in scattering patterns with deformation. It considers the deformation of an individual anisotropic sphere into an ellipsoid of revolution with semimajor axis $\lambda_s R_0$ and semiminor axes $\lambda_s^{-1/2} R_0$, λ_s being the extension ratio of the spherulite. It is assumed that Equations 58 and 59 for the scattering from underformed anisotropic three-dimensional spherulites may also be used for deformed spherulites provided that the definition of the shape factor U is changed to:

$$U = \left(\frac{4\pi R_0 \lambda_s^{-1/2}}{\lambda}\right) \sin(\theta/2)[1 + (\lambda_s^3 - 1)\cos^2(\theta/2)\cos^2\mu]^{1/2} \qquad (79)$$

Although this procedure is not rigorous, it leads to predicted patterns which agree quite well with experiment (Figure 33). As seen in Figure 33, the change in shape of the cloverleaf VH patterns with elongation is the same for the theoretical and the experimental patterns. In both cases the change in the lobe scattering envelope takes the form of an extension of the lobe perpendicular to the tensile axis and a shrinkage of the lobe parallel to the tensile axis.

The model of Samuels considers only the change in the shape of the spherulite upon stretching, but does not take into account modifications which could occur inside the spherulite, e.g., the reorientation of the optic axes. More sophisticated models have been developed by Stein and co-workers, first for two-[42,79] and then for three-dimensional spherulites.[43,79,94] All these models are in qualitative agreement with the experimental observations and with Samuels' theory. However, differences concerning scattering maxima have been noticed.[42] Furthermore, in the three-dimensional case, deformation may lead to the splitting of the theoretical VH pattern into eight lobes, which is not predicted by Samuels' model.

E. SCATTERING BY ANISOTROPIC RODS

To take into account the specific features of rodlike scattering, various types of distribution functions $N(\alpha)$ can be introduced into Equations 60 and 61, for instance[47,95]

$$N(\alpha) = N_0(\epsilon^{-2}\cos^2\alpha + \epsilon^2\sin^2\alpha)^{-1/2} \qquad (80)$$

and

$$N'(\alpha) = N_0(\epsilon^2\cos^2\alpha + \epsilon^{-2}\sin^2\alpha)^{1/2} \qquad (81)$$

where N_0 is the number of rods and ϵ is an orientation or extension parameter. For unoriented rods, $\epsilon = 1$. In such a case, the shape of the SALS patterns depends essentially upon the angle ω_0 between the direction of maximum polarizability and the axis of the rod: VH scattering is of the ×-type if $\omega_0 = 0$ and 90° and of +-type if $\omega_0 = \pm 45°$. Both types of patterns are observed for poly(γ-benzyl-L-glutamate).[61] Figure 34 shows the theoretical VH and VV SALS patterns calculated by Samuels[95] for $\omega_0 = 45°$. These patterns compare well with the experimental patterns shown in Figure 20. For oriented rods, varying ϵ makes it possible to interpret the patterns obtained from stretched films with a rodlike morphology.[47,95]

Similar results are obtained from the model developed by Kawai and co-workers for a

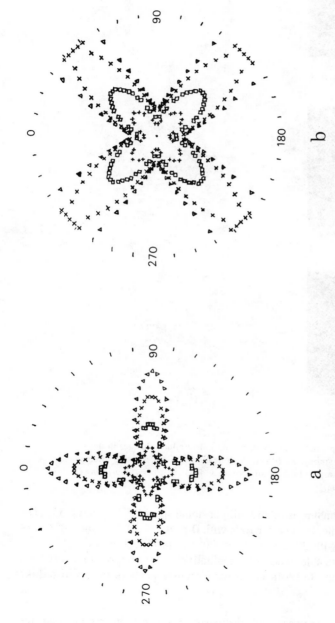

a

b

FIGURE 34. Computed SALS patterns for a random assembly of anisotropic rods in two dimensions. $\omega_0 = 45°$. The polarization direction is vertical; (a) VH and (b) VV. (From Samuels, R. J., *J. Polym. Sci., Part A2*, 7, 1197, 1969. With permission.)

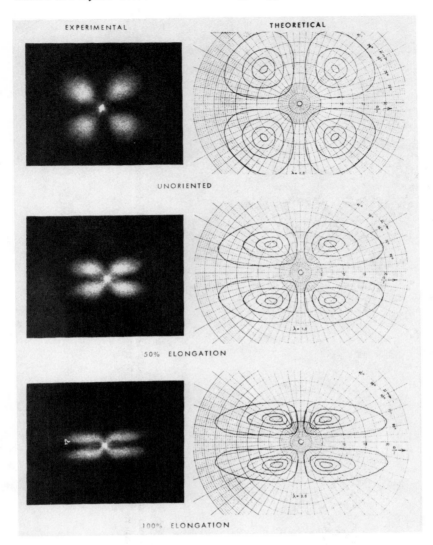

FIGURE 33.   Theoretically predicted and experimentally determined changes in the VH SALS pattern of isotactic polypropylene film with elongation. The polarization direction and the film-stretch direction are vertical. (From Samuels, R. J., *J. Polym. Sci., Part C*, 13, 37, 1966. With permission.)

random assembly of one-dimensional rods in three-dimensional space[48,55,61] (Figure 35). The VH patterns are shown to be of the $\times$-type for rods with $0 \leqslant \omega_0 \leqslant 30° \ 33'$ and $70° \ 07' \leqslant \omega_0 \leqslant 90°$ and $+$-type for those with $30° \ 33' < \omega_0 < 70° \ 07'$ (see Section VI).

Although the above models are able to describe qualitatively the two types of VH rodlike patterns found in experiments, they fail to account for various aspects of observed rodlike scattering[61] (Figure 36):

1.   The measured intensity may increase with decreasing θ much more rapidly than the theoretical intensity (Figure 36A).
2.   The μ-dependence of the observed scattering intensity may not always increase with θ, while the μ-dependence of the calculated intensity always increases with increasing θ (Figure 36A).

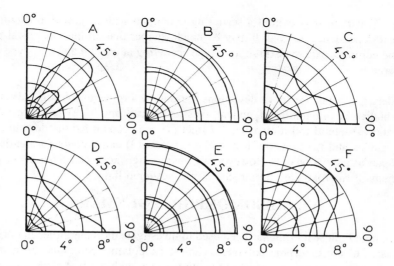

FIGURE 35.   Calculated VH patterns for a random assembly of rods with infinitesimally thin lateral dimensions in three-dimensional space as a function ω_0. (A) $\omega_0 = 0°$, (B) $\omega_0 = 30°$, (C) $\omega_0 = 50°$, (D) $\omega_0 = 55°$, (E) $\omega_0 = 70°$, and (F) $\omega_0 = 90°$. (From Hashimoto, T., Ebisu, S., and Kawai, H., *J. Polym. Sci., Polym. Phys. Ed.*, 19, 59, 1981. With permission.)

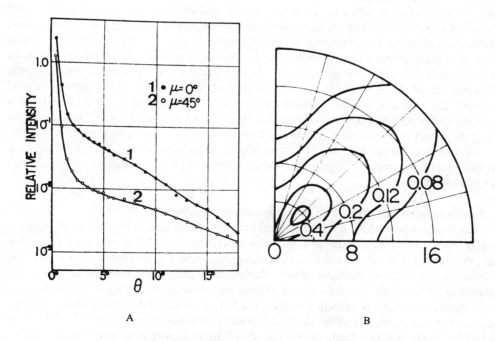

FIGURE 36.   Some experimental deviations from the theoretically predicted VH pattern. (A) VH scattering curves at $\mu = 0$ and 45° showing excess scattering at small angles and decreasing μ-dependence with increasing θ and (B) VH pattern exhibiting a scattering maximum. (From Hashimoto, T., Ebisu, S., and Kawai, H., *J. Polym. Sci., Polym. Phys. Ed.*, 19, 59, 1981. With permission.)

3.   The VH pattern may exhibit a scattering maximum with respect to θ (Figure 36B).
4.   The μ-dependence at small θ may be much sharper than in the theoretical pattern.
5.   Some complex patterns exhibit a + -type scattering at small θ and a × -type scattering at large θ.

In order to account for these deviations various factors were investigated: (1) internal inhomogeneities in orientation of optical axes and anisotropy,[49] (2) polydispersity in rod length,[49] and (3) lateral rod dimension.[58] Effect (1) can account for the deviations at large scattering angles and (2) turns out to be minor. Effect (3) can account for the deviation in μ-dependence as a function of θ. However, all these discrepancies can generally be accounted for by effects of interparticle interference on the scattered light.[61]

VI. MESOMORPHIC POLYMERS

Small-angle light scattering has been used to investigate the supermolecular organization of two classes of mesomorphic polymers: cholesteric polymer solutions and nematic thermotropic main-chain polymers. Most of the studies deal with the first class and concern the organization at rest and under electric, magnetic, and flow fields.

A. SUPERMOLECULAR STRUCTURES AT REST

Thermotropic nematic polymers and cholesteric polymer solutions are considered in this section. Cholesteric polymer solutions are an interesting case since they can have several textures which are not found with other polymers, and also because of their form optical rotation. Cholesterics are formed by a continuous twisting of the mean direction of orientation (or director) along an axis perpendicular to the director (Figure 1). This generates an optical rotation. A plane polarized light going through an optically active medium will see a rotation of its polarization plane by an angle Φ. Φ is a function of the intrinsic property and concentration of the optically active species and of the thickness of the sample. Chiral materials are optically active. Another way for having an optical rotation is to build a succession of planes of uniaxially oriented molecules, with a tilt angle between planes. This is a good model of a cholesteric order and it shows a strong form optical rotation.[96]

As said previously, VH and HV scattering are generally similar, except when some optical rotation is present. Cholesteric solutions are thus a good example which could show this phenomenon.

1. Rodlike Scattering from Polybenzylglutamate Solutions

The first polymer which has been studied in a mesomorphic state,[97] polybenzylglutamate (PBG) gives a cholesteric phase in solution in various solvents.[98] For increasing concentration, the solution undergoes a transition towards a biphasic region where anisotropic spherical domains are among an isotropic phase. Further increasing the concentration brings the solution in a fully cholesteric state with or without well-developed textures. When the solution is too viscous or when not enough time has been left, the solution does not present a specific texture. Several papers describe the SALS pattern obtained in this case.[54,99-104] Most of them[54,99-101] note that two kinds of HV pattern are formed, with a × shape or a + shape, both with the maximum intensity at the center of the pattern. All the authors deduced that this pattern is typical of rodlike scattering (Section V.E), and that the shape is a function of the nature of the solvent. A very detailed and interesting contribution has been done by Hashimoto and co-workers.[102-104] Using the same material, cholesteric PBG solutions, or films, they carefully observed not only the HV and the VV patterns, but also the VH and HH patterns. They also describe theoretically the scattering in terms of rods, and introduced the form optical rotation as Picot and Stein[105] did for spherulites.

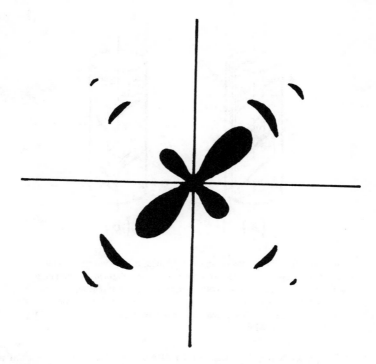

FIGURE 37.   Schematic HV pattern obtained with a PBG cholesteric solution.
Adapted from References 102 to 104.

Figure 37 is a schematic representation of the HV patterns obtained for PBG cholesteric solutions. This representation is a mixture of several experimental light scattering patterns published by Hashimoto and co-workers.[102-104] The scattering intensity of the central figure is very strong compared to the other scattering maxima. Two types ($\times$ or $+$) of HV scattering can be obtained with the same PBG cast from solutions with different solvents. Three questions have been addressed by Hashimoto and co-workers: (1) the origin of the $\times$ or $+$ HV scattering at very low angle, (2) the origin of the anomalous azimuthal dependence of the depolarized patterns, and (3) the presence of multiple-order scattering maxima.

a. $\times$- or +-Type HV Scattering

As said previously, the HV (or VH) central figure can be explained by the scattering of rodlike entities formed by the aggregation of PBG molecules. Several types of anisotropic rods can be imagined (Figure 38).

In the rod model A (Figure 38), the optical axis designated as a unit vector d is confined to a particular plane and makes a polar angle ω_0 with respect to the rod axis. The rod of model B (Figure 38) has a cylindrically symmetric optical axis orientation. Kawai et al.[48,55,61] have shown that the x- or +-type of small-angle HV scattering is a function of the angle ω_0, in three dimensions:

$\times$-type: model A $0° \leq \omega_0 \leq 30°33'$
                                $70°07' \leq \omega_0 \leq 90°$
            model B   all values of ω_0

$+$-type: only found for model A for $30°33' < \omega_0 < 70°07'$

As shown by Watanabe et al.[100] the $\times$-type is obtained for poor solvents while the $+$-

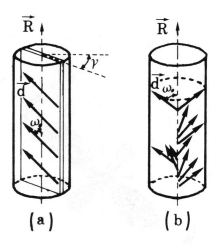

FIGURE 38. Two types of anisotropic rods: (a) in the rod of model A, the optic axis *d* is confined to a particular plane defined by the angle γ and makes a polar angle ω_0 with respect to the rod axis R and (b) in the rod model B, *d* is oriented in cylindrical symmetry with R. (From Hashimoto, T., Ijitsu, T., Yamaguchi, K., and Kawai, H., *Polym. J.*, 12, 745, 1980. With permission.)

type is obtained for good solvents. Hashimoto et al.[102] deduced from optical and electron microscopies that the ×-type obtained for solutions is due to model B with $\omega_0 = 90°$. The supermolecular entity is a rod composed by cholesteric planes, the cholesteric axis being the rod axis. Since it is not possible to obtain a +-type HV scattering with model B, this would mean that the supermolecular entity in this case is different from the one giving the ×-type HV pattern. As will be seen below, the strong optical rotation associated with the cholesteric order transforms the HV pattern in a pattern with no azimuthal dependence, and this makes the experiments not very easy. Using very thin samples (10 μm) in which the cholesteric character is expected to be very weak, clear ×- and +-types are seen depending on the solvent.[100] This does not seem to fit with the model implying the cholesteric rods as suggested by Hashimoto et al.[102] The supermolecular structure is formed by the aggregation of PBG molecules, the size of the aggregates growing with concentration as seen by fluorescence depolarization,[106] and electric birefringence.[107,108] In the cholesteric state, there is no clear picture of the association mode(s) of the PBG molecules; Tsuji and Watanabe[108] found that these aggregates are very large (up to 16 μm large) with a large distribution of size. Such large aggregate length can probably be assimilated to some persistence length of the director orientation. If the origin of the rodlike scattering in the isotropic PBG solution can be due to aggregated molecules at equilibrium in a nonaggregated solution, the scattering in the cholesteric state remains without precise explanation. Moreover, as will be shown below, continuous textures (not involving entities with defined boundaries as the sphere or rod models) can give scattering patterns similar to the ones of spheres or rods. Recently, Thomas et al.[109] have shown that domains with defined boundaries may not exist in nematic polymers.

b. Effect of the Form-Optical Rotation on the Rodlike Scattering

In the following, we will suppose that the ×- or +-HV pattern is due to the scattering of rodlike entities. Hashimoto et al.,[102] using the analysis of Picot and Stein,[105] calculated the effect of an optical rotation on the rodlike scattering. The medium is supposed to have an optical rotation K given by

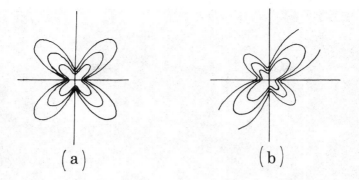

FIGURE 39.   Effect of the form-optical rotation on ×-type HV patterns
for $\omega_0 = 90°$; (a) Kd = 0° and (b) Kd = 30°. (From Hashimoto, T.,
Ijitsu, K., Yamaguchi, K., and Kawai, H., *Polym. J.*, 12, 745, 1980.
With permission.)

$$K = -\pi(\Delta n)^2 P/4\lambda_0^2 \tag{82}$$

where Δn is the birefringence of a layer or of the untwisted material and P is the cholesteric
pitch. The rods are assumed to be randomly oriented in an optically active medium. This
last hypothesis, valid in the original work of Picot and Stein for semicrystalline polymer
spherulites embedded in a cholesteric medium, is more doubtful in the PBG case. The results
are given in Figure 39, which shows the effect of the optical rotation Kd, with d being the
thickness of the optically active medium in which the rods are distributed. The main result
is to clearly show the μ distortion of the scattering pattern. Such a distortion is experimentally
found for PBG solutions and films.

c. Presence of Multiple-Order Maxima

VH and HV patterns have up to four scattering maxima, each one reflecting a single
spacing P, the cholesteric pitch. The Bragg law gives $P \sin(\theta/2) = n\lambda$, with n being an
integer, and θ the scattering angle for the maximum intensity of the order n. Hashimoto et
al.[104] calculated the scattering for the cholesteric rods. For a single cylindrical cholesteric
domain of radius R and length L, with $R \gg \lambda$, and a homogeneous twist, it comes:

$$I_{HV} = K\sin^2 2\mu(I^a_{HV} + I^b_{HV}) \tag{83}$$

$$I^a_{HV} = j_0^2(qL/2) \tag{84}$$

$$I^b_{HV} = \frac{1}{4}\left\{j_0^2\left(\left[q + \frac{4\pi}{P}\right]\frac{L}{2}\right) + j_0^2\left(\left[q - \frac{4\pi}{P}\right]\frac{L}{2}\right)\right\} \tag{85}$$

K is a constant, $q = (4\pi/\lambda)\sin(\theta/2)$, j_0 is the 0th order spherical Bessel function of
the first kind. I^a_{HV} is the small-angle diffuse scattering due to the cholesteric domain as a
whole. I^b_{HV} is the "cholesteric" scattering with the first order at $q = (4\pi/P)$. The line
profile is a function of the domain size L, with a SD $\sigma = \sqrt{6/L}$.

If a random arrangement of such rods exists, only one broad first order diffraction
maximum exists. This is not what is experimentally observed. The authors explain their
results by an inhomogeneous twisting of the PBG cholesteric solutions. If ω_{12} is defined as
the relative twist angle of two directors separated by a distance z, the spatial correlation
function of the twisting g(z) is

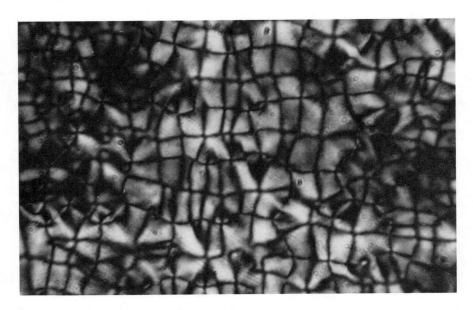

FIGURE 40.   Optical micrograph between crossed polarizers of the polygonal focal conic texture of a cholesteric hydroxypropylcellulose solution. The mean spacing between two black lines is 27 μm.

$$g(z) = <\cos 2\omega_{12}>_z \qquad (86)$$

If the twisting is homogeneous,

$$g(z) = \cos\left(\frac{4\pi z}{P}\right) \qquad (87)$$

If the twisting is inhomogeneous, g(z) is expressed by a Fourier expansion in which the coefficients of the higher-order harmonics are associated with the nonlinearity of the twisting. This generates high-order scattering maxima which match very well the experiments when considering the coefficients of the Fourier expansion of g(z) as adjustable parameters. This is similar to the scattering of ringed spherulites which gives higher angle maxima (Figure 25).

2. Small-Angle Light Scattering from the Cholesteric Polygonal Focal Conic Texture

The cholesteric polygonal focal conic texture is known to occur for cholesteric polymer solutions.[110] It appears between crossed polars as two conjugated squared networks of black lines, attached to the upper and to the lower glass plate, respectively, as shown in Figure 40. Each crossing of the lines corresponds to a dome or a basin formed by successive cholesteric layers. A complete description of the structure has been given by Bouligand.[111] What is interesting in this texture is that there is only a small amount of molecular orientation discontinuity. Most of the structure is continuous. The locus of the dark lines is not the boundary of some domain. This texture has been several times confused with spherulites.[101,112] A recent study[113] has shown that such a texture gives HV and VV patterns typical of anisotropic spheres, despite there is no such entities in the studied solutions. Figure 41 gives an example of the HV pattern obtained for the texture of Figure 40. Preparing different samples with various distances L between domes, and using L as the relevant parameter related to the angular position θ_m of the HV scattering maximum at $\mu \pm 45°$, it comes:

FIGURE 41.   HV pattern obtained from the texture of Figure 40.

$$U = (2\pi L/\lambda)\sin\left(\frac{\theta_m}{2}\right) \tag{88}$$

with U being nearly a constant equal to 2.7. No theoretical description has been given for SALS of this particular texture. These results show that it is very easy to misinterpret the fundamental supermolecular structure responsible for a given SALS pattern when no information is known about this structure. To obtain SALS patterns similar to those of spheres or spherulites is not proof that such entities are present. The above experiments are a good illustration of this. However, when the structure is known, elastic light scattering is a powerful tool which can give valuable information, difficult to obtain with other techniques. The finding of the inhomogeneous twisting of cholesteric PBG solutions is such an example.

3. Small-Angle Light Scattering from Thermotropic Nematic Polymers

Only a preliminary study has been performed by Rojstaczer and Stein.[114] They recorded HV and VV SALS patterns when cooling a thermotropic polymer from the liquid to the nematic state.

In the nematic state, both HV and VV patterns show an azimuthal dependence of the intensity. The HV pattern shows intensity maxima at $\mu = 0$, 90, 180, and 270° while in the VV pattern, maxima are found at $\mu = 45$, 135, 225, and 315°. The origin of these patterns is thought to be the defect structure of the nematic state. One good indication along this line is the decrease of the HV pattern with time, showing that some orientation correlation time is increasing. No precise model has been given by the authors.

B. CHOLESTERIC POLYMER SOLUTIONS UNDER VARIOUS FIELDS (ELECTRIC, MAGNETIC, FLOW)

Several studies have been performed to investigate the structure of PBG cholesteric solutions submitted to electric or magnetic fields.[115-117] As said above, there is no precise description of the structure at rest. This explains that the modifications of this structure under an electric or a magnetic field cannot be clearly interpreted. The major results are that a magnetic or an electric field increases the rodlike character of the scattering, the rods being aligned parallel to the field. Relaxation is peculiar, giving stripes as seen in a polarizing microscope. The same features apply to a flow field. During the flow, a rodlike character appears[118] while the relaxation is complex, several scattering patterns appearing with time.[119]

More work is thus needed to understand the structure, at rest and under various conditions, of mesomorphic polymers. SALS is one of the major tools to reach this goal.

VII. CONCLUSION

SALS is an important tool for studying the supermolecular organization of polymers. The numerous examples given in this chapter are a good illustration. Nevertheless, the technique has some drawbacks. One is that the interpretation of experiments is difficult since it has to be referred to a model based on the geometrical description of the distribution of the scattering elements. An *a priori* knowledge of this distribution is thus necessary. Moreover, as has been shown, several different distributions can give the same SALS patterns, and the interpretation of these SALS patterns is not always unambiguous. It is thus very important to couple SALS with other techniques to be able to draw conclusions about the morphology of polymer supermolecular structures.

ACKNOWLEDGMENTS

We thank G. H. Meeten, B. Ernst, and E. Peuvrel for their very valuable help during the preparation of this chapter.

REFERENCES

1. **Billmeyer, F. W., Jr.,** *Textbook of Polymer Chemistry,* 2nd ed., Wiley-Interscience, New York, 1963.
2. **Flory, P. J.,** *Principles of Polymer Chemistry,* Cornell University Press, Ithaca, NY, 1969.
3. **Flory, P. J.,** *Statistical Mechanics of Chain Molecules,* Wiley-Interscience, New York, 1969.
4. **De Gennes, P. G.,** *Scaling Concepts in Polymer Science,* Cornell University Press, Ithaca, NY, 1979.
5. **Des Cloizeaux, J. and Jannink, G.,** *Les Polymères en Solution: Leur Modélisation et leur Structure,* Editions de Physique, Les Ulis, France, 1987.
6. **Mark, J. E., Eisenberg, A., Graessley, W. W., Mandelkern, L., and Koenig, J. L.,** *Physical Properties of Polymers,* American Chemical Society, Washington, D.C., 1984.
7. **Doi, M. and Edward, S.,** *Theory of Polymer Dynamics,* Clarendon Press, Oxford, 1986.
8. **Ward, I. M.,** *Developments in Oriented Polymers-2,* Elsevier, London, 1987.
9. **Zachariades, A. E. and Porter, R. S.,** *The Strength and Stiffness of Polymers,* Marcel Dekker, New York, 1983.
10. **Chu, B.,** Light scattering studies of polymer solutions and melts, *Polym. J.,* 17, 225, 1985.
11. **Viovy, J. L., Monnerie, L., Veissier, V., and Fofana, M.,** Studies of the depolarization of polymer luminescence: lasers or not lasers?, in *Applications of Lasers in Polymer Science and Technology,* Fouassier, J. P. and Rabek, J. F., Eds., CRC Press, Boca Raton, FL, 1989, chap. 7.
12. **Tsvetkov, V. N.,** Flow birefringence, in *Newer Methods of Polymer Characterization,* Bacon, K., Ed., Interscience, New York, 1964.
13. **Champion, J. V.,** Flow birefringence and the Kerr effect, in *Developments in Polymer Characterisation, Part II,* Dawkins, J. V., Ed., Applied Science Publishers, Essex, England, 1978, 207.
14. **Lenz, R. W.,** Synthesis and properties of thermotropic liquid crystal polymers with main chain mesogenic units, *Polym. J.,* 17, 105, 1985.
15. **Flory, P. J.,** Molecular theory of liquid crystals, *Adv. Polym. Sci.,* 59, 1, 1984.
16. **Blumstein, A.,** *Liquid Crystalline Order in Polymers,* Academic Press, New York, 1978.

17. **Plate, N. A., Freidzon, Ya. S., and Shibaev, V. P.,** Cholesteric and other phases in thermotropic liquid crystalline polymers with side chain mesogenic groups, *Pure Appl. Chem.,* 11, 1715, 1985.
18. **Finkelmann, H. and Rehage, G.,** Liquid crystal side chain polymers, *Adv. Polym. Sci.,* 60/61, 99, 1984.
19. **Chandrasekhar, S.,** *Liquid Crystals,* Cambridge University Press, Cambridge, England, 1983.
20. **Wunderlich, B.,** *Macromolecular Physics,* Vols. 1 to 3, Academic Press, New York, 1980.
21. **Lemstra, P. J., Kirschbaum, R., Ohta, T., and Yasuda, H.,** High-strength/high modulus structures based on flexible macromolecules: gel spinning and related processes, in *Developments in Oriented Polymer-2,* Ward, I. M., Ed., Elsevier, New York, 1987, 39.
22. **Geil, P. H.,** *Polymer Single Crystals,* Wiley-Interscience, New York, 1963.
23. **Born, M. and Wolf, E.,** *Principles of Optics,* Pergamon Press, Oxford, 1983.
24. **Van De Hulst, H. C.,** *Light Scattering by Small Particles,* John Wiley & Sons, New York, 1957.
25. **Chu, B.,** *Laser Light Scattering,* Academic Press, New York, 1974.
26. **Haudin, J. M.,** Optical studies of polymer morphology, in *Optical Properties of Polymers,* Meeten, G. H., Ed., Elsevier, London, 1986, chap. 4.
27. **Meeten, G. H.,** Small-angle light scattering by spherulites in the anomalous diffraction approximation, *Opt. Acta,* 29, 759, 1982.
28. **Feynman, R. P., Leighton, R. B., and Sands, M.,** *The Feynman Lectures on Physics,* Vol. 2, Addison-Wesley, London, 1970.
29. **Mie, G.,** *Ann. Phys.,* 25, 377, 1908.
30. **Wickramasinghe, N. C.,** *Light Scattering Functions of Small Particles,* John Wiley & Sons, New York, 1973.
31. **Meeten, G. H. and Navard, P.,** Small angle polarized light scattering from amorphous spheres, *J. Polym. Sci., Polym. Phys. Ed.,* 22, 2159, 1984.
32. **Meeten, G. H. and Navard, P.,** Small-angle light scattering for spheres, I Theory, *J. Polym. Sci., Polym. Phys. Ed.,* in press.
33. **Meeten, G. H. and Navard, P.,** Small-angle light scattering for spheres. II. Experiments with isotropic spheres, *J. Polym. Sci., Polym. Phys. Ed.,* in press.
34. **Chylek, P.,** Partial-wave resonances and the ripple structure in the Mie normalized extinction cross section, *J. Opt. Soc. Am.,* 66, 285, 1976.
35. **Chylek, P., Kiehl, J. T., and Ko, M. K. W.,** The fine structure of Mie scattering, *J. Colloid Interface Sci.,* 64, 595, 1978.
36. **Meeten, G. H. and Navard, P.,** unpublished results.
37. **Stein, R. S.,** Small angle light scattering from the polymeric solid state, in *Static and Dynamic Properties of the Polymeric Solid State,* Pethrick, R. A. and Richards, R. W., Eds., D. Reidel, Boston, 1982, 109.
38. **Stein, R. S. and Rhodes, M. B.,** Photographic light scattering by polyethylene films, *J. Appl. Phys.,* 31, 1873, 1960.
39. **Champion, J. V., Killey, A., and Meeten, G. H.,** Small-angle light scattering by spherulites, *J. Polym. Sci., Polym. Phys. Ed.,* 23, 1467, 1985.
40. **Samuels, R. J.,** Small-angle light scattering from optically anisotropic spheres and disks. Theory and experimental verification, *J. Polym. Sci., Part A2,* 9, 2165, 1971.
41. **Clough, S. B., Stein, R. S., and Picot, C.,** Low-angle light scattering equations for polymer spherulites, *J. Polym. Sci., Part A2,* 9, 1147, 1971.
42. **Clough, S., Van Aartsen, J. J., and Stein, R. S.,** Scattering of light by two-dimensional spherulites, *J. Appl. Phys.,* 36, 3072, 1965.
43. **Van Aartsen, J. J.,** The scattering of light by deformed three-dimensional spherulites, Office of Naval Research Technical Report No. 83, Project Nr 056-378, Contract No. 3357(01), University of Massachusetts, Amherst, March 1966.
44. **Keijzers, A. E. M.,** Light Scattering by Crystalline Polystyrene and Polypropylene, Ph.D. thesis, Technische Hogeschool, Delft, Netherlands, 1967.
45. **Samuels, R. J.,** Small-angle light scattering and crystallization processes in solid polymer films, *J. Polym. Sci., Polym. Phys. Ed.,* 12, 1417, 1974.
46. **Yoon, D. Y. and Stein, R. S.,** Effect of interference between anisotropic scattering entities on light scattering from polymer films. II. The correlation function approach, *J. Polym. Sci., Polym. Phys. Ed.,* 12, 735, 1974.
47. **Rhodes, M. B. and Stein, R. S.,** Scattering of light from assemblies of oriented rods, *J. Polym. Sci., Part A2,* 7, 1539, 1969.
48. **Murakami, Y., Hayashi, N., Hashimoto, T., and Kawai, H.,** Light-scattering patterns from random assembly of anisotropic rod-like structures with their principal optic axes oriented in cylindrical symmetry with respect to their own rod axes, *Polym. J.,* 4, 452, 1973.
49. **Hashimoto, T., Murakami, Y., Hayashi, N., and Kawai, H.,** Light scattering from polymer having an optically anisotropic rod-like texture in relation to a model of a random assembly of disordered rods, *Polym. J.,* 6, 132, 1974.

50. **Hashimoto, T., Murakami, Y., Okamori, Y., and Kawai, H.,** Origin of a complex light scattering pattern from some crystalline polymer films, *Polym. J.,* 6, 554, 1974.
51. **Hashimoto, T., Murakami, Y., and Kawai, H.,** Light scattering from polymer films having optically anisotropic rodlike texture. A quantitative test of theories, *J. Polym. Sci., Polym. Phys. Ed.,* 13, 1613, 1975.
52. **Adams, G. C. and Stein, R. S.,** Some studies of the crystallization of polychlorotrifluoroethylene copolymer films, *J. Polym. Sci., Part A2,* 6, 31, 1968.
53. **Hashimoto, T., Kawasaki, H., and Kawai, H.,** Crystal and amorphous orientation behavior of poly(chlorotrifluoroethylene) films in relation to crystalline superstructure, *J. Polym. Sci., Polym. Phys. Ed.,* 16, 271, 1978.
54. **Wilkes, G. L.,** Superstructures in polypeptide films as noted by small-angle light scattering, *Mol. Cryst. Liq. Cryst.,* 18, 165, 1972.
55. **Moritani, M., Hayashi, N., Utsuo, A., and Kawai, H.,** Light scattering patterns from collagen films in relation to the texture of a random assembly of anisotropic rods in three dimensions, *Polym. J.,* 2, 74, 1971.
56. **Van Aartsen, J. J.,** Light scattering from a random collection of cylinder symmetrical particles, *Eur. Polym. J.,* 6, 1095, 1970.
57. **Hayashi, N. and Kawai, H.,** Light-scattering patterns from random assembly of anisotropic rods of finite dimensions, *Polym. J.,* 3, 140, 1972.
58. **Matsuo, M., Nomura, S., Hashimoto, T., and Kawai, H.,** Light scattering from a random assembly of anisotropic plates in two- and three-dimensional space, *Polym. J.,* 6, 151, 1974.
59. **Stein, R. S. and Picot, C.,** The effect of interference between anisotropic scattering entities on the light scattering from polymer films. I. Case of no impingement, *J. Polym. Sci., Part A2,* 8, 1955, 1970.
60. **Prud'homme, R. E. and Stein, R. S.,** Light scattering from assemblies of spherulites, *J. Polym. Sci., Polym. Phys. Ed.,* 11, 1357, 1973.
61. **Hashimoto, T., Ebisu, S., and Kawai, H.,** Light scattering from polymer having optically anisotropic rodlike texture. III. Inter-rod interference, *J. Polym. Sci., Polym. Phys. Ed.,* 19, 59, 1981.
62. **Hashimoto, T., Todo, A., and Kawai, H.,** Light scattering from crystalline superstructure in tubular extruded poly(1-butene) films. II. Analysis of interparticle effect based upon a paracrystal model, *Polym. J.,* 10, 521, 1978.
63. **Plaza, A. and Stein, R. S.,** The scattering of light from polyethylene films at low angles, *J. Polym. Sci.,* 40, 267, 1959.
64. **Rhodes, M. B., Keedy, D. A., and Stein, R. S.,** The use of a laser as a light source for photographic light scattering from polymer films, *J. Polym. Sci.,* 62, S73, 1962.
65. **Samuels, R. J.,** Small angle light scattering from deformed spherulites. Theory and its experimental verification, *J. Polym. Sci., Part C,* 13, 37, 1966.
66. **Samuels, R. J.,** *Structured Polymer Properties,* Wiley-Interscience, New York, 1974.
67. **Stein, R. S. and Stidham, S. N.,** Effect of birefringence on light scattering from oriented systems, *J. Polym. Sci., Part A2,* 4, 89, 1966.
68. **Stein, R. S. and Chu, W.,** Effect of birefringence on low-angle light scattering patterns from oriented polymer films. II, *J. Polym. Sci., Part A2,* 8, 489, 1970.
69. **Stein, R. S. and Keane, J. J.,** The scattering of light from thin polymer films. I. Experimental procedure, *J. Polym. Sci.,* 17, 21, 1955.
70. **Keijzers, A. E. M., Van Aartsen, J. J., and Prins, W.,** Light scattering by shock-cooled isotactic polypropylene films, *J. Appl. Phys.,* 36, 2874, 1965.
71. **Van Antwerpen, F. and Van Krevelen, D. W.,** Light scattering method for investigation of the kinetics of crystallization of spherulites, *J. Polym. Sci., Polym. Phys. Ed.,* 10, 2409, 1972.
72. **Pakula, T. and Soukup, Z.,** New apparatus for studies on supermolecular structures of thin polymer films by small-angle light scattering, *J. Polym. Sci., Polym. Phys. Ed.,* 12, 2437, 1974.
73. **Chu, W. H. and Horne, D. E.,** Automated light scattering apparatus for studying polymer films, *J. Polym. Sci., Polym. Phys. Ed.,* 15, 303, 1977.
74. **Wasiak, A., Pfeifer, D., and Stein, R. S.,** Application of an optical multichannel analyzer to small-angle light-scattering studies, *J. Polym. Sci., Polym. Lett. Ed.,* 14, 381, 1976.
75. **Russel, T. P., Koberstein, J., Prud'homme, R., Misra, A., Stein, R. S., Parsons, J. W., and Rowell, R. L.,** A calibration procedure for a low-angle light-scattering apparatus, *J. Polym. Sci., Polym. Phys. Ed.,* 16, 1879, 1978.
76. **Tabar, R. J., Stein, R. S., and Long, M. B.,** A two-dimensional position-sensitive detector for small-angle light-scattering, *J. Polym. Sci., Polym. Phys. Ed.,* 20, 2041, 1982.
77. **Stein, R. S.,** Optical studies of the morphology of polymer films, in *Structure and Properties of Polymer Films,* Lenz, R. W. and Stein, R. S., Eds., Plenum Press, New York, 1973, 1.

78. **Stein, R. S. and Wilson, P. R.,** The scattering of light by anisotropic spheres and discs, Office of Naval Research Technical Report No. 35, Project NR 356-378, Contract No. 33 57 (00), University of Massachusetts, Amherst, August 1961.

79. **Stein, R. S., Erhardt, P., Van Aartsen, J. J., Clough, S., and Rhodes, M.,** Theory of light scattering from oriented and fiber structures, *J. Polym. Sci., Part C*, 13, 1, 1966.

80. **Hashimoto, T., Prud'homme, R. E., and Stein, R. S.,** Dynamic light scattering of polyethylene. II. Results and interpretation, *J. Polym. Sci., Polym. Phys. Ed.*, 11, 709, 1973.

81. **Motegi, M., Oda, T., Moritani, M., and Kawai, H.,** Light-scattering patterns from polyethylene films in relation to spherulitic crystalline texture, *Polym. J.*, 1, 209, 1970.

82. **Stein, R. S. and Picot, C.,** Small-angle light scattering by random assemblies of incomplete spherulites. II. Truncated spherulites, *J. Polym. Sci., Part A2*, 8, 2127, 1970.

83. **Prud'homme, R. E. and Stein, R. S.,** Scattering from truncated spherulites, *J. Polym. Sci., Polym. Phys. Ed.*, 11, 1683, 1973.

84. **Tabar, R. J., Wasiak, A., Hong, S. D., Yusa, T., and Stein, R. S.,** Small-angle light scattering by random assemblies of incomplete spherulites. System with partial degree of volume filling, *J. Polym. Sci., Polym. Phys. Ed.*, 19, 49, 1981.

85. **Tabar, R. J., Stein, R. S., and Rose, D. E.,** The effect of spherulitic truncation on small-angle light scattering, *J. Polym. Sci., Polym. Phys. Ed.*, 23, 2059, 1985.

86. **Picot, C., Stein, R. S., Motegi, M., and Kawai, H.,** Small-angle light scattering by random assemblies of incomplete spherulites. I. Sheaflike textures, *J. Polym. Sci., Part A2*, 8, 2115, 1970.

87. **Misra, A. and Stein, R. S.,** Light scattering of the early stages of the crystallization of poly(ethylene terephthalate), *J. Polym. Sci., Part B*, 10, 473, 1972.

88. **Hashimoto, T., Todo, A., Murakami, Y., and Kawai, H.,** Light scattering from crystalline superstructure in tubular-extruded polybutene-1 films, *J. Polym. Sci., Polym. Phys. Ed.*, 15, 501, 1977.

89. **Prud'homme, R. E., Yoon, D., and Stein, R. S.,** Scattering of light from spherulitic polymers: effect of internal structure, *J. Polym. Sci., Polym. Phys. Ed.*, 11, 1047, 1973.

90. **Stein, R. S. and Chu, W.,** Scattering of light by disordered spherulites, *J. Polym. Sci., Part A2*, 8, 1137, 1970.

91. **Yoon, D. Y. and Stein, R. S.,** A lattice theory of light scattering from disordered spherulites, *J. Polym. Sci., Polym. Phys. Ed.*, 12, 763, 1974.

92. **Yang, R. and Stein, R. S.,** Deformation of polybutene-1 spherulites, *J. Polym. Sci., Part A2*, 5, 939, 1967.

93. **Weynant, E., Haudin, J. M., and G'Sell, C.,** In-situ observation of the spherulite deformation in polybutene-1 (Modification I), *J. Mater. Sci.*, 15, 2677, 1980.

94. **Van Aartsen, J. J. and Stein, R. S.,** Scattering of light by deformed three-dimensional spherulites, *J. Polym. Sci., Part A2*, 9, 295, 1971.

95. **Samuels, R. J.,** Solid-state characterization of the structure and deformation behavior of water-soluble hydroxypropylcellulose, *J. Polym. Sci., Part A2*, 7, 1197, 1969.

96. **De Vries, H.,** *Acta Crystallogr.*, 4, 219, 1951.

97. **Robinson, C. and Ward, J. C.,** Liquid crystalline structures in polypeptides, *Nature*, 4596, 1183, 1957.

98. **Uematsu, I. and Uematsu, Y.,** Polypeptide liquid crystals, *Adv. Polym. Sci.*, 59, 37, 1984.

99. **Frenkel, S.Ya., Baranov, V. G., and Volkov, T. I.,** A study of the formation of supermolecular order in polymeric solutions by means of low-angle scattering of polarized light, *J. Polym. Sci., Part C*, 16, 1655, 1967.

100. **Watanabe, J., Imai, K., and Uematsu, I.,** Light scattering studies of poly(γ-benzyl L-glutamate) solutions and films, *Polym. Bull.*, 1, 67, 1978.

101. **Tanaka, T., Mori, T., Tsutsui, T., Ohno, S., and Tanaka, R.,** Poly(L-glutamic acid)-poly(oxyethylene glycol) systems: perpetuated cholesteric liquid crystal structures in noncrosslinked and crosslinked solid films, *J. Macromol. Sci. Phys.*, B17(4), 723, 1980.

102. **Hashimoto, T., Ijitsu, T., Yamaguchi, K., and Kawai, H.,** Supramolecular structure of polypeptides in concentrated solutions and films. I. Origin of rod-like scattering and effect of form-optical rotation, *Polym. J.*, 12, 745, 1980.

103. **Hashimoto, T., Ebisu, S., Inaba, N., and Kawai, H.,** Supramolecular structure of polypeptides in concentrated solutions and films. II. Small-angle light scattering from cholesteric mesophase, *Polym. J.*, 13, 701, 1981.

104. **Hashimoto, T., Inaba, N., Ebisu, S., and Kawai, H.,** Supramolecular structure of polypeptides in concentrated solutions and films. III. Inhomogeneous twisting of cholesteric mesophase as clarified from multiple-order light-scattering maxima, *Polym. J.*, 13, 897, 1981.

105. **Picot, C. and Stein, R. S.,** Effect of optical rotation on low-angle light scattering patterns, *J. Polym. Sci., Part A2*, 8, 1491, 1970.

106. **Ushiki, H., Tanaka, F., and Mita, I.,** Study of the fluorescence depolarization of a chromophore attached to polymer chain end, *Eur. Polym. J.*, 22, 827, 1986.

107. **Pyzuk, W. and Krupkowski, T.,** High electric birefringence and association modes of poly-γ-benzyl-L-glutamate solutions, *Makromol. Chem.,* 178, 817, 1977.

108. **Tsuji, K. and Watanabe, H.,** Molecular association of polypeptides in concentrated solutions investigated by the electrooptical method, *J. Colloid Interface Sci.,* 62, 101, 1977.

109. **Thomas, E. L., Hudson, S. D., and Lenz, R. W.,** Imaging of textures and effects of thermotropic liquid crystalline polyesters by electron microscopy, paper presented at the Int. Conf. on Liquid Crystal Polymers, Bordeaux, France, July 20 to 24, 1987.

110. **Livolant, F.,** Cholesteric liquid crystalline phases given by three helical biological polymers: DNA, PBLG and xanthan. A comparative analysis of their textures, *J. Phys.,* 47, 1605, 1986.

111. **Bouligand, Y.,** Recherches sur les textures des états mésomorphes. II. Les champs polygonaux dans les cholestériques, *J. Phys.,* 33, 715, 1972.

112. **Onogi, Y., White, J. L., and Fellers, J. F.,** Structural investigations of polymer liquid crystalline solutions: aromatic polyamides, hydroxypropylcellulose and poly(γ-benzyl-L-glutamate), *J. Polym. Sci., Polym. Phys. Ed.,* 18, 663, 1980.

113. **Meeten, G. H. and Navard, P.,** Cholesteric hydroxypropylcellulose solutions: microscopy and small-angle light scattering, *J. Polym. Sci., Polym. Phys. Ed.,* 26, 413, 1988.

114. **Rojstaczer, S. and Stein, R. S.,** Domain growth in thermotropic liquid crystalline polymers by small angle light scattering, paper presented at the ACS Meeting, New Orleans, August to September 1987.

115. **Matsuo, M., Kakei, K., Nagaoka, Y., Ozaki, F., Murai, M., and Ogita, T.,** A light scattering study of orientation of liquid-crystalline rodlike textures of poly(γ-benzyl-L-glutamate) in an electric field by saturation electric birefringence method, *J. Chem. Phys.,* 75, 5911, 1981.

116. **Iizuka, E.,** Properties of liquid crystals of polypeptides, *Adv. Polym. Sci.,* 20, 80, 1976.

117. **Iizuka, E., Keira, T., and Wada, A.,** Light scattering by liquid crystals of poly-γ-benzylglutamate in electric field, *Mol. Cryst. Liq. Cryst.,* 23, 13, 1973.

118. **Hashimoto, T., Takebe, T., and Suehiro, S.,** Apparatus to measure small angle light scattering profiles of polymers under shear flow, *Polym. J.,* 18, 123, 1986.

119. **Navard, P.,** Formation of band textures in hydroxypropylcellulose liquid crystals, *J. Polym. Sci., Polym. Phys. Ed.,* 24, 435, 1986.

Chapter 3

APPLICATION OF LASERS IN THE SCATTERING FOR THE STUDY OF SOLID POLYMERS

Marian Kryszewski

TABLE OF CONTENTS

I. INTRODUCTION

It is a well-established fact that the properties of polymers depend on the chemical constitution of macromolecules and on the physical structure of the system. It follows that the properties can be modified by changing the molecular structure, e.g., by choosing appropriate monomers and their copolymerization in specific conditions as well as by changing the supermolecular structure of the system, i.e., the spatial packing of macromolecules and phases arising in the material as a result of different physical processes that take place during processing. Another possibility which has received increased attention is the modification of the properties by using mixtures of polymers of different chemical structure. In this way one can obtain not only multicomponent systems but also multiphase ones, when one of the components can undergo crystallization.

Modification of the physical properties of polymers is connected with the possibility of obtaining systems of definite orientation of molecules and crystallinity, as well as with definite geometrical arrangement of the ordered phases. This makes modification of properties relatively simple in the case of polymers which easily undergo crystallization or form anisotropic structures already during synthesis, e.g., polymeric liquid crystals. At the opposite pole from crystalline materials are amorphous systems but their orientation, e.g., due to external forces, may also change the mechanical properties of such systems.

The basic structural units in crystalline polymers are more or less perfect crystalline lamellae which may exhibit distinct correlation of orientation and arrangement forming higher-order structures, e.g., spherulites, where the lamellae are radially arranged. The knowledge of the overall crystallinity degree is not sufficient for description of the properties of such complex systems. It is indispensable to have information about the shape, size, and perfection of these basic crystalline entities as well as some information concerning the correlation of orientation and arrangement in spherulites, and finally, data are needed about size and size distribution of such structures. All these different structures are influencing the mechanical properties of polymers as well as the permeability to gases and vapors, optical properties, e.g., clarity, etc.

The above remarks show that many studies, which cannot be cited here devoted to the establishment of the relations between properties and structure are connected with determination of nonhomogeneities that scatter electromagnetic radiation.

X-ray scattering is used to obtain basic information about the size and shape of molecules and molecular assemblies. It is due to the fact that the spatial distribution of the scattering pattern is related to the size of the scattering object; thus, X-rays scattered or diffracted at high angles give information about the structure of molecules or crystals at resolutions where it is possible to distinguish individual atoms. Small-angle X-ray scattering (SAXS) provides information on the overall shape and size of the scattering object and on its packing within other objects, density, etc.

The information obtainable from SAXS from disordered systems is often limited and the interpretation ambiguous. When SAXS are used in conjunction with other methods, they can be very powerful first of all in the case when sufficient information is available to provide the starting point for the analysis. It is possible to interpret SAXS data for the systems in a variety of states and to follow dynamically the changes in structure as a function of temperature, stress, etc. To follow rapid structural changes in a scattering pattern the data needs to be collected quickly. This requires a high photon flux on the samples. Thus, besides the bright X-ray sources the synchrotron radiation (SR) source is extremely useful for these experiments.

SAXS is extensively used to study the size and shape of crystalline elements in semi-crystalline polymers as well as to solutions to measure molecular dimensions and to amorphous bulk polymers where electron density fluctuations exist due to phase separation, e.g., in polymer blends, block, or graft polymers.

Quite novel results have already been obtained by time-resolved SAXS studies of structure development during the early stages of isothermal crystallization as well as of crystal annealing above the crystallization temperature. Reorientation of lamellar crystals accompanying the stretching of samples as well as the initial stages of microcrack formation so-called crazing which involves the creation of voids has been monitored *in situ*. The progress of phase separation has also been investigated in amorphous polymers. SR is particularly useful in liquid crystal research to investigate the temperature, flow, or field-induced relaxation processes related to orientation or phase transitions.

It is not possible here to illustrate the type of current research on polymers using SR but as typical examples concerning crystalline polymers one can mention annealing of polymer crystals, lamellae thickening during isothermal crystallization, and chain-folded crystallization in very long n-alkanes. Before synchrotron radiation was available many of such rapid changes could not be observed, thus, these studies resulted in much interesting information about the relative role of molecular diffusion mechanism. The same remark is valid for isothermal crystallization (the correlation between the fold period i and supercooling ΔT) and transient states in chain folding. To take advantage of synchrotron radiation ancillary equipment is required; for example, time resolution of time-induced processes is limited not only by the photoncounting period but also by the limited range of temperature. Attempts are being made to record time-resolved SAXS patterns while simultaneously measuring the changes in other physical properties; for example, a combination of scanning calorimetry (DSC) and scattering techniques is possible.

Because this paper is devoted to a discussion of some aspect of progress in the use of lasers as bright monochromatic sources of light for studies of polymer morphology and related phenomena, the ample literature related to SAXS will not be quoted here and only some remarks given above aim to show that X-ray scattering studies provide important information about the mechanism of structure changes in a very short time scale. The use of SR will surely involve development of new powerful sources and high-count rate detectors with ancillary equipment for controlling the specimen conditions. It means, however, that this technique will be still more expensive and available to a limited number of research groups.

Similar experimental results on a larger dimensional scale can be obtained by small angle light scattering (SALS). The usually much longer wavelength of light causes some limitation with respect to the size of investigated heterogeneities. The method of light scattering at different angles has been used for 40 years for studies of polymer solutions and later to investigate various polymeric solids. Since the pioneering work of Stein[1] in which earlier papers related to this subject are cited, the theory and experiment of SALS on various morphological structures in crystalline polymers and heterogeneous polymeric systems develop steadily. A real turning point of these studies was the application of lasers as a source of light. They provide intense, parallel monochromatic, polarized light permitting scattering patterns to be obtained in times less than 1 ms, thus, changes accompanying fast time-dependent processes may be followed. With proper immersion it is possible to study the internal structure of films as well as fibers. In this paper I will consider mainly the SALS on solid synthetic polymers and polymer blends as well on structure formation in polymer blends and in polymers which are able to crystallize. Experiment and theory of SALS on morphological structures in crystalline polymers and heterogeneous polymeric systems develops steadily. Since the early 1960s and basic papers of Stein, several advances have been made in the theoretical understanding of this process and in its practical application. Distorted scattering patterns from stretched samples and different morphological structures have also been analyzed. These results will be shortly reviewed in the first part of this paper including the generalizations of the theory of scattering from films containing rod-like aggregates of crystals and from films having correlated orientation of crystalline aggregates. Details of several of these studies are published, thus, only the assumptions and results will be briefly described.

Properties of polymer blends are believed to depend on their structure at supermolecular level. The insight on fundamental molecular parameters affecting the structure-property relationship of polymer blends is gained by the time-dependent light scattering technique coupled with other methods, e.g., a temperature-jump method. Because these studies are of particular interest, the theoretical and experimental achievements in that field will be discussed taking into consideration demixing structures in solvent cast films of copolymers and polymer mixtures as well as in mechanical mixtures.

Modern polymer research using laser light scattering will be included but overloading with theory as well as experimental data will be avoided. Many tables and figures which can be found in the cited papers have not been included. I wish to present some results which, according to my personal view, are of interest concerning the progress in this area. I have concentrated on application of the technique by some examples and on information on the scope of the method and its potentialities for polymer research.

Some problems and works which have contributed to the present status of polymer studies using light scattering techniques are undoubtedly omitted.

It is beyond the scope of this paper to discuss biological macromolecules in solution and in the solid state; however, important progress is to be noted in the study of these materials. Only a few remarks on light scattering in these systems will be given.

II. SMALL ANGLE-LIGHT SCATTERING FROM POLYMERIC SOLIDS

A. BASIC CONCEPTS

Polymers scatter light as a result of fluctuation in density, refractive indices, and orientation in correlated domains. Electron and light microscope investigation show that an ordered structure exists in semicrystalline polymers over many thousands of angstrom units. These structures consist of crystallites which are of the order of few angstroms or from other correlated structural elements. The above-mentioned structural correlations in the areas comparable with the wavelength of visible light are responsible for specific light-scattering effects. The angular dependence of scattering is related to these fluctuations and the intensity of scattered light is connected with the local refractive index variation. The polarization of the scattering is defined by optical anisotropy of the fluctuating regions. Thus the analyses of the intensity and of angular dependence of SALS give important information about the complex morphology of crystalline polymer samples or about the structure of heterogeneous polymeric systems consisting of components characteristic of different optical properties. In that respect light scattering complements optical and electron microscopy but also leads to much new quantitative information on the structures arising in polymers. It seems interesting to note that relatively thick samples can be investigated under controlled strain, orientation, and temperature without damage of the sample. Investigations in the presence of solvent can also be carried out. To study orientation of fluctuations, scattering intensity is measured as a function of the polarization direction of the scattered and incident light. One distinguishes two different scattering patterns: H_v SALS in which the analyzer has its plane of polarization perpendicular to that of the incident beam and V_v pattern in which the analyzer has its plane of polarization parallel to that of the incident beam.

The first theoretical foundations of light scattering on density fluctuations in heterogeneous media was laid by Debye and Bueche. The main contribution to theoretical and experimental development of SALS has been made by Stein[1] and collaborators. The theory of SALS patterns interpretation and many examples of its application have been also described in monographs devoted to studies of polymer structure and morphology.[2-4] Later on, other groups have contributed to this technique. It is not the aim to cite here even the most important reviews and original papers; thus, we will only mention important achievements of groups of workers headed by Benoit, Samuels, Kawai, and Hashimoto. In our laboratory we have investigated different morphological structures using SALS and constructed various experimental arrangements permitting structural studies taking advantages of all aspects of this method.

B. EXPERIMENTAL TECHNIQUE

Light scattering at wide angles can be studied by a conventional scattering apparatus with good angular resolution. Film samples are used most frequently. They are investigated between glass plates with a fluid of matching refractive index in order to minimize surface scattering. Usually it is necessary to apply correction factors for reflection, refraction, and secondary scattering.

For SALS less than 10° from the incident beam, special equipment has been constructed. The incident beam should be exactly parallel in order to make accurate measurements of scattering angles and the light source must be monochromatic, polarized, and stable. This is the reason why lasers are usually used as light sources, e.g., He-Ne laser. Increasing the power of lasers may provide heat destruction of the polymer sample which in the case of highly crystalline films the presence of spherulite structures markedly affects the character of destruction.[4,5]

Various kinds of SALS instruments have been described in many papers and developed to high standards. It does not seem necessary to cite this work here because their basic elements are similar to that described by Stein.[1] The apparatus must be equipped with two polarizers in the incident and scattered beam. They should be precisely rotated. Because the scattered intensity for oriented films depends on the orientation of the film with respect to the light beam, usually the samples are mounted in a goniometer, permitting the variation of two spherical coordinates. Special cells for controlling sample temperature immersing in liquids and for stretching polymer films are available. For measuring the intensity of scattering as a function of angle one can apply photographic plates or photometric systems containing a photomultiplier mounted in a goniometer. Electron scanning systems containing an optical multichannel analyzer or rapid scan spectrometer have also been constructed. The detection system should have a high resolution at low scattering angles and high sensitivity and accuracy at various scattering angles. Sometimes neutral filters are used for attenuation of the intensity of scattered light. Calibration of apparatus has to be made for absolute intensity measurements. The use of careful immersion makes it possible to study the internal structure of fibers and fibers imbedded in another medium.

To follow rapid changes in the scattered light intensity, e.g., during cyclic deformation, special equipment has been constructed in several laboratories. Construction and principle of operation of SALS apparatus for studies of supermolecular structures in static and dynamic range in thin polymer films were described by Pakuła and Soukup.[7] This system can be employed for time-dependent experiments which require rapid recording of a variety of scattering pattern parameters. A stabilized He-Ne laser emitting polarized light of wavelength $\lambda = 0.6328$ μm was used. Luminescence from the laser tube is removed by monochromatic filter and the laser beam passes through a quarterwave plate which is so adjusted to achieve circular polarization. This system permits an unrestricted choice of polarization plane by adjustment of polarizer at an appropriate angle relative to vertical. The light falls on the sample placed in the heating chamber or in a stretching device. The scattered light from the sample passes through an analyzer. Without going into detail of the description of this apparatus it must be mentioned that a special goniometer makes it possible to carry out the measurements at high recording speed at a defined angle μ or along a circle of any radius must be applied. The coupling of the stretching device with the optical system makes it

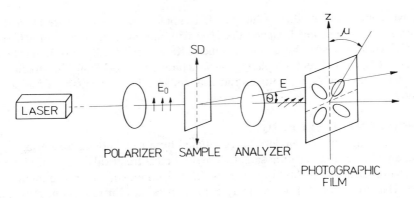

FIGURE 1. Scheme of a basic equipment for SALS studies.

possible to study the deformation at a constant rate or at sinusoidal deformation. With this equipment, including a heating chamber for the samples, many kinds of time-dependent experiments can be made, e.g., studies of isothermal and nonisothermal crystallization and changes of structures due to deformation at constant rate or at periodic deformation. Appropriate analysis of scattered light distribution makes it possible to determine the real and imaginary part of the strain compliance and the real and imaginary parts of dynamic rheooptical coefficient. Such complex and diverse applications to the study of the polymer supermolecular structure and its changes could be made only using laser as an intense, polarized light source.

C. OUTLINE OF THE THEORETICAL BACKGROUND AND OF EXPERIMENTAL RESULTS

The advances in application of SALS have been reviewed in many papers. The basic theoretical and experimental results of SALS including the pioneering works of Stein and collaborators are collected in the symposium volume on small angle scattering from fibrous and partially ordered systems and in many monographs describing the structure-property relationship of semicrystalline-oriented polymers, e.g., see Reference 3.

Interpretation of SALS can be made using two approaches: the model and the statistical approach. The first method assumes a particular form of the scattering entity and the experiments are made to determine the parameters introduced in the model. For spherulitic samples the scattering pattern is analyzed in terms of scattering from a homogeneous anisotropic sphere in an anisotropic medium.

The equations for V_v and H_v patterns have been derived. The intensity of the scattered light for I_{V_v} and I_{H_v} is shown to be a function of the volume of this sphere, of λ — the wavelength of light in the medium, of Θ — polar scattering angle, of μ — azimutal scattering angle, and of polarizabilities α_t, α_r, and α_s being tangential and radial polarizabilities of the scattering sphere and polarizability of surroundings (see Figure 1). With the assumptions that α_t, α_r, and α_s are close (λ is not changed appreciably by passing the sphere) Rayleigh-Gans approximation was used. The V_v and H_v SALS patterns calculated using these equations can be compared with experimental results and the polarizabilities are selected in such a way that the theoretical and experimental patterns are in correspondence. In dense spherulitic samples, e.g., in polyethylene and polypropylene, each spherulite is surrounded by other spherulites and in the simple theory it is assumed that the surrounding has a polarizability about equal to the average polarizability of the spherulites. The position of maxima in the well-known ''four leaf clover'' in H_v pattern makes it possible to calculate the radius of spherulite from the equation:

$$U = (4\pi R/\lambda)\sin(\Theta/2)$$

and $U = 4$.

The anisotropy of spherulites results principally from preferred orientation of anisotropic crystallites and from their orientation with respect to radial orientation of lamellae. Thus the birefringence of spherulites depends on crystallites orientation and on the orientation of the amorphous phase which is always included in such structures.

Evidently this morphological structure is related to the stereoregularity of investigated polymer and conditions of crystallization.

The simple theory describes well the scattering patterns at small angles. For higher angles the predicted intensity values are too small because the simple model does not describe the heterogeneities over small distances in a correct way. Several refinements have been made by Stein to take into consideration the internal heterogeneity within the spherulites, it can be of random or of an ordered nature, e.g., regularly ordered orientation of crystals in spherulites which gives rise to the ringed structures frequently detected using a polarizing microscope, as well as the transmission of light by films with spherulites of a lower degree of crystalline order. Further developments concerned the theory of scattering of light from films having distorted spherulites as a consequence of deformation of spherulites to elipsoids which were described by Stein et al. and Samuels in the papers published in the above-mentioned symposium as well as the SALS pattern from anisotropic rodlike structures. The random and nonrandom orientation fluctuations have been considered too.

An alternate approach is the statistical approach which takes into consideration the correlation function to the calculation of scattering intensity. It can be done for homogeneous sphere and in the case of internal heterogeneities which modify the correlation function. The correlation function itself can be obtained by inversion of the experimental data (Fourier inversion).

In the last 20 years many problems connected with polymer texture have been studied using SALS.

Some of the obtained results are mentioned in order to show the wide range of issues which have been solved in various laboratories, particularly by the groups of Stein, Kawai, and Hashimoto. We have contributed to a small extent to elucidate some of the open problems. Generally these results can be divided into the development of the methods of SALS pattern analysis or/and into studies of morphological structure changes due to external forces or processing and preparation of films from various polymers.

The most typical morphological structures found in crystalline polymers during melt crystallization are spherulites. Depending on polymer chain structure and crystallization conditions other types of morphological structures may arise. The theory of SALS for such morphological entities like various hedrites, ovoids, and spiral ovoids was elaborated and experimentally verified in our laboratory.[8a,8b,9] It shows that SALS is very powerful for the analysis of such complex structures arising, for example, in polyoxymethylene.

The origin of the complex SALS patterns from some crystalline polymers was shown to be related to internal disorder and specific orientation of the optic axis with respect to spherulite radii and specific fine structure of spherulites in poly (tetrafluoroethylene) and in polypropylene-polyethylene A-B and A-B-A block copolymers. These specific SALS patterns exhibit spherulitic scattering at small angles and rod-like scattering at high scattering angles.[10]

Some polymers crystallize in ringed spherulites which give unusual ring scattering patterns, e.g., poly(decamethylene terephtalate) crystallized in the range from 0 to 80°C. The analysis of H_v and V_v scattering patterns for this material shows that the observed deviations from usual SALS patterns may have resulted from light scattering from ringed spherulites with the optical axis inclined at about 45° to spherulite radius.[11] These two examples show clearly the possibility of SALS method to detect fine differences in spherulitic structure.

Different other textures arising in polymers as well as many detailed studies of the specific SALS pattern attracted much research. Some examples given below shall give the feeling of past and actual trends in these investigations.

In an exact analysis of SALS patterns it is necessary to introduce corrections concerning the retardation of the incident and scattered beams[12] especially for studies of uniaxially deformed two- and three-dimensional spherulites. The application of SALS for studies of deformed spherulites, since early works of Stein and Samuels cited in the collected symposium volume,[13] were continued by other groups. Dimensional changes of polymer spherulite with uniaxial stretching were investigated by Kawai et al.[13] showing that Samuel's calculations are valid and that the spherulites deform in the affine fashion while keeping their volumes constant during stretching the bulk specimen up to 50% elongation. The particular scattering patterns depends on the kind of polymer investigated, e.g., different brands of polyethylenes.[14] Elongation ratio λ_s of deformed spherulites is always an interesting parameter characterizing the changes of morphology of spherulitic samples and micromechanics of deformation. A simplified function of intensity distribution in H_v pattern being a good approximation to Stein's equation has been proposed and a new method of calculating the elongation ratio of deformed spherulites was elaborated.[15] This method was verified for deformed spherulites in polyethylene and it is shown to be useful for studies of affine deformation despite simplifications introduced in calculations. Further development along this line was the application of the mechanical model of deformation of two-dimensional spherulites.[16] Assuming two-phase structure and proper internal correlation of orientation of crystalline elements as well as mechanical properties of phases, we have calculated the distribution of intensity in SALS patterns. A good description of the observed effects was found, e.g., deviation from affine deformation of spherulites in polyoxymethylene at smaller stress than noted before for polyethylene.

The study of the deformation of PE spherulites by wide angle light scattering has been shown useful to elucidate the change in helicoidal twisting of lamellae with applied uniaxial stretching.[17] Spherulitic samples do not possess spherulites of uniform size and usually the mean spherulite size is not sufficient characterization of the sample. The mean radius cannot be used to relate the properties, e.g., mechanical, with the structure. Complete description of the sample on the supermolecular level requires the determination of distribution of spherulite sizes. For this aim, microscopic techniques can be used but they are laborious and not easily applicable to small spherulites. It has been shown both theoretically and experimentally that SALS can be also used for this purpose.[18] When special corrections of experimental data are introduced to achieve great accuracy in the determination of scattered light intensity, the estimation of spherulite size distribution can be easily made, e.g., for low density polyethylene samples obtained in various crystallization conditions. The possibility of the calculation of number-average spherulite size R_s, volume average size R_v, and the maximum size R_{max} allows computation of the variances of the volume distribution which, in turn, can be approximated by $(R_{max} - R_v)$ being a good characteristic of the dispersion of volume distribution of spherulite sizes.

We will not discuss here the theoretical and experimental results concerning the effects of interspherulitic interference, nonrandom orientation fluctuation, and the change of light-scattering patterns when the sample is tilted, because it seems more important to characterize briefly the application of SALS to other textures arising in polymers as well as to study the morphology of various polymers depending on their processing.

The high intensity of laser beams make it possible to record the scattering patterns very rapidly, e.g., by a TV camera video-type recording system. Therefore, this technique is especially useful to investigate rapid growth of superstructures accompanied by crystallization and time-dependent deformation of arising morphological entities, e.g., in blown polybuten-1 films.[19] The analysis of such results is usually not simple and different effects have to be taken into consideration, e.g., interparticle effects. It was suggested that the paracrystal model[20] of polymeric solids can be useful for this purpose. Morphological studies of oriented crystallization of a high-density polyethylene using SAXS and SALS patterns as well as wide-angle X-ray diffraction patterns have shown that the molecular orientation in the melt has an important influence on the final morphology.[21] The scattering patterns clearly show that the draw ratio influences the nucleation parameters during oriented crystallization, e.g., bulk-free energy and the kinetic factor characterizing fraction of the chain segments being attached to growing nuclei. It seems interesting to note that such subtle, but important, effects can be detected by scattering techniques. Calendar manufacturing of polymers, e.g., of polyethylene, results also in structures depending on processing conditions. Such morphology consists of rod-like units oriented preferentially at some angles with respect to the machine direction. The formed network of superstructure can be correlated with the specific SALS patterns[22] and accounts for the whitening of the specimens on stretching along the machine direction.

Independently of polyolefins, polyamides, and other commodity polymers, various polymers have been investigated using SALS, e.g., collagen films,[23] which in analogy to natural cellulose, polytetrafluoroethylene,[24] and poly chlorotrinitrofluoroethylene[25] exhibit rod-like structure. This texture is characteristic of an assembly of anisotropic rods in three dimensions and depends to high extent on crystallization conditions. Sometimes spherulitic morphology appears too but the deformation of such morphology is not simple owing to strong interaction of crystallites in the spherulites.

The above-mentioned results of morphological studies of various polymers are only some selected examples of the application of SALS for elucidation of such structures. They have merely been possible because of the advances in the theory of light scattering from rod-like entities of different optical anisotropy, dimensions, length, and various arrangement in the investigated materials, e.g., see papers of Kawai's research group.[26,27]

From the methodical and application point of view the SALS studies of various assemblies of anisotropic plates of different dimensions in two- and three-dimensional space are very important.[28] An interesting application of the light scattering by anisotropic discs was the study of single crystals.[29] It has been shown theoretically that many parameters characterizing such a system, e.g., the sign of anisotropy as well as the growth of homo- and copolymer single crystals directly in solution, can be evaluated from SALS patterns. Differences in the rate of growth and platelet aggregation have been found for homo- and copolymers.[29] At low crystal concentration these measurements are difficult but the authors have suggested that polymer single crystal orientation by a weak electric field might be useful. The combination of various techniques offers a powerful method for studies of single crystals at low concentrations, e.g., at the very beginning of the crystallization process.

With the growing interest in reinforced polymers and composites the light-scattering method was also successfully applied to such complicated systems. The basis for this analysis was the study of light scattering from dielectric cylinders, e.g., glass fibers in polymers. The results were discussed as Mie and Rayleigh-Gans-Born scattering theories from an isolated and infinitely long cylinder.[30] An approximate but rapid and convenient method of evaluating the size and refractive index of the fiber was elaborated. The same analytical treatment can be used to study coaxial cylinders, e.g., E-glass coated and uncoated with polystyrene fibers.[31] These theoretical results were further widely extended and applied for investigation of various fiber reinforced plastics and composites.[32]

Another interesting application of SALS is the study of helically twisted fibers.[33] These synthetic fibers have an internal structural twist which is similar to that which occurs in natural fibers which were obtained by specific melt-spinning process consisting of drawing and rotating the take-up element.[34] The SALS pattern was calculated for fibers with various pitch of a helix using appropriate models and corrections for refraction. The theoretical predictions were compared with the experimental results obtained from the above-mentioned fiber and such filaments in which helical orientation was obtained by cold twisting of the normal cylindric fibers around their axes up to the desired twist. The agreement with the theory is satisfactory, especially for cold-twisted fibers. For the melt-spun fibers with various twists there is a specific modulation (periodic reflexes) which seems to be correlated with internal inhomogeneities; for example, twisting of the fiber does not occur exactly around the geometrical axis. The SALS shows inhomogeneities of helical orientation which have not been discovered by either optical or electron microscopy or X-ray examination.

The versatility of SALS as applied to the study of the structure of solid polymers could be demonstrated on other examples, e.g., the investigations of craze formation in glassy polymers.[35] Light-scattering techniques have been applied for years for characterizing polymeric lyotropic liquid crystalline systems, e.g., concentrated solutions of polypeptides in helicogenic solvents, in order to determine the sense of twisting in cholesteric liquid crystals, orientation correlation, and the variation of their properties with the solvent, concentration, and applied electric and magnetic fields. Interesting results have been obtained for concentrated solutions and solid films prepared by casting.[36-39] Supramolecular structures found in these systems were analyzed in terms of elucidation of rod-like scattering patterns effects, of form-optical rotation, existence of mesophases, inhomogeneities in twisting of mesophase, etc. These studies also lead to the development of the theory of scattering in such systems, e.g., the application of Rayleigh-Gans-Debye theory, considering the fact that the scattering dominantly arises from orientation fluctuations of optically anisotropic scattering elements rather than from concentration and density fluctuations.[37,40] These are only a few remarks on some structures found in liquid crystalline systems.

The application of light scattering for investigation of organized liquid crystalline systems was expected, because 20 years ago the swelling-induced structures in cellulose gels (in water and in solvent-exchanged cellulose gels) have been investigated by Marchassanlt et al.[41,42] It is beyond the scope of this paper to discuss in detail these types of studies; also, important progress is recently noted. Regardless of the advances in theory and experiments many problems remain to be answered.

Scattering from partially crystalline polymers preferably swollen in amorphous regions or in porous matrix (imbibition technique) is used to study texture and morphology as well as their changes due to various treatments: annealing, selective extraction, and formation of various interpenetrating networks. These studies can be made under equilibrium conditions, various states of swelling, as well as under deformation in static and dynamic conditions. Since the pioneering work of Stein these studies have been carried out in many laboratories including ours (see Reference 43) and resulted in deeper insight in the morphological structure and its influence on diverse properties showing how extended is the scope of scattering technique. The investigations of structure and morphology of deformed polymers by SALS, particularly of such samples which were subjected to dynamic stresses, are usually complemented with rheooptical studies of deformation mechanism, i.e., by dynamic birefringence, dynamic SALS, and dynamic (e.g., see Reference 44) X-ray diffraction. The number of papers related to the micromechanics of crystalline and amorphous polymeric systems steadily increases through the years since the early works of Stein. They are mentioned here to show that such type of studies, carried out in many laboratories using various experimental techniques and adequate theoretical interpretations, leads to better understanding of various types of mechanical dispersion and to proper correlations with details of morphology and crystalline or/and amorphous phases arrangement in supramolecular structured units.

It is not possible to discuss here the application of SALS to study the structure of numerous semicrystalline polymers but it is proper to mention the investigations on ionomers, e.g., the results obtained for carboxylated perfluorinated ionomer membranes. The application of SAXS, wide-angle X-ray diffraction, and SALS studies revealed supermolecular structures at various levels ranging from a few nanometers to a few micrometers. The size of ionic clusters, spherulite size, and the amount of water uptake have been correlated with equivalent weights and with functional groups as well as the morphological structure changes induced by swelling and drying processes.[45]

Before closing this incomplete review of the use of lasers as a light source for SALS from various polymeric solids, let us mention the application of lasers as a source of heat impulse (spot heating) in structural studies. Such application of laser is evidently very different from the better known uses of laser for studying several aspects of polymer degradation.

Investigations of the structure of crystalline polymers have been supplemented in recent years by studies on thermal conduction. Usually macroscopic samples are used and characterized by macroscopic thermal conductivities, thus, only limited information on structure influence can be achieved. It appeared interesting to study the phenomenon of heat propagation in flat, radial spherulite, spot heated by an impulse of laser at a defined place of such structure.[46] The anisotropy of thermal conduction in a single spherulite was calculated for a two-phase model. The temperature distribution due to cooling of a spherulite impulse heated at one point was computed leading to isothermal contour evaluation for a wide range of thermal conduction anisotropy, cooling time, and position of hot spot along the radius of spherulite. Experiments with narrow laser pulse of heat (point source) were carried out on rather large spherulites formed in thin films of isotactic polypropylene (i-PP) and isotactic polystyrene. The isotherms recorded using a thin coating of heat-sensitive indicator substance could be compared with theoretical isotherm patterns. The coincidence of theoretical isotherms with the experimental ones was very satisfactory making it possible to estimate conduction anisotropy of i-PP. Additional measurements of heat-conduction coefficients of polymer samples characteristic of different crystallinity made possible the calculation of the thermal conductivity of the amorphous phase as well as the conductivities of crystallites (constituting the lamellae) along and perpendicular to the chains were determined. We have discussed here these results to some extent because they show that impulse laser beams can be used in quite sophisticated structural investigations.

Nowadays the application of SALS for solid polymer studies is extended by SAXS and particularly by small angle neutron scattering (SANS) technique which leads to novel information on polymeric systems at different dimension scale as compared with SALS. This fact is clearly seen looking at the programs of numerous symposia devoted to discussion of scattering techniques as applied to the study of polymers, e.g., see Symposium on New Developments in Small-Angle Scattering from Polymers Organized by Division of Polymer Chemistry of American Chemical Society (ACS)[47] the Symposium on ACS Awared in Polymer Chemistry honoring Richard S. Stein,[48] whose concepts, creativity, and steadily increasing contributions to the theory and wide application of SALS have to be highly estimated. The same remark concerns many other more recent symposia at ACS meetings as well as specialized meetings devoted to the discussion on various levels of structure of solid polymers.

By combining measurements of angular distribution of absolute scattered intensity using visible light, X-rays, and SANS results with the Rayleigh line width determination, we can learn a great deal on polymer chain configuration and dynamics as well as on polymer structures.

However, it seems that SALS, being the simplest technique at least from the experimental point of view provided lasers as light sources, will retain its importance. It is due to its actual state and further possible developments both from the theory and experimental aspects particularly for the studies of multiphase and multicomponent systems in the solid state and their evolution with time. This aspect of SALS will be discussed in the next section.

III. MORPHOLOGY OF MULTICOMPONENT POLYMER SYSTEMS AS STUDIED BY LIGHT SCATTERING

A. INTRODUCTORY REMARKS

The concept of heterogeneous polymeric systems comprises various high molecular weight materials in which separation of phases consisting of components of different chemical constitution occurs, e.g., block and graft copolymers, interpenetrating networks (IPN), and polymer blends called polymer alloys. This wide class of materials also includes materials in which different phases exist due to the presence of inclusion of various substances. These materials are obtained by specific processes yielding a variety of filled polymers or polymer composites. Such systems consist of a polymer matrix and various particulates, platelets, fibers, etc. of a different chemical nature and physical structure.

Composites are characteristic of internal structure which can be varied freely in order to match their performance to the most demanding structural roles. The literature related to filled polymers and polymer composites is today innumerable concerning monographs, reviews, symposia proceedings, original papers, and specialized journals. Taking into consideration the general aim of this paper, i.e., the application of lasers to study the structures arising in polymers, we have to restrict the discussion of composites to some remarks presented in the first chapter, e.g., considerations of the scattering on glass fibers imbedded in polymer matrix, being aware of their incompleteness.

In the general term of heterogeneous polymeric systems one has to include one component system in which separation of phases is related to crystallization. The application of the light-scattering technique, using lasers as light sources for characterizing various structures arising during crystallization, has also been discussed in the first section. We will consider polymer blends as a combination of two or more polymers resulting from various processing procedures and to some extent block and graft copolymers. Multicomponent polymeric systems, block and graft polymers, IPNs, and polymer blends are now one of the main current subjects of polymer research and application. Their potential technological importance and many open problems connected with the formation and properties of such heterogeneous systems have attracted many academic and industrial research groups.

Pursuit of the understanding of the physical properties (particularly mechanical properties) has revealed new principles and refined earlier fundamental concepts as well as uncovered further opportunities for research and practical applications. In contrast to copolymers, where the components are linked by strong covalent bonds, the components in polyblends adhere only through van der Waals forces, dipole interactions, or hydrogen bonding. In polyblends usually one component makes a matrix in which another polymer is imbedded but at a certain concentration a phase inversion may occur. The distinguishing of polyblends into four types which can be obtained considering rigid matrix and soft disperse phase or vice versa is illustrative and makes it possible to predict the improvement of some properties. It is certainly not adequate for a real understanding of structure-property relationship.

The great deal of interest in the studies of polymer alloys and other multicomponent systems especially since the early 1970s is documented by a large number of monographs, reviews, and original papers as well as symposia proceedings, e.g., see References 49 to 51 and 53 to 58. We have to note important contributions collected in three volumes edited by Klempner and Frisch.[59] Let us mention that the major areas of study on polymer blends are along three main lines. The first is related to formation of specific structures including all aspects of phase separation compatibility, etc. The second are the studies on interactions in multicomponent systems and the last one is connected with structure-property relationship investigations.[60]

The modification of glassy polymers with dispersed rubber phase was the first major effort to obtain useful polymer blends. Later came the detailed studies of toughening theory as well as on fracture phenomena and ultimate properties.[61]

The connections between various fields of study on polymer blends and heterogeneous block and graft copolymers are evident considering their mechanical properties[6] or permeation behavior.[63,64] Because of the heterogeneous nature of most polymeric blends their physical properties can be treated as microcomposites. It has been shown in fact that many theories dealing with elasticities of composite materials are applicable to polymeric alloys. The morphology has been again shown to have a profound influence on the mechanical properties.

Mixtures of polymers exist in a very large range of morphological states from coarse to very fine. It is evident even from the first look at IPNs and rubber blends. However, aside from the existence of unusual size, shape, and geometrical arrangements of dispersed phase, more subtle complexities are possible and especially the intrinsic arrangement of amorphous and crystalline phases makes polymer blends particularly flexible in property modification.

A number of experimental techniques are used to characterize the disposition of molecules and phases on submicroscopic range; thus, it is not possible to consider all of them in a chapter devoted to a short discussion of structures in heterogeneous polymeric systems as revealed by application of optical methods — one has to select a limited number of topics.

Because of the large body of literature available on this subject, the treatment must therefore necessarily be illustrative rather than exhaustive. Thus our discussion will be restricted to basic comments on the structure of polymeric alloys in the solid state and to a discussion of the dynamics of structure formation due to phase separation. The last subject seems to be of particular interest due to the advances in experimental techniques and theory of phase separation in complicated polymeric systems.

B. STRUCTURE OF MULTICOMPONENT POLYMERIC SYSTEMS

Because of the generally incompatible nature of polymers, polymeric alloys and block copolymers show micro- and macrophase separation. The techniques most frequently used for observation of structures being formed are electron microscopy (EM) and SAXS. In EM studies staining agents are applied, e.g., osmium tetraoxide, which reacts with double bonds or other specific groups of the component. They are responsible for the necessary contrast for EM. These microstructures obtained in solvent cast samples are not necessarily equilibrated ones but apparent ones. They can sometimes be changed to equilibrated structures by proper thermal treatment.

Although the essential morphologies of spheres, lamellae, and cylinders are well documented by EM, they are usually not very regular but under appropriate conditions of specimen preparation it is possible to observe long-range order in block and graft copolymers. The nearly perfect long-range order extends over large areas. In some cases imperfections in this order may appear as grain boundaries normally observed in metallic systems; thus, the analogy to metallic multicomponent systems is sometimes evidenced. The detailed discussion of these structures is out of the scope of this paper; thus, we will present some comments on optical properties of polymer blends and their relation to the morphology.

C. OPTICAL PROPERTIES AND STRUCTURE OF MULTICOMPONENT POLYMERIC SYSTEMS

A very comprehensive and penetrating treatment of optical properties of polymer blends is published by Stein[65] some years ago. It is always topical because of the perfect description of the state-of-art which makes it possible to follow further developments concerning to structure determination by means of optical methods.

Optical properties of polymer blends depend on the scattering of light, e.g., turbidity, but the studies of birefringence and dichroism lead to valuable information on structure too.

The birefringence of a multicomponent system can be expressed as a sum of the contributions of components characterized by their volume fraction, orientation functions, and intrinsic birefringence as well as by form birefringence and refractive indices of components. The studies of birefringence are usually carried out with conventional light sources with appropriate filters. The use of lasers sometimes simplifies the experimental conditions especially in the case of dynamic birefringence measurements but it does not change the power of this technique in a principal way. This is the reason why we will close these remarks saying only that the birefringence study results in valuable information on the dispersion and arrangement of components but less direct than in the case of scattering techniques. It does not mean that the results concerning the correlation between composition, orientation, and different bearing of load fraction for samples subjected to stress are not useful for understanding the role of morphology which may depend upon the formation conditions, for example, by determination of stress optical coefficients in a static and dynamic way.

The interpretation of birefringence is sometimes ambiguous because it is a composite property. IR dichroism has a potential selectivity to measure the orientation of particular phases. The dichroism of a given phase (selecting the wavelength) is characteristic of its particular orientation; thus, the measurements of IR dichroism have been often applied for determination of crystallinity, orientation function of crystallites, and amorphous phases in deformed spherulitic samples as well as for characterization of segmented polyurethanes and some other materials. There are limited examples of application of this technique to the study of blends in which the relative orientation of components or/and the compatibility was also determined.

Before going into the discussion of results of polymer-blend structure determination using the light-scattering technique let us mention the increasing use of polarized Raman scattering technique to the selective study of orientation of components in polymer heterogeneous systems. Due to very intense, monochromatic, and polarized light emitted by lasers at various wavelengths, this method steadily increases in importance also for elucidation of polymer blend morphology and for evaluation of interactions at the interface and/or in the interphases. The application of this technique is not simple because of a very large amount of Raman scattering on the heterogeneities. It would be appropriate to devote more attention to the above mentioned technique from the viewpoint of structural studies but it is not possible in the frames of this chapter. However, various aspects of this method may be discussed in other parts of this book.

The problem of analysis of scattering effects from heterogeneous systems under discussion is somehow similar to that presented briefly in Chapter 1. It consists in proper calculation of the form factor which is directly correlated with the structure of the scattering system. Here again one can apply the model approach which is based on evaluation of the angular dependence of scattering intensity for a properly selected model of structure or the statistical approach. It has been shown that the model approach applied in terms of Rayleigh-Gans-Debye theory can be used for polymer blends consisting of components with appreciably different refractive indices and spherical, ellipsoidal, and other shaped particles. For larger particles the model approach applies the Mie theory. Some procedures have been elaborated to solve Mie's equations and to calculate the variation of specific turbidity with particle diameter. This model approach was often used to study various lattices and dispersions and less frequently to investigate the polymer blend structure. Usually in this treatment the form of particle size distribution is assumed including properly chosen parameters which are evaluated in such a way that the minimum deviation between experimental and calculated intensities is reached.

All the model theories are in reality single scattering theories which are a good approximation for films of low turbidity and thickness. In a number of papers theoretical corrections are discussed in order to adapt the calculations to thicker films but usually thin samples in appropriate immersion are used. The high intensity and monochromacity of laser light is of real importance for exact measurements of spacial distribution of the scattered light thus making possible the reliable comparison of experimental results with theoretical predictions.

The statistical analysis of the scattered light intensity from polymer blends and other heterogeneous polymeric systems seems to be the most universal; however, one encounters several difficulties. The scattering is described in terms of two parameters: $\overline{\eta}^2$ which defines the mean-squared deviation of the scattering power from the average and $\gamma(r)$ which describes the spatial extent of fluctuations. The last parameter may be obtained from Fourier inversion of the scattered intensity leading to determination of correlation distance. This type of treatment is known and used for elucidation of the intensity distribution in SAXS experiments in correlation with particles size and their distribution. The reader will find valuable information about these problems in exhaustive monographs.[66-69]

The correlation function technique was used for studies of various polymer blends, e.g., poly (methylmethacrylate) and polystyrene, styrene-isoprene diblock copolymers with polyisoprene, and polystyrene, and of many other systems employing both SAXS and SALS. It is not possible to cite all of the polymer blends and copolymers investigated. The earlier works are reviewed in the monograph on polymer blends, e.g., see Reference 60. These papers show that a variety of morphological structures may be obtained depending on chemical constitution of the components and preparation conditions including the kinetics of phase separation which will be characterized later. The same remarks are evidently valid for heterogeneous block and graft copolymers for which mostly SAXS and SANS techniques are used due to the existence of heterogeneities usually much smaller than the wavelength of visible light.

It seems important, however, to mention the analysis of the domain boundary thickness which is an important parameter characterizing polymer blends and block copolymers. There are similar problems which discuss the lamellae thickness in crystalline polymers for which the less ordered surface layers plays an important role in all aspects of structure-property correlations. Block copolymers composed of incompatible components of blocks A and B form in general in a pseudo-two-phase structure in the solid state as a consequence of microphase separation process. There exists an interfacial boundary region of a certain thickness in which these components are mixed. The origin of such interphase has recently been investigated on the basis of statistical thermodynamics by Meier,[70] Helfand,[71] and others. A large number of works on mechanical properties of block and graft copolymers also have evidenced the existence of the interphase. Many aspects of its role were described in the proceedings of specialized symposia concerning polymer blends as well as polymer composites, e.g., see References 72 and 73. The experimental technique mostly used to study the thickness of the interphase in SAXS[74] and recently the SANS technique, making it possible to test the existing, current theories of domain information and to reach some correlations between structure and mechanical properties. It has been shown that some block copolymers and their blends with corresponding homopolymers exhibit many interesting effects, e.g., strain-softening, clearly shown in the studies of cyclic tensile stress-strain behavior and directly in the change of SAXS patterns with stretching. This strain-induced plastic-to-rubber transition is related to various structural changes (starting with lamellar one) in function of stress which are partially reversible in the released samples[75] and in many other papers of Kawai's and Hashimoto's groups including segmental interdiffusion,[75] molecular weight,[76] solubilization,[77] annealing,[78] deformation,[79] etc. These are only some examples of the progress in the field of block copolymer studies. Many other groups have contributed to our understanding of the fine structure of these systems and their changes with time and external conditions.

There are some concerns with the advances in the theory of scattering of electromagnetic radiation and SANS from polymer solutions of higher concentrations and from the multi-component polymer solution, e.g., see, Reference 80. The aim of these works is, among others, the evaluation of the generalized interaction parameter χ_{ij} as function and correlated with it the excluded volume. The interaction parameter χ is always of basic importance for understanding the behavior of polymer chains in rather concentrated solutions and evidently in heterogeneous systems in which phase separation may occur.

During the last years many groups are involved in the study of structure of polymer blends. The use of light-scattering technique was shown to be a very useful method to study the final morphology, e.g., see Reference 81. The above remarks concerned mostly non-crystalline systems. Special attention shall be given to scattering studies of polymer blends in which one component may crystallize. The formation of crystallites and other morphological structures leads to additional scattering also arising from anisotropy differences between the crystalline and amorphous regions. Depending on the concentration of crystalline component its morphology may dominate that of blend. A clear example of such behavior is the system consisting of poly(vinyl chloride) and poly(coprolactone).[65,82] Poly(ϵ-caprolactone) is known to blend with a variety of polymers as well as other polycaprolactones and copolymers containing this monomer. In that respect the review of Stein[64] and Koleske[81] deserve special attention. They show particular possibilities of modification properties of blends containing these crystallizable components in composition with other amorphous and crystalline materials. The morphological structure formed in these systems as revealed by SALS by Stein[64] and other researchers cited by him may be a good example of the power of this technique. A crystalline blend may be regarded as three-phase system consisting of one crystalline phase and possibly of two amorphous phases. The experimental and theoretical values of the intensity of scattered light may be rationalized by either assuming mixed amorphous phases or assuming the existence of diffuse boundary when the components are incompatible in amorphous phase. Similar behavior was found, e.g., crystallinity at a certain composition, for the blends consisting of poly(vinylidene fluoride) and poly(methyl methacrylate), Noryl type blends: poly(2.6-dimethyl-1.4 phenylene oxide) (PPO) with isotactic polystyrene and poly(ethylenterephtalate) with poly(butylene terephtalate) for which the structure is volume filled with spherulites when one of the components is present in excess. At intermediate compositions a nonspherulitic mixed crystalline structure has been found. Evidently SALS permits the characterization of polymer structure at the level compared with the wavelength of light; thus, SAXS may be used for the determination of the internal arrangement of crystals within the spherulite. It does not seem necessary to cite other results, as they are similar to specific differences connected with various chemical constitution, compatibility, and crystallization ability of the components.

Rather important progress in understanding the dynamics of phase separation and segmental diffusion in multicomponent polymeric systems was reached with development of high-speed measurement of SAXS profiles with a position-sensitive detector.[83] The dynamic SAXS technique of course can be applied to study the dynamics of polymer deformation, crystallization, etc. The results of time-dependent structural changes of a few angstroms by the SAXS undoubtedly provided a new source of experimental evidence to that obtained by other techniques, e.g., rheooptical; thus, many research groups started to apply it in order to establish new information in the study of the structure-property relationships in polymer blends. A large number of papers appeared describing the results obtained using SALS for studies of several systems which cannot be discussed here. However, these investigations did not consider the light scattering technique. This shows an important revival in the method of time-resolved light scattering studies which will be shortly reviewed in Section IV.

IV. MORPHOLOGY FORMATION IN MULTICOMPONENT POLYMER SYSTEMS

For thermodynamic reasons most polymer pairs are immiscible; thus, phase separation occurs. The degree of compatibility may vary widely for different systems; thus, the studies of phase separation are the subject of numerous theoretical and experimental investigations. This research is also significant to understand nonequilibrium phenomena in general. In polymers the nonequilibrium phenomena involve the studies of cooperative phenomena and self-organization of systems consisting of long-chain molecules because the constitution of macromolecules is a specific physical factor which influences the analysis of nonequilibrium phenomena as compared with low molecular weight liquids. The knowledge of the mechanisms of phase separation and stabilization also is of basic importance for the arising morphology which together with properties of the components define the final behavior of these systems.

It is not possible and may be not our purpose to characterize here even the most important theories and experimental results obtained during the last years applying various experimental techniques.

The basic concepts as well as experimental results of phase separation in polymer blends are clearly presented in the review of Kawai and Wang.[84] It is well known that one can calculate both spinodal and binodal curves for low molecular weight binary mixtures employing simple thermodynamic approach. The application of thermodynamic equilibria to describe the phase separation in polymer systems is more complicated. One has to find a suitable expression for the free energy of mixing and to take into consideration the polydispersity of polymer molecules. It follows from the simple Flory-Huggins theory that spinodal composition and critical composition are dependent on volume fractions, on chain length, and interaction parameter χ_{12}. Much work has been done to analyze the effect of polydispersity and to find a correct expression for the free energy of mixing modifying Flory's theory. The interested reader should consult literature on polymer solutions, thermodynamics, and properties.

When a homogeneous mixture undergoes phase separation upon lowering of temperature the cloud point curve is associated with upper critical solution temperature (UCST). This curve is convex upward and its maximum point shifts to higher temperatures with increasing molecular weight of polymer. For various polymer-solvent and polymer-polymer mixtures a decrease of solubility is also observed with temperature increase. In that case the cloud point curve is convex downward and the critical temperature is called the lower critical solution temperature (LCST), although it occurs for many systems, actually at a higher temperature than UCST. The LCST shifts to a lower temperature with increasing molecular weight.

The UCSTs reported in the literature are usually concerned with polymers of modest molecular weight. The systems exhibiting LCST are better known and many factors influencing the LCST behavior have been investigated, e.g., molecular weight thermal expansion coefficients, pressure, interaction parameters, χ_{12}, etc. We shall be more interested in the kinetic aspect of liquid-liquid phase transformation in binary mixtures and in the arising morphologies than in the deeper analysis of the thermodynamic conditions describing phase separation. A concise review on the theoretical treatment of the problem and on the arising morphologies is presented by McMaster.[85]

Because the energy state of homogeneous mixture responds to composition changes in various ways the rate process leading to phase transformation inside spinodal and binodal ranges are different. Inside spinodal because of instability to infinitesimal fluctuations there is no barrier to phase growth; thus, the separation occurs continuously and spontaneously. This is called spinodal decomposition (SD). In the metastable region however the small fluctuation tends to decay. The separation can proceed only by overcoming the barrier with large fluctuation in composition. This fluctuation is called a nucleus. When a nucleus is formed it grows by diffusion; thus, this process is called nucleation and growth mechanism. The nucleation and growth mechanism is less known in detail because of difficulties in simple theoretical treatment. It seems that classical theory developed for systems of mon-oatomic substances can lead to its qualitative description but extensions for polymeric systems are needed.

SD is, to a first approximation, treated in terms of diffusional problem which make possible some predictions of the arising structure, particularly for early stages of decom-position, linearized Cohn's theory. As the amplitude of composition of the new phase increases, the nonlinear effect becomes important but also some qualitative agreement with experiment was reached. Some remarks related to these problems will be presented later. Before discussing the experimental results and before going into some aspects of theoretical predictions concerning the phase separation, let us mention some pecularities of the formed structures and various techniques applied to the studies of phase separation.

There are several experimental methods to identify phase domains in heterogeneous polymers: light and electron microscopy, light and X-ray scattering, thermal expansion, heat capacity, and dielectric or mechanical measurement as well as pulse magnetic resonance. Taking into consideration the main scope of this paper we will concentrate evidently on light scattering which provides particularly wide possibilities to the insight into the structure formation in polymer blends. It does not mean that other methods are of smaller significance, e.g., determination of Tg and NMR measurements. The description of phase composition in a solid polymer mixture by Tg is useful but it is really applicable when two sharp glass transitions are observed. This method encounters a high degree of complexity when one component is crystallizable. The pulsed NMR offers advantages because two relaxation times appears as phase separation occurs and can be easily detected and resolved. In recent time an important development of the NMR technique has to be noted and many research groups are involved in the studies of phase separation and characterization of the arising structures in polymer blends.[86-88]

It seems necessary, before discussing the kinetics of phase separation, to give some general remarks on the morphology of polymer blends. Connectivity is an important mor-phological feature of SD. It depends on the procedure applied for the formation of specimens with various morphologies as well as on the specific characteristics of the components. The first qualitative information may be reached by phase contrast or/and polarizing microscope. The differences between the above-mentioned ways of phase separation can be sometimes easily seen. When phase separation occurs rapidly and the precipitated phases are intercon-nected with each other, the phase separation occurs by SD. Further this pattern is finer at higher decomposition temperature and at the latter stage of phase separation the phase domains seem to increase in size while maintaining their interconnectivity and eventually they break into small spheres or merge into macrospheres.

In the range between spinodal and binodal, different morphologies are usually observed. The phase separation process occurs slowly and the precipitated particles are spherical in shape. The dimensions of these spheres are small as compared with average spacing of the phase patterns in the binodal range. This characteristics seems to be directly related to the nucleation and growth (NG) mechanism. The distinction between these mechanisms is however not as simple because of the important influence of temperature and dynamics of this process. More conclusive information can be surely obtained by the simultaneous studies of kinetics and morphology. A very interesting up-to-date review on structure formation in polymer mixtures by SD was recently published by Hashimoto.[89] This review considers both a contemporary discussion of the theory of phase separation in polymeric systems and the experimental results obtained for many systems. We wish, however, to follow his treatment particularly from the point of view of the application of SALS as an important experimental technique for studies of kinetics of transformations and for characterizing the arising structures. This is due to the fact that the control of the structures or morphology determines all the properties of the system. On the fundamental level it is also important to obtain some insight on molecular parameters affecting the structure-property relationship of polymer blends. The role of the molecular parameters on the structure and properties is evident because they influence both the equilibrium and kinetic aspects of the phase separation, e.g., binodal and spinodal curves and mechanism of phase separation, NG vs. SD. The kinetics of phase separation depends on molecular dynamics in thermodynamically metastable and unstable regimes. Although the equilibrium structure is identical regardless of the mechanism of the phase separation, the intermediate structures formed, especially in the early stage of phase separation, may be quite different for the two above-mentioned mechanisms. Both depend on the growth of the concentration fluctuations of one component. If the early SD stage periodic concentration fluctuations with a certain wavelength are built up in the whole space and their amplitude increases with time, the corresponding wavelength is nearly constant. This wavelength depends on thermodynamic conditions, e.g., on supercooling or supersaturation, and the kinetics of phase separation is related to the amplitude of these fluctuations.

In the case when NG dominates, the nuclei have the composition close to that at equilibrium and they grow by thermal activation process. In analogy to simple crystallization process the nuclei, larger than critical ones, spontaneously grow. The average size of the nuclei is determined by kinetics of the phase separation process and the amplitude of fluctuations is also determined by the quench depth. In this way size and amplitude of fluctuations in NG range is different.

Usually the experimental investigations are concerned with very early stages of SD in various polymeric ternary solutions, polymer 1, polymer 2, and solvent in block copolymers with solvent and in binary solutions of two polymers. Some papers have been devoted to determine molecular parameters characterizing the later stage of SD too.

Early stages of SD were recently studied using time-resolved light-scattering technique. In that case the growth of each Fourier component of the concentration fluctuation, with wave number q can be analyzed by real time detection of light intensity at the scattering vector q. In real space $q = 2\pi/\Lambda$ and in the reciprocal space $q = (4\pi/\Lambda) \sin \frac{\Theta}{2}$ where λ is the wavelength of light and Θ is the scattering angle in the medium. The use of the light scattering technique is limited for the small-q-regime, i.e., $q R_0 \ll 1$ or $\Lambda \gg R_0$ where R_0 is the unperturbed chain dimension of the polymer. We cannot go into the details of the analysis of the conditions which have to be fulfilled to consider a small-q-regime and we cannot analyze the deviations from the linear SD theory proposed at first by Cohn. They may be of some importance for polymeric systems at latter stages of SD. The same remark is valid for a more pertinent discussion related to translational diffusion of the center-of-mass of polymer chains in entangled solutions or in the bulk solid state as described in terms of reputation process.

A. TIME-RESOLVED LIGHT SCATTERING AND RESULTS OF MORPHOLOGICAL STUDIES
1. Amorphous Polymer Blends

The use of the laser made it possible to study the demixing process by the time-resolved light scattering technique on the sample rapidly quenched from the thermodynamically stable state into the metastable or unstable state. An automated step-scanning laser-light scattering photometer was constructed by Hashimoto et al.[90] Here we will describe its main parts. It consists of 20 mW He-Ne gas laser emitting an incident light beam (Figure 2). Its polarization

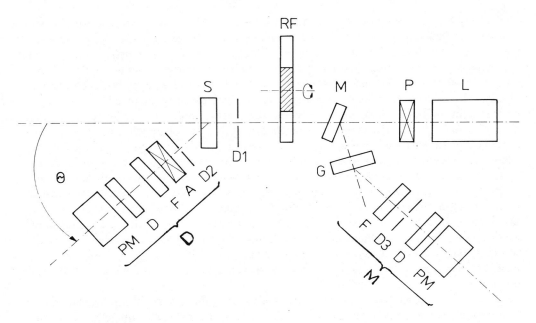

FIGURE 2.   Scheme of the time-resolved SALS apparatus. L — He-Ne gas laser, P — rotating polarizer, M — glass slide, RF — rotating circular neutral density filter, D — detector, D_1 — aperture to eliminate parasite scattering from the optics, S — sample, D_2 — aperture determining the angular resolution of the scattering profile, F — neutral density filter, D — opaque glass, PM — photomultiplier, Θ — measuring angle, and M — monitor. G, F, D_3, and D are the monitoring system with a photomultiplier PM. (From Hashimoto, T., Kumaki, J., and Kawai, H., *Macromolecules,* 16, 641, 1983. With permission.)

direction could be rotated by a polarizing rotator. A part of the incident beam was reflected by a glass slide (M) scattered by an opaque glass plate and detected by monitoring phototube (PM). The rest of incident beam was passed through a rotating circular neutral filler (RF) to adjust correctly the light intensity of the incident beam and an aperture (D1) to eliminate parasite scattered light. The scattered beam intensity from the sample (S) was detected by a photomultiplier (PM) through an aperture (D) which determines the angular resolution of the scattering profile, an analyzer (A), and neutral density filter (F).

The scanning of the PM is driven by a pulse motor and the scanning modes are controlled by a microprocessor. The scattered intensity data and times at which the intensity data were sampled are stored in a microcomputer. The time data are important because they tell the time after which the isothermal phase separation has started or/and they define the scattering angles at which the intensity data are sampled during the step- or continuous-scanning operation of PM.

This technique was used for investigation of dynamic liquid-liquid phase separation of a polymer blend consisting of polystyrene and poly(vinyl methyl ether). Both regimes of NG and SD have been investigated. It was found that at early stages of SD the intensity at a given q increases exponentially with time after the initiation of the isothermal phase separation induced by a temperature jump from the temperature well below the binodal point. The relaxation rate 2R(q) of the intensity increases the growth of fluctuations in the linear change in the SD regime, a function of q such that $R(q)/q^2$ linearly decreases with q^2. This is in accordance with the linear theory proposed by Cahn and extended by de Gennes for polymers. It shows well the power of this technique. The spinodal temperature was determined from the dynamics measured as a function of temperature in the linear SD regime. It seems interesting to note that, in the late stage of SD, the intensity increases with time and starts to deviate from exponential behavior. The scattering maximum shifts to smaller q, corresponding to the onset of the coarsening process. The higher the superheating, as expected, the earlier the range where the coarsening starts. In the NG regime the intensity was found to increase nonexponentially with time.

We have devoted more time to the discussion of the last paper because it shows clearly the potentialities of this technique. It seems that time-resolved SALS has certain advantages over the NMR technique. However, it will be inadequate to neglect the recent progress made in many laboratories concerning the use of pulsed NMR for structure formation in polymer blends, e.g., see References 86a and 86b, and many other papers of Nishi's group in which they describe many aspects of phase separation and structure of the interphase being formed. The investigations of Kitamaru's group are similar. They have investigated phase separation and behavior of interphases mostly in polyolefins, e.g., see Reference 87. In this paper a sample literature review and monographic literature are indicated. It is appropriate to mention here the new progress in the application of digital image analysis to pattern formation in polymeric systems.[88] This method makes it possible to extract various types of information from the image of a high-order structure in the polymer system observed under a microscope. Due to various numerical operations in real and wave-number space is it possible to extract physical information which is not simply obtained from light scattering. Let us continue further discussion on the results obtained using time-resolved light scattering.[91-93] They show its wide application for phase separation and phase dissolution in several polymer blends. The development of much simpler equipment for this purpose has to be noted too. Hashimoto et al.[91] describe a simpler setup for recording the scattering patterns. A plane-polarized He-Ne laser was used as a source of light. The polarization direction of the beam can be rotated by a polarization rotor one-half wavelength retardation plate. A special sample cell made of quartz is located in a temperature-controlled chamber, and placed on horizontal stage. The scattered light intensity was recorded through an analyzer onto film or by a TV camera system composed of a silicone vidicona, videotape recorder, and a TV monitor capable of recording 60 frames per second. The time of insertion of the sample cell into a constant temperature chamber was exactly measured. We will omit the more detailed description of the sample heating system making it possible to regulate the temperature or to carry out the experiment in temperature jump conditions.

This method makes it possible to study the evolution of structure with time after the temperature drops from above the cloud points to those below them and to investigate the structural changes as a function of the quench depth ΔT and polymer composition. The study carried out for polystyrene (PS), polybutadiene (PB), and toluene as well as for block copolymer PS/PB and solvent have shown that the early stage of SD (a few tens of seconds) can be described by the linearized theory proposed by Cahn. It has been also demonstrated that the time-dependent scattering profiles depend in an important way on the phase separation mechanism SD vs. NG. It is interesting to note that the phase-separated structures observed in solvent cast films are similar to those which were obtained at a certain concentration during the solvent evaporation process and reflect very much the course of SD and its coalescence process occurring in solution. These studies made it possible to extend our knowledge of the arising structures but also they have promoted the extension of the theories of phase separation, e.g., by coupling the linearized Cahn's theory with a pseudobinary approximation and mean-field theory approximation for the ternary solutions. The above-mentioned theoretical extensions seem to be very appropriate for describing the early stages of SD. The experimental support of these theoretical progresses does not mean a full success, because the discovered deviations can be attributed to a specific mean-field model for ternary solution and one also shall consider some experimental errors arising from the data evaluation at the primary stages of phase separation.

Recently this time-resolved SALS method was extended to the studies of phase dissolution in binary mixtures of PS and poly(vinyl methylether)[93] in liquid state. Time-resolved light scattering studies were applied for the mixture, which shows a LCST. This mixture of PS with poly(vinyl methylether) was demixed in dynamically metastable or unstable state for a short period in order to generate concentration fluctuations characteristic of early stage of phase separation. By the temperature jump technique this mixture was rapidly transferred to a thermodynamically stable state and the decay rate R(q) of the concentration fluctuation was investigated as a function of q (wave number of the fluctuations). It has been shown that in early stage of dissolution in small q-region the concentration fluctuations decay approximately exponentially; R(q) is proportional to q^2. These types of studies also make it possible to determine the apparent diffusion constants D_{app} and further conclusions of its dependence on temperature and its correlation with thermodynamic driving force for phase dissolution. The above-presented characteristics summarizes the results obtained using various systems. Let us give some information on structural and molecular parameters which can be obtained from these studies. There are, for example, the average radius of the dispersed domain R_s which may be estimated approximately from

$$4\pi(R_s/\lambda)\sin\Theta_m/2 = 5.768$$

Assuming that the maximum arises from interparticle interference, the average distance D_s may be calculated:

$$2D_s\sin(\Theta/2) = \lambda$$

It seems necessary, however, to note that the observed regularity of interdomain distance is connected probably with the basic aspect of structure formation during the solvent evaporation process. When concentration increases, the phase separation grows according to the mechanism of either NG or SD. The structure formed in solution is frozen in some stage of phase separation at which concentration and viscosity both become high enough so that the system cannot attain equilibrium with a given time scale of solvent evaporation process. This frozen-in structure in solution remains and it is observed in final films. Evidently this structure depends on temperature or more exactly on ΔT and time. There is a kind of "structure memory" developed in solutions and it can be retained somehow in bulk. In ternary solutions the situation is more complex. Usually for elucidation of the decomposition process one assumes pseudobinary approximation but the detailed discussions are beyond the scope of the present paper in spite of the fact that they are of interest in both academic and industrial viewpoints. The analyses based on linear theory of SD and on special mean-field model for the mixtures make it possible to predict in what regime the NG or SD will dominate whereby the time plays an important role. A very lucid and pertinent analysis of the time scale where the linear theory is applicable is given in Hashimoto's review[89] where the interested reader may find all necessary theoretical background as well as pertinent citations.

Summarizing the results presented here one shall say that the time evolution of scattering intensity is affected by the random thermal force less than that of small-molecular systems, despite of the fact that the effect at small q is more significant in polymers than in small molecules. It is probably due to the quench depth employed which in that case is much larger than usually used for small molecules.

The present knowledge on the demixing kinetics is much better for the early stage; thus, evidently efforts shall be made to understand the late stages of demixing especially with regard to the aspects of polymer demixing mechanisms which are different from that concerning low molecular systems.

The above-presented discussion concerned mainly the amorphous binary or ternary systems. An important part of blends consists of one crystallizable component. In that case

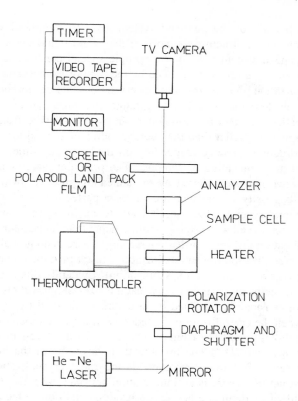

FIGURE 3.   Simplified schematic diagram of the time-resolved apparatus.
(From Hashimoto, T., Sasaki, K., and Kawai, H., *Macromolecules*, 17,
2812, 1984. With permission.)

time-resolved SALS has been shown to be very useful for studies of kinetics of phase separation and for morphology control.

2. Crystallizable Polymer Blends

Polymer blends containing at least one constituent which is able to crystallize are of special practical interest. The structure of such systems is controlled not only by the arrangement of separated components during cooling but also by the crystalline morphology of the second phase which usually exists in form of more or less clear spherulitic entities.

The SD of molten polymer blend involves demixing of the constituents according to SD. Rapid cooling and crystallization of the demixed system results often in a morphology of space-filling spherulites containing "structure memory" of the liquid-liquid phase separation. These effects and structures deserve in my opinion a special consideration; thus, I wish to discuss in some detail a very interesting paper of Hashimoto et al.,[94] devoted to this problem. The model system investigated consisted of polypropylene (PP) and of ethylene-propylene random copolymer (EPR). These particular polymers are highly immiscible and consequently they will demix in the liquid state above the melting temperature of the constituting polymers. Using time-resolved SALS one can investigate in some detail the various stages of SD and subsequently coarsening processes as well as the crystallization effects induced rapid quenching. The system used in this investigation is similar to that described in Reference 91 (Figure 3). The phenomena which take place are not simple because demixing causes spatial periodic concentration fluctuations and coarsening which in turn causes spatial variations of melting temperatures and results in different crystallization rates. Rapid crystallization of the demixed system leads to morphology of space filling spherulites, characteristic of structure memory of the liquid-liquid phase separation of PP and EPR. This organized structure is due to periodic spacial concentration fluctuations of defined wavelength which in turn is controlled by the time of isothermal demixing process. During rapid crystallization process formation of crystallites and of morphological structures takes place in PP-rich domains without a long-range rearrangement of PP molecules connected with segregation of the EPR component from crystallizing fronts of PP. When, however, the domains are smaller than the average radius of spherulites the morphological entities retain the structure memory, depending on the demixing time. Interesting effects have been observed after annealing of samples. The principal features of the spherulites, e.g., its diameter, have not been altered by the annealing process; thus, one can say that the annealing results mostly in the increase of their perfection, i.e., in the increase of crystallinity, by secondary crystallization. It leads to an interesting conclusion that L-L demixing memory is conserved within the spherulite rather than in the interspherulitic regions and that the crystallization prevents further coarsening. There is no global rearrangement of PP or EPR during crystallization (the crystallization is faster than the rate of global diffusion of polymers which would be required for segregation). Evidently this is due to the specific quench conditions. It seems on purpose to say that the sizes of spherulites are not controlled directly and their dimensions are defined by structure established by L-L demixing. This situation changes when the crystallization is conducted at higher temperatures. This problem as well as the morphology control of binary polymer was discussed in the next paper on crystalline morphology and SD.[95] The investigations have shown that the bimodular, bicontinuous, and periodic structure conserved during rapid crystallization is a particular example of blend structure. There is an interrelation between these two types of structures which may be modified by demixing and crystallization conditions. One has observed such structures in which the spherulite radius is smaller than the width of the domain, as in the previous case, but it is also possible to observe the opposite situation by modification of demixing time before crystallization. The study of slow crystallization, which is not diffusion limited, may define the conditions or establish a criterion in which the structure memory of liquid-liquid demixing is conserved during the crystallization process. The time-resolved SALS studies made it possible to define the diffusion-limited crystallization (detailed studies of H_v and H_v patterns) as well as to determine the spherulite growth rates which depending on temperature are linear or nonlinear with time. This last effect is easily interpreted as a consequence of the long-range rearrangement of modulated structure and segregation of EPR on the crystallization front. During long-time crystallization the fraction of EPR at the crystallization front increases with time, resulting in suppressing of the rate of the spherulite growth. The time-resolved SALS investigations also made possible the studies of anisotropy fluctuations of H_v scattering from spherulites which exhibit certain peculiarities, e.g., H_v spinodal rings, at the same angle as the spinodal ring. At the end of this characteristic of the structure formation in the crystallizable blends it is proper to add that this investigation made also possible the observation of various ''macrostructures'' being formed. One shall include here samples with dispersed spherulites as well as ''spherulite villages'' separated by domains rich in amorphous polymer. These structures depend on time of heating above the melting point of both components and/or quench procedure which determines the crystallization rates.

These interesting results show how useful are the time-resolved studies of SALS. It is not out of place to say that all these studies were carried out assuming homogeneous nucleation in the PP phase. Usually polymers like PP and EPR contain impurities which can be hardly separated even by a very careful purification process. These impurities act as external heterogeneous nucleating agents and their migration, in the molten state, from one component into the another one may change the crystallization condition in a very pronounced way. These effects have been investigated in our laboratory.[67,68,96,99]

It has been shown for several miscible and immiscile polymeric blends that the driving force for the migration is the difference of the interfacial free energy of the impurity with respect to the melt of the components of the blend. The detailed discussion of these phenomena is out of the scope of this discussion but they have to be considered in order to reach a real picture of the phenomena controlling the final morphology.

V. CONCLUDING REMARKS

The aim of this review was to show, on the basis of some selected examples, the wide possibilities of the application of laser light scattering technique to solve many important problems related to structure determination in solid polymers. Both static SALS as well as time-resolved SALS investigations furnish information on the existing structures or on their formation in various external conditions. Other optical techniques frequently used for the investigation of structure-property relationships, e.g., birefringence studies and rheooptical investigations, have also profited to a large extent from high intensity and polarization of laser light.

The above-mentioned examples of studies of polymer structure using mostly SALS are evidently not exhaustive but they give to the reader a feeling on the potentialities of these techniques.

It is worth saying that the advantages of SALS and other optical methods mentioned above are limited and cannot give the completely satisfactory answer to all problems encountered in the study of solid polymers. Let us mention some of these limitations. They are mostly related to the resolution power which is determined by the wavelength of light. For the investigations of order-disorder transitions characteristic for block copolymers, e.g., induced by a temperature jump, time-resolved SAXS[100,101] combined with small angle neutron scattering seem to be more appropriate.[102] These techniques make it possible to study transient structural responses at a time interval on the order of seconds, the kinetics of such transitions as well as the equilibria in function of concentration. The kinetics studies resulted in evaluation of the effective translational diffusivities for the center of mass motion of the block copolymers in the solution and in determination of the activation energies related to this diffusion process. The problem of the molecular conformation of a block copolymer chain in the separated domain can be directly solved by SANS while SAXS measurements make possible to separate the scattering arising from the microdomain structure. The further advantage of these combined techniques is the determination of perturbed state of macromolecules. SAXS studies[104] are also extremely important to construct a model of oriented lamellar microdomains in diblock copolymer films. The comparison with one-dimensional paracrystal model with uniaxial orientation has been shown particularly useful including the role of domain boundary thickness.[103] I hope that the reader will realize the basic differences between the problems briefly treated above and the macroscopic structural studies carried out using SALS.

Let us finish this discussion with indication of some problems which are not finally solved as yet but are of basic importance for the understanding of structure formation in polymer blends. They concern the studies of the late stages of decomposition of binary polymer mixtures by SALS. The aim of these investigations is not only to learn about the mechanism of the decomposition at long time period[104,105] but also to reach some conformity with general laws found for the dynamics of phase transition in other mixtures: metallic alloys and inorganic glasses. Evidently it is necessary to establish also the specific characteristic associated with connectivity inherent to polymer chains. This approach is very important because the physics of polymers, in spite of its specificity, shall reach many basic common points with general physics of condensed state matter.

SALS has also been recently used for rheooptical studies of polymers under shear flow[106] or for liquid crystalline systems, as it was mentioned before. The most interesting seems the study of the evolution of structure in polymer blends in which one component is a liquid crystalline material.[107] This review does not give a whole account of the achievements obtained using lasers and particularly SALS for structure determination of structures arising in polymers. The theoretical and experimental framework which has been built during the last 2 decades to describe the morphology of various polymeric systems using light-scattering techniques is rather satisfactory and many structure-property relationships already established are very useful for the present and for the future uses of heterogeneous polymeric systems. The application of these techniques is fraught with theoretical and experimental difficulties. In particular more adequate methods are needed for determination of domain sizes, phase composition and interphases, or/and interphases which are surely of primary importance. They define the adhesion between phases and affect mechanical properties, fracture diffusion, etc. not only in usual polymer blends but also in IPNs and in polymer composites which are now the main subject of interest of numerous research groups.

The separation of phases via NG or SD is principally understood, but much work remains to be done to control the phase transformations in polymeric systems. From the theoretical point of view studies on advanced statistical thermodynamics are of principal importance including the experimental justification of the predicted rules. The same concerns the solid-state transitions in blends as well as the aging phenomena which also are now more and more recognized as main factor governing the long-term uses.

I am convinced that in spite of rapid development of more sophisticated and more expensive techniques, the versatility and simplicity of SALS are the main factors which will preserve its well-deserved place among many other methods used for evaluation of complex structure and morphology of polymeric materials. Imagination and experimental skillfulness surely will lead to the further extension of this method.

The author is grateful to his collaborators for their enthusiastic contribution to this work.

In addition, I am indebted to Professors R. Stein, H. Benoit, Dale J. Meier, H. Kawai, and T. Hashimoto for many competent and insightful discussions during several years of our scientific contacts.

I wish also to express my gratitude to the Japan Society for Promotion of Science in Tokyo which provided a research grant for visiting in Japan which made it possible to write a part of this work.

This work was supported by the Polish Academy of Science in the frames of Project 01.14.

REFERENCES

1. **Stein, R. S.**, Optical methods of characterizing high polymers, in *Newer Methods of Polymer Characterization*, Ke, B., Ed., Wiley-Interscience, New York, 1964, chap. 4.
2. **Samuels, R. J.**, *Structured Polymer Properties: The Identification, Interpretation and Application for Crystalline Polymer Structure*, Wiley-Interscience, New York, 1974.
3. **Ward, I. M.**, *Structure and Properties of Oriented Polymers*, Applied Science Publishers, London, 1975.
4. *Small Angle Scattering from Fibrous and Partially Ordered Systems*, Marchessault, R. H., Ed., J. Polymer Sci. C. 13, 1966.
5. **Kryszewski, M., Krasnikowa, N. P., and Milczarek, P.**, On the interaction of an impulse laser beam with supermolecular structural elements in crystalline polymers, *J. Polymer Sci. Polym. Lett.*, 2, 813, 1971.
6. **Kryszewski, M., Galęski, A., and Milczarek, P.**, Destruction of polymers with crystalline inhomogeneities under the influence of intense laser beams, *J. Polym. Sci. Polym. Lett.*, 14, 365, 1976.
7. **Pakula, T. and Soukup, S.**, New apparatus for studies on supermolecular structures of thin polymer films by small angle light scattering, *J. Polym. Sci. Polym. Phys. Ed.*, 12, 2437, 1974.

8a. **Gałęski, A. and Kryszewski, M.,** Small-angle light scattering from hedrites, ovoids and spiral ovoids. I. Theoretical consideration, *J. Polym. Sci. Phys. Ed.,* 12, 455, 1974.

8b. **Gałęski, A. and Kryszewski, M.,** Small angle light scattering from square and hexagonal hedrites, *J. Polym. Sci. Phys. Ed.,* 14, 181, 1976.

9. **Gałęski, A. and Kryszewski, M.,** Small angle light scattering from hedrites, ovoids and spiral ovoids. II. Experimental, *J. Polym. Sci. Phys. Ed.,* 12, 471, 1974.

10. **Hashimoto, T., Murakami, Y., Okamori, Y., and Kawai, H.,** Origin of complex light scattering pattern from some crystalline polymer films, *Polym. J.,* 6, 554, 1974.

11. **Daniewska, I., Hashimoto, T., and Nakai, A.,** Laser-light scattering from ringed spherulites in poly(decamethylene)terephtalate: observation and interpretation of unusual "ring scattering" patterns, *Polym. J.,* 16, 49, 1984.

12. **Montegi, M., Moritani, M., and Kawai, H.,** Effect of birefringence on light scattering from deformed spherulitic polymer films. II, *J. Polym. Sci.,* 8, 499, 1970.

13. **Oda, T., Montegi, M., Moritani, M., and Kawai, H.,** On the dimensional change of polymer spherulite with uniaxial stretching of bulk specimen. II. Observation of polyethylene by means of light scattering technique, *Kobunshi Kagaku,* 25, 639, 1968.

14. **Kryszewski, M., Gałęski, A., and Pakuła, T.,** Verification of the small-angle light scattering from undeformed spherulites and the relation between spherulite deformation and stretching of polyethylene films, *Eur. Polym. J.,* 7, 241, 1971.

15. **Pakuła, T. and Kryszewski, M.,** A study of the elongation ratio of deformed spherulites, *J. Polym. Sci. C.,* 38, 87, 1972.

16. **Pakuła, T., Gałęski, A., and Kryszewski, M.,** Studies of single spherulite deformation. I, *J. Polym. Sci. C.,* 42, 753, 1973.

17. **Todo, A., Hashimoto, T., and Kawai, H.,** Deformation of polyethylene spherulites as observed by wide angle light scattering, *Polym. J.,* 11, 59, 1979.

18. **Pakuła, T., Kryszewski, M., and Soukup, Z.,** Determination of spherulite distribution by small-angle light scattering, *Eur. Polym. J.,* 12, 41, 1976.

19. **Hashimoto, T., Todo, A., Murakami, Y., and Kawai, H.,** Light scattering from crystalline superstructure in tubular-extruded poly(butene-1) films, *J. Polym. Sci. Phys. Ed.,* 15, 501, 1977.

20. **Hashimoto, T., Todo, A. and Kawai, H.,** Light scattering from crystalline superstructure in tubular-extruded poly(1-butene) films. II. Analysis of interparticle interference effect based upon a paracrystal model, *Polym. J.,* 10, 521, 1978.

21. **Hashimoto, T., Ishido, S., Kawai, H., and Ziabicki, A.,** Morphological studies of oriented crystallization of high-density polyethylene, *Macromolecules,* 11, 1210, 1978.

22. **Hashimoto, T., Nagatoshi, K., Todo, A., and Kawai, H.,** Superstructure of high density polyethylene film crystallized from stressed polymers as observed by small angle light scattering, *Polymer,* 17, 259, 1976.

23. **Moritani, M., Hayashi, N., Utso, A., and Kawai, H.,** Light scattering patterns from collagen films in relation to the texture of a random assembly of anisotropic rods in the three dimensions, *Polym. J.,* 2, 74, 1971.

24. **Hashimoto, T., Murakami, Y., Hayashi, N., and Kawai, H.,** Light scattering from polymer films having an optically anisotropic rod-like texture in relation to a model of a random assembly of disordered rods, *Polym. J.,* 6, 132, 1974.

25. **Hashimoto, T., Kawasaki, H., and Kawai, H.,** Crystal and amorphous orientation behaviour of poly(chlorotrifluoroethylene) films in relation to crystalline superstructure, *J. Polym. Sci. Phys. Ed.,* 16, 271, 1978.

26. **Murakami, Y., Hayashi, N., Hashimoto, T., and Kawai, H.,** Light scattering from random assembly of anisotropic rod-like structure with principal optic axes oriented in cylindrical symmetry with respect to their own rod axes, *Polym. J.,* 4, 128, 1973.

27. **Hashimoto, T., Yamaguchi, K., and Kawai, H.,** Light scattering from films having optical anisotropic rodlike texture. II. Principle to estimate its size, *Polym. J.,* 9, 271, 1977.

28. **Matsuo, M., Nomura, S., Hashimoto, T., and Kawai, H.,** Light scattering from a random assembly of anisotropic plates in two and three dimensional space, *Polym. J.,* 6, 151, 1974.

29. **Picot, C., Weill, G., and Benoit, H.,** Light scattering by anisotropic discs. Application to polymer single crystals, *J. Colloid Interface Sci.,* 27, 360, 1968.

30. **Uemura, Y., Fujimura, M., Hashimoto, T., and Kawai, T.,** Application of light scattering from dielectric cylinder based upon Mie and Rayleigh-Gans-Born theories to polymer systems. I. Scattering from a glass fiber, *Polym. J.,* 10, 341, 1978.

31. **Uemura, Y., Hashimoto, T., and Kawai, H.,** Application of light scattering from dielectric cylinder based upon Mie and Rayleigh-Gans-Born theories to polymer systems. II. Scattering from a coaxial and circular cylinder, *J. Soc. Fiber Sci. Technol., Jpn.,* 34, T 481, 1978.

32. **Uemura, Y., Hashimoto, T., and Kawai, H.,** Application of light scattering from a dielectric cylinder based upon Mie and Rayleigh-Gans-Born theories to polymer systems. III. An application to polymer composites, *Polym. J.,* 11, 413, 1979.

33. **Gałęski, A., Pakuła, T., and Kryszewski, M.,** Small-angle light scattering SALS from helically twisted fibers, *J. Polym. Sci. Polym. Symp.,* 61, 35, 1977.

34. **Pakuła, T., Morawiec, J., Świętosławski, J., and Kryszewski, M.,** Melt-spun polypropylene fibers with helical orientation. I. Morphology and mechanical properties characterization, *J. Polym. Sci. Polym. Symp.,* 58, 323, 1977.

35. **Uemura, Y., Tsujiei, K., Hashimoto, T., and Kawai, H.,** Application of light scattering from dielectric cylinder based upon Mie and Rayleigh-Gans-Born theories to polymer systems. IV. An application to craze formation in glassy polymers, *Polym. J.,* 11, 955, 1979.

36. **Hashimoto, T., Ebisu, S., and Kawai, H.,** A novel technique for characterizing the sense of twisting in cholesteric liquid crystals, *J. Polym. Sci. Polym. Lett. Ed.,* 18, 569, 1980.

37. **Hashimoto, T., Ijitsu, T., Yamaguchi, K., and Kawai, H.,** Supermolecular structure of polypeptides in concentrated solutions and films. I. Origin of rod-like scattering and form-optical rotation, *Polym. J.,* 12, 745, 1980.

38. **Hashimoto, T., Eibisu, S., Inaba, N., and Kawai, H.,** Supermolecular structures of polypeptides in concentrated solutions and films. II. Small-angle light scattering from cholesteric mesophase, *Polym. J.,* 13, 701, 1981.

39. **Hashimoto, T., Inaba, N., Ebisu, S., and Kawai, H.,** Supermolecular structure of polypeptides in concentrated solutions and films. III. Inhomogeneous twisting of cholesteric mesophase as clarified from multiple-order light scattering maxima, *Polym. J.,* 13, 897, 1981.

40. **Hashimoto, T.,** Laser light scattering from cholesteric liquid crystals, *J. Appl. Polym. Sci. Appl. Polym. Symp.,* 41, 83, 1985.

41. **Beebe, E. V., Coalson, R. L., and Marchessault, R. H.,** Characterization of cellulose gels by small angle light scattering, *J. Polym. Sci. C.,* 13, 103, 1966.

42. **Coalson, R. L., Marchessault, R. H., and Peterlin, A.,** Small angle scattering by a three-component system. Testing of the theory for light scattering, *J. Polym. Sci. C.,* 13, 123, 1966.

43. **Milczarek, P., Pluta, M., and Kryszewski, M.,** SALS investigations of internal spherulitic order and *N*-heptane penetration in styrene modified low density polyethylene, *Colloid Polym. Sci.,* in press.

44. **Pakuła, T., Kryszewski, M., and Pluta, M.,** Mechanical and rheo-optical in low-density polyethylene, *Eur. Polym. J.,* 13, 141, 1977.

45. **Hashimoto, T., Fujimura, M., and Kawai, H.,** *Structure of Sulfonated and Carboxylated Perfluorinated Ionomer Membranes,* in *Perfluorinated Ionomer Membranes,* (ACS Symp. Ser., Vol. 180), Eisenberg, A. and Yeager, H. L., Eds., American Chemical Society, Washington, D.C., 1982, chap. 11.

46. **Gałęski, A., Milczarek, P., and Kryszewski, M.,** Heat conduction in a two dimensional spherulite, *J. Polym. Sci. Phys. Ed.,* 15, 1267, 1977.

47. Symposium on New Developments in Small Angle Scattering for Polymers, Division of Polymer Chemistry, American Chemical Society, Washington, D.C., August 1983, *Polym. Prepr.,* 24, Vol. 2, 213.

48. Symposium on ACS Award in Polymer Chemistry Honouring R. S. Stein, Division of Polymer Chemistry American Chemical Society, Seattle, March 1983, *Polym. Prepr.,* 24, Vol. 1, 180.

49. **Platzer, N. A. J., Ed.,** *Multicomponent Polymer Systems,* (Adv. Chem. Ser., Vol. 99), American Chemical Society, Washington, D.C., 1971.

50. **Platzer, N. A. J., Ed.,** *Copolymers, Polyblends and Composites,* (Adv. Chem. Ser., Vol. 142), American Chemical Society, Washington, D.C., 1975.

51. **Aggarwal, S. L., Ed.,** *Block Polymers,* Plenum Press, New York, 1970.

52. **Molau, G. E., Ed.,** *Colloid and Morphological Behavior of Block and Graft Copolymers,* Plenum Press, New York, 1971.

53. **Manson, J. A. and Sperling, L. H., Eds.,** *Polymer Blends and Composites,* Plenum Press, New York, 1976.

54. **Sperling, H. L., Ed.,** *Recent Advances in Polymer Blends, Grafts and Blocks,* Plenum Press, New York, 1974.

55. **Kuleznev, V. N.,** *Mnogokomponentnye Polimernye Sistemy,* (in Russian), Khymya, Moscow, 1974.

56. **Shen, M. and Kawai, H.,** Properties and structure of polymeric alloys, *ATChE J.,* 24, 1, 1978.

57. **Martuscelli, E., Palumbo, R., and Kryszewski, M., Eds.,** *Polymer Blends — Processing, Morphology and Properties,* Vol. 1, Plenum Press, New York, 1980.

58. **Kryszewski, M., Gałęski, A., and Martuscelli, E., Eds.,** *Polymer Blends — Processing, Morphology and Properties,* Vol. 2, Plenum Press, New York, 1984.

59. **Klempner, D. and Frisch, K. C., Eds.,** *Polymer Alloys — Blends, Blocks, Grafts and Interpenetrating Networks,* Polym. Sci. and Technol., Vol. 20), Plenum Press, New York, 1983.

60. **Paul, D. R. and Newman, S., Eds.,** *Polymer Blends,* Vol. 1 and 2, Academic Press, New York, 1978.

61. **Bucknall, C. B.,** *Toughend Plastics,* Applied Science Publishers, London, 1977.
62. **Nielsen, L. E.,** *Mechanical Properties of Polymers and Composites,* Marcel Dekker, New York, 1974.
63. **Hopfenberg, H. B., Ed.,** Permeability of plastic films and coatings to gases, vapours and liquids, *Polymer Science and Technology,* Vol. 6, Plenum Press, New York, 1974.
64. **Corfagna, C., Apicella, A., Drioli, E., Hopfenberg, H. B., Martuscelli, E., and Nicolais, L.,** *Polymer Blends — Processing, Morphology and Properties,* Martuscelli, E., Palumbo, R., and Kryszewski, M., Eds., Plenum Press, New York, 1980, 383.
65. **Stein, R. S.,** Optical behaviour of polymer blends, in *Polymer Blends,* Vol. 1, Paul, D. R. and Newman, S., Eds., Academic Press, New York, 1978, chap. 9.
66. **Kakudo, M. and Kasai, N.,** *X-Ray Diffraction by Polymers,* Elsevier, New York, 1972.
67. **Hosemann, R. and Bagchi, S. N.,** *Direct Analysis of Diffraction by Matter,* North-Holland, Amsterdam, 1962.
68. Physical optical of dynamic phenomena and processes, in *Macromolecular Systems,* Sedlacek, B., Ed., Walter de Gruyter, Berlin, 1985.
69. **Alexander, L. E.,** X-ray diffraction methods, in *Polymer Science,* John Wiley & Sons, New York, 1969.
70. **Meier, D. J.,** *Am. Chem. Soc. Polym. Prepr.,* 15, 174, 1974.
71. **Helfand, E.,** Block copolymers, polymer-polymer interfaces and the theory of inhomogeneous polymers, *Acc. Chem. Res.,* 8, 295, 1975.
72. **Ishida, H. and Koenig, J. L., Eds.,** *Composite Interfaces,* Elsevier/North-Holland, New York, 1986.
73. **Sedlacek, B., Ed.,** *Polymer Composites,* Walter de Gruyter, Berlin, 1986.
74. **Kawai, H. and Hashimoto, T.,** Microdomain structure and some related properties of block copolymer, in *Contemporary Topics in Polymer Science,* Vol. 3, Shen, M., Ed., Plenum Press, New York, 1979, 245.
75. **Hashimoto, T., Tsukahara, Y., and Kawai, H.,** Analysis of segmental interdiffusion in phase-separated block polymers and polymer blends by small-angle x-ray scattering, *J. Polym. Sci., Polym. Lett. Ed.,* 18, 582, 1980.
76. **Hashimoto, T., Shibayama, M., and Kawai, H.,** Domain-boundary structure of styrene-isoprene block copolymer films cast from solutions. IV. Molecular weight dependence of lamellar microdomains, *Macromolecules,* 13, 1660, 1980.
77. **Hashimoto, H., Fujimura, M., Hashimoto, T., and Kawai, H.,** Domain-boundary structure of styrene-isoprene block copolymers cast from solutions. VII. Quantative studies of solubilization of homopolymers in sphericol domain systems, *Macromolecules,* 14, 844, 1981.
78. **Pakuía, T., Saijo, K., Kawai, H., and Hashimoto, T.,** Deformation behaviour of styrene-butadien triblock copolymer with cylindrical morphology, *Macromolecules,* 18, 1244, 1985.
79. **Pakuía, T., Saijo, K., and Hashimoto, H.,** Structural changes in polystyrene-polybutadiene-polystyrene block polymers caused by annealing in highly oriented state, *Macromolecules,* 18, 2037, 1985.
80. **Benoit, H., Strazielle, C., and Benmoua, M.,** An evaluation of the flory X parameter as a function of concentration using the polymer solution scattering theory, *Acta Polymerica,* 39, 75, 1988.
81. **Walsh, D. J., Higgins, J. S., and Moconnachie, A.,** *Polymer Blends and Mixtures,* (NATO AST Series), Series E 89, Martinus Nijhoff, The Hague, 1985.
82. **Koleske, J. V.,** *Blends Containing Poly (ε-Caprolactone) and Related Polymers in Polymer Blends,* Vol. 2, Paul, D. R. and Newman, S., Eds., Academic Press, New York, 1978, 369.
83. **Hashimoto, T., Suehiro, S., Shibayama, M., Saijo, K., and Kawai, H.,** An apparatus for high speed measurements of small angle x-ray scattering profiles with a linear position sensitive detector, *Polym. J.,* 13, 501, 1981.
84. **Kawai, T. and Wang, T. T.,** Phase separation behaviour of polymer-polymer mixtures, in *Polymer Blends,* Vol. 1, Paul, D. R. and Newman, S., Eds., Academic Press, New York, 1978, 141.
85. **McMaster, L. P.,** Aspects of liquid-liquid phase transition phenomena, in *Multicomponent Polymeric Systems,* (Adv. Chem. Ser. No. 142), American Chemical Society, Washington, D.C., 1975.
86a. **Tanaka, H. and Nishi, T.,** New type of separation behaviour during crystallization process in polymer blends with phase diagram, *Phys. Rev. Lett.,* 55, 1102, 1985.
86b. **Tanaka, H. and Nishi, T.,** Direct determination of the probability distribution function of concentration in polymer mixtures undergoing phase separation, *Phys. Rev. Lett.,* 59, 692, 1987.
87. **Kitamaru, R., Flori, F., and Murayama, K.,** Phase structure of lamellar crystalline polyethylene by solid state high resolution ^{13}C NMR. Detection of the amorphous crystalline interphase, *Macromolecules,* 19, 636, 1986.
88. **Tanaka, H., Hayashi, T., and Nishi, T.,** Application of digital image analysis to pattern formation in polymer systems, *J. Appl. Phys.,* 59, 3627, 1986.
89. **Hashimoto, T.,** Structure formation in polymer mixtures by spinodal decomposition, in *Current Topics in Polymer Science,* Ottenbrite, R. M., Utracki, L. A., and Inone, S., Eds., Hanser Publishers, Munich, 1987.
90. **Hashimoto, T., Kumocki, J., and Kawai, H.,** Time resolved light scattering studies on kinetics of phase separation and phase dissolution in polymer blends. I. Kinetics of phase separation of a binary mixture of polystyrene and poly(vinyl-methyl-ether), *Macromolecules,* 16, 641, 1983.

91. **Hashimoto, T., Sasaki, K., and Kawai, H.,** Time resolved light scattering studies on the kinetics of phase separation and phase dissolution of polymer blends. II. Phase separation of ternary mixtures of polymer A, polymer B, and solvent, *Macromolecules,* 17, 2812, 1984.

92. **Sasaki, K. and Hashimoto, T.,** Time resolved light scattering studies on the kinetics of phase separation and phase dissolution of polymer blends. III. Spinodal decomposition of ternary mixtures of polymer A, polymer B and solvent, *Macromolecules,* 17, 2818, 1984.

93. **Kumaki, J. and Hashimoto, T.,** Time-resolved light scattering studies on kinetics of phase separation and phase dissolution of polymer blends. IV. Kinetics of phase dissolution of a binary mixture of polystyrene and poly(vinyl-methyl-ether), *Macromolecules,* 19, 763, 1986.

94. **Inaba, N., Sato, K., Suruki, S., and Hashimoto, T.,** Morphology control of binary polymer mixtures by spinodal decomposition. I. Principle of method and preliminary results on PP/EPR, *Macromolecules,* 19, 1690, 1986.

95. **Inaba, N., Yamada, T., Saruki, S., and Hashimoto, T.,** Morphology control of binary mixtures by spinodal decomposition. II. Further studies on PP/EPR, *Macromolecules,* in press.

96. **Bartczak, Z., Gałęski, A., Martuscelli, E., and Janik, H.,** Primary nucleation behaviour polypropylene/ ethylene/propylenerandom copolymer blend, *Polymer,* 25, 1843, 1985.

97. **Bartczak, Z. and Gałęski, A.,** Changes in the interface shape during crystallization in two-component polymer system, *Polymer,* 27, 543, 1986.

98. **Bartczak, Z. and Martuscelli, E.,** Spherulite primary nucleation in binary miscible blend of poly(ethyleneoxide) with poly(methyl-methacrylate), *Macromol. Chem.,* 188, 445, 1987.

99. **Bartczak, Z., Gałęski, A., and Krasnikowa, N. P.,** Primary nucleation and spherulite growth rate in isotactic polypropylene-polystyrene blends, *Polymer,* 28, 1627, 1987.

100. **Hashimoto, T., Kowasaka, K., Shibayama, M., and Suehiro, S.,** Time-resolved small angle X-ray scattering on the kinetics of the order-disorder transitions of block copolymers. I. Experimental technique, *Macromolecules,* 19, 750, 1986.

101. **Hashimoto, T., Kowasaka, K., Shibayama, M., and Kawai, H.,** Time resolved small angle X-ray scattering studies on the kinetic of the order-disorder transition of block polymers. II. Concentration and temperature dependence, *Macromolecules,* 19, 754, 1986.

102. **Hasayawa, H., Hashimoto, T., Kawai, H., Lodge, T. R., Amis, E. J., Glinka, E. J., and Han, Ch. C.,** SANS and SAXS studies on molecular conformation of a block polymer microdomain space, *Macromolecules,* 18, 67, 1985.

103. **Shibayama, M. and Hashimoto, T.,** Small-angle X-ray scattering analysis of lamellar microdomains based on a model of one-dimensional paracrystal with uniaxial orientation, *Macromolecules,* 19, 740, 1986.

104. **Hashimoto, T., Itakura, M., and Hasagawa, H.,** Late stage spinodal decomposition of a binary polymer mixture. I. Critical test of dynamic scaling on scattering function, *J. Chem. Phys.,* 85, 6118, 1986.

105. **Hashimoto, T., Itakura, M., and Shimidru, N.,** Late stage spinodal decomposition of a binary polymer mixture. II. Scalling analyses on $Q_m(\tau)$ and $I_m(\tau)^a$, *J. Chem. Phys.,* 85, 6773, 1986.

106. **Hashimoto, T., Takabe, T., and Suahivo, S.,** Apparatus to measure small-angle light scattering profiles of polymer under shear flow, *Polym. J.,* 18, 123, 1986.

107. **Nakai, A., Shiwaku, T., Hasagowa, H., and Hashimoto, T.,** Spinodal decomposition of polymer mixture with a thermotropic liquid crystalline polymer as one component, *Macromolecules,* 19, 3008, 1986.

Chapter 4

LASER SPECTROSCOPY IN LIFE SCIENCES

Angelika Anders and Marita Knälmann

TABLE OF CONTENTS

I. LASER PROPERTIES — THEIR USE IN SPECTROSCOPY

Lasers became indispensable tools in many fields of spectroscopy because of their great advantages over conventional light sources regarding spectral, spatial, and time resolution.[1,2] They opened a new era not only in physics and chemistry but also in biology and medicine.[3-9,12] Laser methods of spectroscopy are being widely used in basic biological and medical research. Laser investigations increase our knowledge about structures and interactions of biomolecules as well as about kinetics and dynamics of various biochemical processes. In medicine, the laser is gradually being introduced as a tool for treatment and diagnosis.

The many possibilities for using lasers in the investigation of biomolecules are based on their unique properties, which are also described in detail in Chapter 1, Volume I. Table 1 summarizes the properties of lasers with typical examples of their application.

High intensity and monochromaticity, resulting in a high spectral intensity, are ideal prerequisites for spectroscopic investigations especially for fluorescence measurements with low quantum yields, for the study of multiphoton processes and excited states as well as for Raman spectroscopy. For example, important biomolecules such as nucleic acids have an extremely low fluorescence quantum yield at room temperature. The Raman spectroscopy grew explosively with the availability of reliable laser sources. Laser Raman spectroscopy provides high selectivity and sensitivity also for questions of biochemical importance. The spectroscopic investigation of cells with lasers is another important field in biology (cyto-fluorimetry, see Section V). An application in photomedical research is the measurement of spectroscopic properties of thick specimens like human tissue. In photomedicine a spectral narrow-band irradiation of skin lesions and tumors can be obtained (photoradiation therapy, see Section V).

The development of tunable lasers enormously extended the capabilities of all kinds of spectroscopic research. Furthermore, the tunability of the wavelength over great spectral ranges is a prerequisite for the induction of selective biochemical effects or the determination of action spectra. Tunable lasers may perform a selective excitation of practically any quantum state of atoms or molecules in the wavelength range from about 200 nm to 20 μm (Section II). Raman spectroscopy, too, could not have been developed to its present utility without tunable lasers. The advantage mainly lies in the ability to obtain selectively the vibrational spectrum of a chromophore in a complex biological system. Action spectra (biological effects in dependence on the irradiation wavelength) of photobiological or pho-totherapeutical reactions can be investigated with high accuracy (Sections II and V).

The extremely short pulses of special laser systems give the possibility of observing the short lifetimes and very fast reactions of biomolecules. With the technological advances in pico- and subpicosecond laser spectroscopy the mechanisms of various physical, chemical, and biological processes can be evaluated. Ultrafast laser pulses opened the study of processes which are not currently amenable to be measured by any other method.

The very small focus due to the small divergence of laser light permits exposure of very selected sites within cells or tissues, using a laser focused through a microscope (microbeam experiments). Spots of about 0.01-μm diameter could be produced (Section V.E).

Quite another field of application in medicine is the thermal use of lasers in therapy, surgery, and coagulation.

II. LIFE SCIENCES — GENERAL ASPECTS

A. BIOLOGICAL PHENOMENA

Laser technology has opened a new era in the life sciences biology and medicine.[10,11] Biological phenomena which may be investigated with lasers are UV and visible radiation

TABLE 1
Laser Properties and Examples of Applications in Life Sciences

Laser properties	Typical applications
High intensity and monochromaticity (i.e., high-spectral intensity)	Spectroscopy of biomolecules, cells, and tissue
	Spectral narrowband irradiation in phototherapy and photochemotherapy
Tunability	Selective biological effects
	determination of action spectra
Short pulses	Time-resolved spectroscopy
	fast reactions
Small focus	Irradiation of selected parts in cell or tissue
Coherence	Scattering spectroscopy, Doppler spectroscopy

effects on cells, photosynthesis, vision, photomovement, photomorphogenesis, photoactivation of enzymes; bioluminescence, reactions of, for example, heme proteins, nucleic acids, and enzymes; cellular energy production, molecular motion, or photosensitization of cells. The receptor molecules, for example, of photosynthesis (chlorophyll and accessory pigments), of vision in vertebrates (rhodopsin), and of photomorphogenesis (phytochrome) are known. The genetic information is carried by the DNA (deoxyribonucleic acid).

With the formation of pigments absorbing visible light such as chlorophylls organisms succeeded — during the evolution of life — in harvesting the solar energy and transforming it into chemical energy in the process of photosynthesis. High energy compounds (biomass) are produced from low-energy precursors in this way. These phototrophic organisms also generate the oxygen in our atmosphere by using a water-splitting system. The development of visible light-absorbing pigments was necessary after the formation of a stratospheric ozone layer which filters the damaging UV radiation below 300 nm out of the solar spectrum (see Figure 2). Only during the initial phase of the evolution when no oxidizing atmosphere existed, short UV radiation reached the surface of this planet and could cause changes in the already existing molecules.

There are processes in which energy is derived from light such as in photosynthesis (energy fixation) and processes in which light is used as sensory signal. This group comprises effects of light on development (photomorphogenesis) and on orientation of plants (phototrophy) and microorganisms (photomovement) as well as on sensory systems of animals and men.

Lasers in life sciences are used to examine different types of processes; the study of biological processes which naturally require light, such as photosynthesis and vision, or lasers are used to investigate the structure of biomolecules, e.g., by Raman spectroscopy. Another class of applications are relaxation methods, where a biological system is perturbed away from equilibrium by means of laser pulses and the following recovery of the system is observed by laser spectroscopy. Systems which have been investigated very extensively with the last method are the heme proteins, typified by the proteins hemoglobin (Hb) and myoglobin (Mb). The role of Hb is to transport O_2 while that of Mb is to store it.

Biomolecules like chlorophyll, rhodopsin, and Hb can be investigated by visible excitation because of their visible electronic absorption bands. However, other biomolecules such as nucleic acids have electronic absorption bands only in the UV region. Action ranges of phototherapy and photochemotherapy of special dermatosis also lie in the UV range (Section V). Figure 1 gives a survey about the absorption ranges of some characteristic biomolecules and fields of action in medicine as well as about the most important types of lasers required for spectroscopic use in dependence upon the wavelength in the visible and UV region. The possibilities of using lasers in photobiology and in photomedicine depend

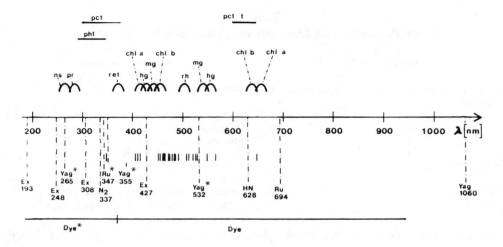

FIGURE 1.   Survey of absorption ranges of some characteristic biomolecules and fields of action in photomedicine (upper) and important lines and wavelength ranges of lasers (lower) in the visible and UV-wavelength region (middle). The absorption spectra are only drawn in their main maxima (see also Sections II.A and III). ns: nucleic acids, pr: proteins, ret: 11-*cis* retinal, rh: rhodopsin, hg: hemoglobin, mg: myoglobin, chl: chlorophyll a resp. b, pht: phototherapy of dermatoses (e.g., psoriasis), pct: photochemotherapy of dermatoses (e.g., psoriasis), pct t: photochemotherapy of tumors with hematoporphyrin, Ex: excimer laser, YAG: Nd:YAG laser (Neodym:yttrium/aluminum garnet), N₂: nitrogen laser, HN: helium neon laser, Ru: ruby laser, l: different argon and krypton laser lines, Dye: dye laser (with different pumping sources), and *: laser wavelengths obtained by frequency-doubling or frequency-mixing techniques from the ground laser wavelength.

on the development of laser technology (Section III) particularly on whether the laser properties required (such as intensity, spectral bandwidth, pulse or continuous wave (CW) operation, pulse length, and pulse repetition rate) are available in the wavelength ranges needed. In some cases it is sufficient to use fixed frequency lasers to solve the problems. The use of a fixed wavelength is, of course, only possible when it coincides with the absorption or action spectra of the biomolecules to be investigated. This is, for example, the case with a frequency-doubled Nd:YAG laser (530 nm) and Hb. In addition, in the investigation of nucleic acids an Nd:YAG laser (fourth harmonic) can be applied; having a wavelength of 265 nm, it lies within the absorption maximum of nucleic acids. However, in biomolecules the dependence of certain processes on the excitation wavelength is of particular interest. Therefore, the use of tunable lasers (especially pulsed and CW dye lasers) for investigating biomolecular processes is of special importance. Besides that, a remarkable range of biological events consists of ultrafast processes and require pulsed lasers to study them. Table 2 presents a summary of fast processes in biology.[9]

For the discussion of spectroscopic problems in photobiology that impact on photomedicine, the wavelength region near 300 nm is of special interest (Figure 2). The long wavelength absorption edges of the nucleic acids are situated in this wavelength range. The maximum of the erythema action spectrum (action spectrum of sunburn) is to be found also near 300 nm. However, this is also the region where mutations in DNA are induced, under natural conditions from sunlight on the surface of the earth, because shorter wavelengths are absorbed in the ozone layer of the atmosphere, as mentioned above. UV light may cause mutations, inhibition of cell mitosis, of DNA-, RNA-, and protein-synthesis as well as sunburn and skin cancer in men. During evolution, organisms have developed mechanisms (e.g., photoreactivation) enabling them to repair such UV-induced lesions.

UVA and UVB are also wavelength regions in which phototherapy and photochemotherapy of skin diseases are applied (Section V).

TABLE 2
Examples of Important Fast Processes in Biology — Measured by Short-Pulse Techniques Using Lasers

Biological system	Fast process
Photosynthesis	Energy transfer, electron transfer
Vision	Isomerization, proton transfer
Nucleic acids	Chain dynamics, photodamage, energy transfer
Heme proteins	Ligand dynamics
Hemoglobin	Heme and protein dynamics
Myoglobin	Allosteric effects
Proteins	Side-chain motions, chain dynamics, collective motions
Amino acids and polypeptides	Vibrational dynamics
Membranes	Molecular reorientation, transport
Bacterial membranes	Isomerization, proton transfer, ion transfer
Phototaxis	Proton transfer

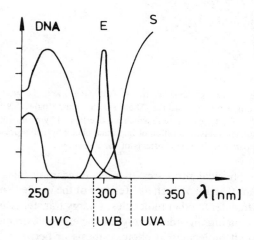

FIGURE 2.   DNA absorption (DNA), erythema effectiveness curve (E), and typical terrestrial solar spectrum (S).

B. PHOTOPHYSICAL PROCESSES

Considering biomolecules one is normally concerned with complex polyatomic molecules. One example is the DNA consisting of 10^3 to 10^6 monomers. Figure 3 shows the conventional kinetic scheme for electronic excitation of a polyatomic molecule and summarizes the important photophysical processes (see also Chapter 2, Volume I).

After light absorption and electronic excitation, the absorbed energy can be released either in form of heat (radiationless processes) or dissipated by radiation (fluorescence or phosphorescence). The direct transition from S_0 to T is a "forbidden" transition. From excited singlet states the following radiationless deactivation processes may occur: vibrational relaxation (VR) from higher vibrational levels to the lowest one, by internal conversion (IC) to the ground state, or intersystem crossing (ISC) to triplet states. The probability of radiationless processes is strongly enhanced in polyatomic molecules; e.g., a high IC is found in nucleic acids (Section IV.C). Radiative deactivation processes are the emission of photons by fluorescence or phosphorescence. Furthermore, the excited molecule may undergo a photochemical reaction (Chem. R.) or transfer its energy (E.T.) to other molecules (M.). Photoproducts, e.g., highly reactive species like free radicals, may be produced in a secondary reaction.

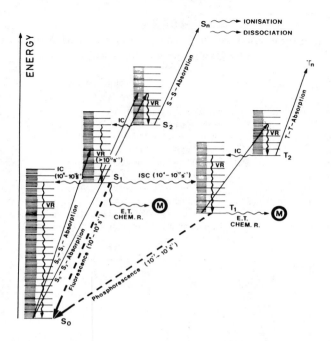

FIGURE 3.   Energy levels of the excited states above the ground state (S_0) for a polyatomic molecule. There are electronic singlet (S_1, S_2, etc.) and triplet (T_1, T_2, etc.) states, which are overlapped by many vibrational and rotational levels. Indication of important photophysical processes (see text); in parentheses: rate constants of the transitions.

Many of the photophysical processes in such large molecules are ultrafast within a picosecond time domain. (Figure 5) such as lifetimes of the excited singlet states, vibrational relaxation, and energy transfer to other molecules. Energy transfer is an important phenomena in biomolecules, used in nature in order to migrate the energy extremely fast and efficiently, e.g., between chlorophyll molecules in photosynthesis or between bases of nucleic acids. Therefore, the need for picosecond laser pulses arises for time-resolved investigations of such processes. Triplet states are important for photochemical reactions because of their relatively long lifetimes, e.g., in the reaction of nucleic acid with psoralens (Section V). All deactivation processes compete with each other; therefore, molecules having strong radiationless processes show an extremely low quantum yield of fluorescence as in the case of nucleic acids (Section IV.C).

Typical photochemical reactions which also occur in biomolecules are dissociation, ionization, dimerization, or photoinduced isomerization (see Figure 4). Thus, photodissociation of the sulfur bond has been observed in the amino acid cystine to be most likely the dominating photochemical reaction of proteins excited by UV radiation. As the cystine absorption is rather low in the UV, intramolecular energy transfer processes from other amino acids must be important for this reaction. The formation of a dimer between two adjacent thymine bases on the same DNA strand constitutes the predominant biological effect of UV radiation. Having its structure changed in such a way, the DNA loses its ability to transcribe the information at this side. Intramolecular energy transfer processes along the DNA to thymine which has the lowest triplet state of all bases play an important role in this reaction. An essential naturally occurring photoinduced isomerization is the *cis-trans* reaction associated with the primary reaction in vision. Photoionization is only rarely observed, because the ionization energies are normally too high to be supplied by a single UV photon (see Figure 8).

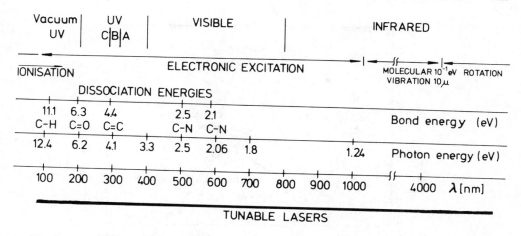

FIGURE 4. UV, visible, and IR regions of the electromagnetic spectrum with indication of photophysical processes and wavelength regions of tunable lasers.

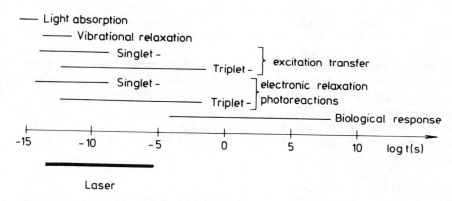

FIGURE 5. Time scale of photophysical and photobiological processes with the indication of the range of pulsed lasers.

Figures 4 and 5 survey the wavelength ranges with respect to the time scale of molecular processes as far as they concern photobiological questions. Furthermore, the regions attainable with lasers are indicated. There are lasers with pulse widths between the femtosecond and CW regimes. Wavelengths between 200 to 300 nm inactivate whereas wavelengths between 300 to 700 nm may activate and reactivate biological systems. The biological response after light exposure of cells takes place from milliseconds (e.g., photoreaction) to years (e.g., sunlight induced human skin cancer). Visible light is absorbed by pigmented cells and even slightly by unpigmented cells when cytochrome and flavoproteins are present. In the presence of pigments like chlorophyll in the cell the pigments serve as primary receptor molecules for a photobiological reaction.

The effects of visible light are strongly increased by the addition of dyes complexing with biomolecules and acting as photosensitizers in the cell (Section V). UV light, as mentioned above, is absorbed by nucleic acids and proteins and may damage the cells. IR light is absorbed by water, and cells are composed to about 80% of water. However, the energy absorbed from natural light is too low to raise the temperature in the cells.

In complex biological systems, the biological effect does not necessarily appear at those molecules which primarily absorbed the light but at the ones to which the energy is transferred. Therefore, the action spectrum (Section I) of a photobiological reaction is often not

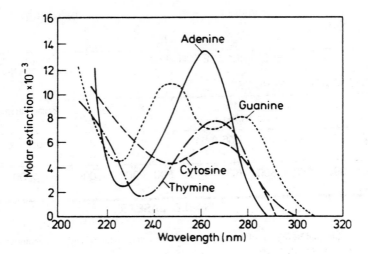

FIGURE 6.   Absorption spectra of DNA bases. (From Hillenkamp, F.,
Pratesi, R., and Sacchi, C. A., Eds., *Lasers in Biology and Medicine*,
Plenum Press, New York, 1980. With permission.)

similar to the absorption spectrum of the primary light receptor molecules. Furthermore,
photoreceptor molecules interact with neighbor molecules. The altered photochemical properties are reflected by changes in the absorption spectra. The absorption maxima of biomolecules *in vivo* are often shifted by many nanometers as compared to those *in vitro* in
solution. Therefore, the problems in photobiology are more complicated because in sufficiently diluted solutions there are no interactions between molecules, a fact which cannot
be neglected in biological systems.

C. SELECTIVE EXCITATION

Tunable lasers with their high spectral intensity selectively excite single quantum states
in atoms and molecules in a wide range of the visible, UV, and IR part of the spectrum,
as mentioned above. Up to now, selective laser action was successfully applied in photophysics and photochemistry.[2,12,13] The application on biomolecules could achieve great importance for the control of complicated biochemical reaction in the future.[14]

Selective laser stimulation acts mainly in two ways: (1) the excitation of special molecules
in a mixture of different molecules as probed for isotope separation in photophysics and (2)
the excitation of single energy levels in one kind of atoms or molecules, in order to control
the photochemical reaction of such a selectively excited molecule.

The selective excitation in complex biological molecules does, however, present considerable difficulties because of the wide and overlapping absorption bands and because of
the very fast relaxation and intramolecular energy transfer processes. One example for the
overlapping absorption spectra of nucleic acid bases (Section IV.C) is given in Figure 6. A
selective excitation of only one kind of base is only possible at the long wavelength absorption
edge, e.g., of guanine.

Therefore, because of few differences in the absorption spectra, differences of absorption
cross sections of the excited states, lifetimes of the excited states, or differences in rates of
photoinduced chemical reactions must be used. Furthermore, to enhance the selectivity,
apart from a one-step electronic excitation two- or more-step excitations have to be carried
out. In case of an IR-UV excitation, the first step must be done in picoseconds in order to
obtain the selectivity for the second step, because of the very fast vibrational relaxation.
Considering future experiments in living cells the problem of IR absorption by the water in
the cells has to be solved, too.

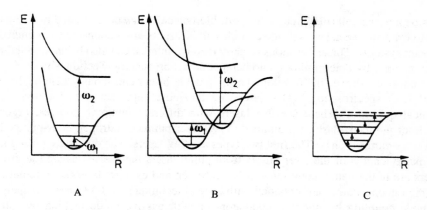

FIGURE 7. Typical molecular potential functions with types of selective photoexcitation (from left): (A) two-step IR-UV excitation, (B) two-step excitation via an electronic state, and (C) multiphoton IR excitation.

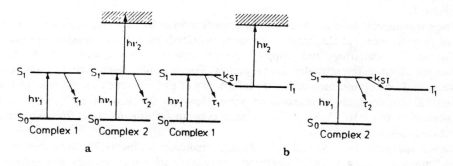

FIGURE 8. Electronic state of complexes 1 and 2: (a) selective ionization of complex 2 and (b) selective ionization of complex 1. (From Andreoni, A., Cubeddu, R., De Silvestri, and Svelto, O., *Lasers in Photomedicine and Photobiology,* Springer-Verlag, Heidelberg, 1980. With permission.)

The principle of the two- or multiphoton excitations method is shown in Figure 7. An example of a selective two-step excitation taking advantage of differences in the lifetimes of excited states is given in Figure 8[15] (see also Section IV.C). Two complexes with different lifetimes τ_1 and τ_2 ($\tau_1 < \tau_2$) are irradiated with two laser pulses. The second pulse is delayed with τ_D against the first pulse. If $\tau_1 < \tau_D < \tau_2$ the second pulse (frequency ν_2) will find a smaller singlet state population of complex 1 due to its faster decay.

Thus, complex 2 can be excited with the second laser pulse to higher energy levels, e.g., complex 2 can be selectively ionized. If $\tau_D < \tau_T$ (τ_T: triplet lifetime), complex 1 can be selectively influenced. Since $\tau_1 < \tau_2$, the ISC rate and therefore the triplet-state population will be larger for complex 1 than for complex 2. In the figure it is assumed that the ground-state absorption cross section and the triplet-state lifetimes τ_T are the same for the two complexes.

III. METHODS

Investigations using laser spectroscopy mostly require sophisticated technology including complex laser and accessory analytical equipment, e.g., instruments like streak cameras which enable a high speed of detection.[16-19]

First, biomolecules can be examined *in vitro* in solution where conditions should resemble those inside a living cell (pH 7, etc.). Second, biomolecules can be irradiated in living cells. In order to irradiate special sites inside a single cell, a laser beam is connected to a microscope (microbeam system). Third, questions of photomedical interest can also be answered directly on patients, e.g., by determining phototherapeutic action spectra (Section V).

As mentioned above, the development of tunable lasers enormously benefit the application of laser spectroscopy.[20] Dye lasers are the most important ones for investigating biochemical questions. There are dye lasers with different pumping sources: argon and krypton laser-pumped, flashlamp-pumped, YAG laser-pumped, nitrogen laser-pumped, and excimer laser-pumped ones. The first two types are CW lasers. The others are pulsed lasers. Typical pulse lengths of the laser pulses from pump lasers themselves as well as from the dye lasers are in the nanosecond range (e.g., nitrogen and excimer lasers). Extremely short pulses (picosecond range) are obtained with special techniques in CW lasers, namely with active mode coupling by internal modulators or with passive mode locking by saturable absorbers. Optical frequency doubling, nonlinear mixing, or Raman scattering allow the extension of spectral ranges of the laser output. An excimer laser-pumped dye laser, for example, is tunable from about 320 to 970 nm using various organic dyes and from 218 to 320 nm with frequency doubling in addition. For details the reader is referred to Chapter 1 in Volume I.

The experimental laser methods for the investigation of biomolecules cover a wide field such as fluorescence and absorption measurements (flash photolysis, transient spectra)[20-24] and Raman,[25-28] scattering, and Doppler spectroscopy.[29-31] The various methods deliver information about different molecular properties. For example, fluorescence, fluorescence lifetimes, and absorption measurements inform about excited states of biomolecules and biochemical reactions; the generation of pulses down to the pico- (1 ps $= 10^{-12}$s) and femtosecond (1 fs $= 10^{-15}$ s) range is of great concern for time-resolved fluorescence and absorption measurements.[32-35]

Raman spectroscopy [CW and pulse Raman, resonance Raman (RR), and coherent anti-Stokes Raman scattering (CARS)] and fluorescence correlation spectroscopy give knowledge mainly about structural alterations, conformation, and motions of biomolecules. Polarized light excitation and scattering techniques also yield information about molecular dynamics.[22,34] In investigating complex events, often a combination of different methods has to be used. For example, the observation of changes in the Raman spectrum occurring after photodissociation can simultaneously deliver information about structural and dynamic processes.

Special techniques on the basis of fluorescence, absorption, and scattering parameters are developed for cell diagnosis of medical concerns: flow cytometry, cell analysis, and sorting.[36]

Figure 9 gives an example of an experimental arrangement for picosecond emission measurements in the region of single-photon events; a single-photon sensitive synchroscan streak camera is used for fluorescence detection. In this example an argon laser which pumps a mode-locked CW ring dye laser is used and an internal frequency doubling technique is integrated additionally to generate UV laser radiation. This experimental configuration, for example, is applied to measure fluorescence dynamics of nucleic acid bases.[19]

Different transient absorption instruments are described by Holtom.[17] In general a sample is excited with a pulsed laser and the following spectral transients are observed; for this an excitation laser beam and another one to probe the sample are necessary. In Figure 10 a setup to measure transient spectra in the subpicosecond range is shown. In this case the amplified pulses from the laser are split and half of them are sent into a cell of H_2O to generate a continuum beam for probing the sample. The residual pulses pass through a variable delay line and are focused to excite the sample. The passively mode-locked CW dye laser and a subpicosecond amplifier system are described in Reference 5. This system

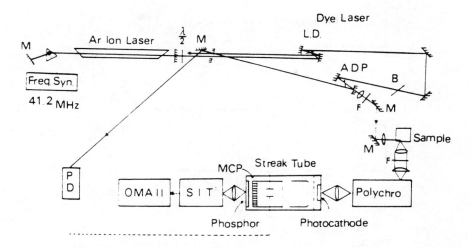

FIGURE 9.   Schematic diagram of an experimental arrangement for picosecond fluorescence detection (M: mirror, LD: Laser dye, ADP: ADP-crystal for frequency doubling, and PD: photodiode). (From Kobayashi, S., Yamashita, M., Sato, T., and Muramatsu, S., *IEEE J. Quantum Electron.*, 12, 1383, 1984. With permission.)

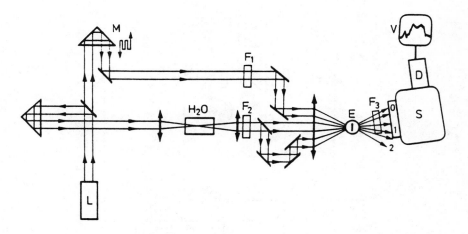

FIGURE 10.   Schematic of experimental setup for measuring transient spectra in the subpicosecond range (L: dye laser, E: sample, 2: excitation beam, 0,1: continuum beam, and S,D,V,: spectrograph and recording systems). (From Pratesi, R. and Sacchi, C. A., Eds., *Lasers in Photomedicine and Photobiology*, Springer-Verlag, Heidelberg, 1980. With permission.)

was applied to investigate photodissociation processes of hemoglobin complexes (see Section IV.D).

An example of UV Raman spectroscopy is given in Figure 11. The fourth harmonic wavelength (266 nm) of a Nd:YAG laser is taken as excitation wavelength and further UV lines are generated with the Raman scattering technique using a hydrogen-filled Raman shifting cell. For example, this technique was used for UV resonance spectroscopy of nucleic acid components (see Section IV.C).[37]

In Figure 12 the principle of a two-step excitation is shown which is used to photoionize polynucleotide-dye complexes (Section IV.C). In this example a nitrogen-pumped dye laser excites the first singlet state and the wavelength of the nitrogen laser itself is taken for the second excitation step.[15]

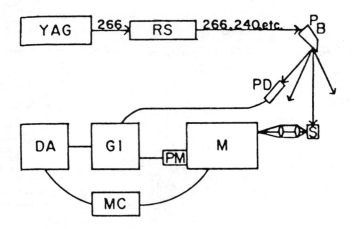

FIGURE 11.   Experimental setup for UV resonance Raman spectroscopy)
YAG:Nd: YAG-laser, RS: Raman shifting cell, S: sample, M, PM, DA:
detection unit (monochromator, photomultiplier, and computer). (From
Fodor, S. P. A., Rava, P. R., Hays, T. R., and Spiro, T. G., *J. Am.
Chem. Soc.*, 107, 1520, 1985. With permission.)

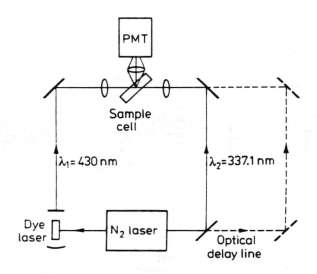

FIGURE 12.   Schematic diagram for a two-step excitation and fluores-
cence observation with a photomultiplier (PMT). (From Andreoni, A.,
Cubeddu, R., De Silvestri, S., Laporta, P., and Svelto, O., *Phys. Rev.
Lett.*, 45, 431, 1980. With permission.)

In cell biology, laser microbeam systems are applied to expose special parts in cells or
tissues. An example is given in Figure 13. In this case an excimer laser pumps a dye laser.
The dye laser beam is directed into the fluorescence illumination path of a microscope.[38]

One of the most important areas in cell biology that has resulted in the development of
commercial instruments is flow cytometry. By this method a great amount of cells can be
analyzed and also selectively separated on the basis of fluorescent signals. Figure 14 shows
an example of a flow cytometer system that employs laser excitation orthogonal to the flow
axis of the samples.[36] For example, different fluorescence parameters of dye-stained cells
can be detected. There are also multi-wavelength systems incorporating three to five different
laser excitation beams capable to record many different data simultaneously.

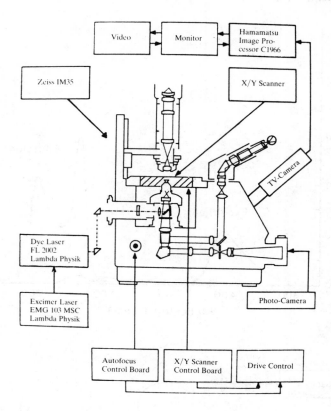

FIGURE 13. Schematic diagram of a laser microbeam system with the basic components: dye laser pumped by an excimer laser, inverted microscope, and the elements for documentation. (From Wiegand, R., Weber, G., Zimmermann, K., Monajembashi, S., Wolfrum, J., and Greulich, K.-O., *J. Cell Sci.*, 88, 145, 1987. With permission.)

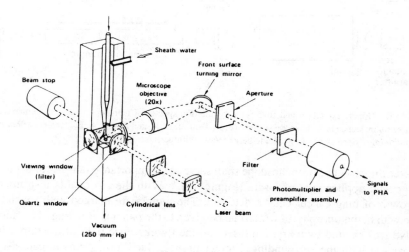

FIGURE 14. Flow cytometer with orthogonal axes of sample flow, laser beam illumination, and fluorescence detection. (From van Dilla, M. A., Dean, P. N., Laerum, O. D., and Melamed, M. R., Eds., *Flow Cytometry: Instrumentation and Data Analysis,* Academic Press, London, 1985. With permission.)

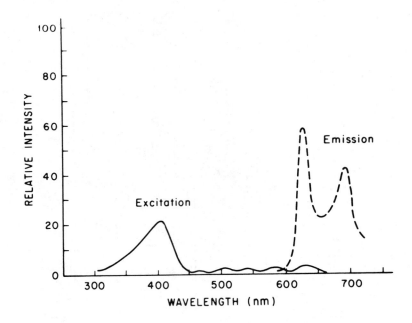

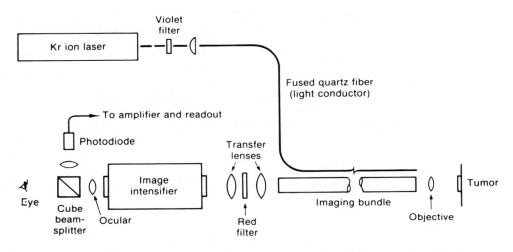

FIGURE 15. Upper: excitation and fluorescence emission spectra of hematoporphyrin in human serum and lower: laser fluorescence bronchoscope system. (From Doiron, D. R. and Profio, A. E., *Lasers in Photomedicine and Photobiology*, Springer-Verlag, Heidelberg, 1980. With permission.)

A laser fluorescence bronchoscope shall serve as an important example for the therapeutic and diagnostic applications of lasers (Figure 15).[39] With the aid of this instrument a very early growth of lung cancer can be detected. The tumors are marked with special dyes, in this case with hematoporphyrin which can be given to the patients as a drug. The fluorescence of the dye was excited by a krypton laser and the fluorescence light of the tumor is imaged in comparison with the surrounding dimmer tissues. Such an instrument can also be used for treatment — for the photochemotherapy of tumors (Section V).[40]

IV. PHOTOBIOLOGY

In the following sections some principal remarks shall be made about fundamental biological systems and processes and we present a few examples which shall illustrate the

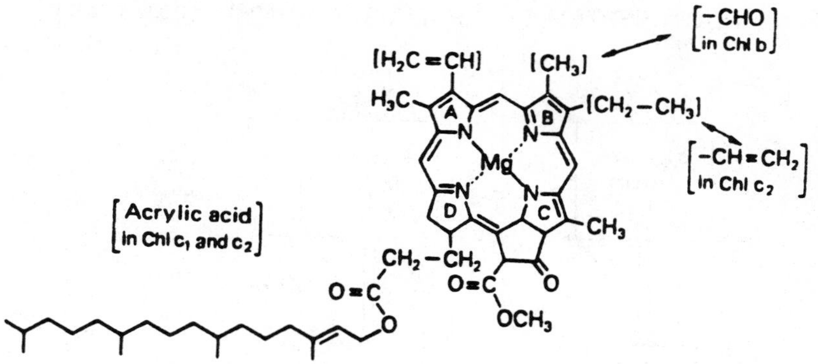

FIGURE 16. Chemical structure of chlorophyll a, b, c_1, and c_2. (From Häder, D. P. and Tevini, M., *General Photobiology*, Pergamon Press, Elmsford, NY, 1987. With permission.)

advantages and often the necessity of laser spectroscopy application for the investigation of such processes.

A. PHOTOSYNTHESIS

During the photosynthesis of green plants the energy of visible light is absorbed by pigments and transduced to chemical energy.[11] Organic compounds are synthesized from the simple inorganic molecules carbon dioxide and water. The radiation energy is stored in carbohydrates. The reaction is summarized by the following equation:

$$6CO_2 + 12H_2O \xrightarrow{\;h\nu\;} C_6H_{12}O_6 + 6H_2O + 6O_2$$

Chlorophylls are the most important light absorbing pigments in the process of photosynthesis (Figure 16). The chemical structure of chlorophyll is a tetrapyrole with porphyrin ring structure and a central magnesium atom. In almost all photosynthetic systems carotinoids are also present and act as accessory pigments which absorb energy and transfer this excitation energy to the chlorophylls. The pigment molecules with their associated proteins, in which they are located, are found in specific organelles in the plants, the chloroplasts. Within the chloroplasts there are closed membranous structures, the thylakoids, in which the pigments are incorporated in a lipid matrix as pigment-protein particles.

There are two light reactions (Figure 17): photosystem I and II. The primary steps include absorption by aggregates of pigments and a transfer of the absorbed energy to the reaction centers of photosystem I and II. The excitation and action of the two reaction centers are responsible for the initiation of the chemical processes in higher plants. The reaction centers are surrounded by large antennae of pigment molecules. The center and the surrounded pigment molecules with their associated proteins are called photosynthetic unit. The antennae increase the effective absorption cross section funneling the absorbed energy to the reaction center. The reaction centers consist of two pigment molecules, e.g., a chlorophyll dimer. Both of the reaction centers have different functions which are connected with complicated reactions, I: the reduction of the coenzyme $NADP^+$ (nicotinamide-adenine-dinucleotide phosphate), which is necessary to reduce CO_2 for the carbohydrates synthesis and II: the photolysis of water. Both photosystems are connected by an electron transport chain.[11]

For the investigation of photosynthesis with laser spectroscopy it is of special interest to examine its first steps: light absorption, energy migration to the reaction centers, and the transformation of the energy in the reaction centers. The reaction rates of these processes

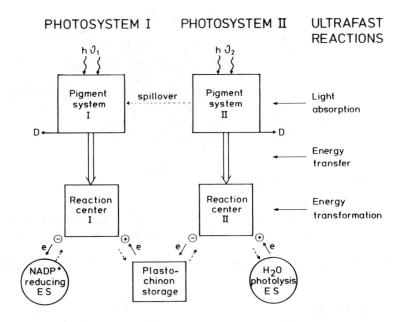

PHOTOSYSTEM I   PHOTOSYSTEM II   ULTRAFAST REACTIONS

FIGURE 17.   Schematic presentation of the functional organization of the quantum transformation in photosynthesis. D: dissipation of electronic excitation energy, ES: enzyme system, e: main reaction pathways of electrons ($\rightarrow$), and possible reverse reactions (--$\rightarrow$).

are partial in the femtosecond range. They can be monitored by observing the temporal behavior of the fluorescence or absorption from different molecular species after pico- or subpicosecond light excitation.[48,49]

1. Energy Transfer

In 1974 Seibert and Alfano[41] got evidence of energy transfer by time-resolved fluorescence spectroscopy. They investigated the fluorescence of isolated spinach chloroplasts at room temperature. They used a frequency-doubled, mode-locked Nd:glass laser and an optical Kerr gate with an instrument resolution time of 10 ps. The excitation took place by 4 ps pulses at 530 nm and the fluorescence was observed at 685 nm (Figure 18). With this time resolution, one observes two maxima in the fluorescence emission kinetics with a delay between them.

The two peaks could be associated with components of photosystem I (lifetime of 10 ps) and photosystem II (lifetime of 210 ps). The 90-ps delay between the two peaks is related to the energy transfer between accessory pigments. These interpretations were based on theoretical models. The best conformity with the experimental results was delivered by a model which considers two independent absorbing species; one is fluorescing and the other transfers energy to a third fluorescing species. This suggests that the 90-ps delay of the second peak is due to an absorbing species at 530 nm (a carotenoid) which passes energy to the photosystem II chlorophyll.

Another example of energy transfer experiments is that of Biennie et al.[42] who investigated the energy transfer between dimeric chlorophyll a and β-carotene *in vitro*. They excited the fluorescence spectra with 4-ns light pulses from a tunable excimer laser-pumped dye laser. They present an energy level diagram with indication of the levels which are candidates for energy transfer and they propose a sequence of the events. Furthermore, they estimated the lifetimes of the first excited states of chlorophyll a from fluorescence decay curves (excitation at 436 nm) with about 6 ns for the monomer and about 5 ns for the dimer.

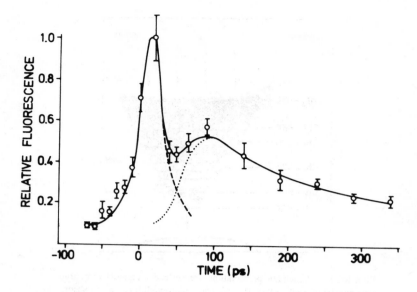

FIGURE 18. Time dependence of fluorescent emission from spinach chloroplasts at 685 nm after excitation by a 4-ps pulse at 530 nm. Chlorophyll concentration: 35 μg/ ml. (From Seibert, M. and Alfano, R. R., *Biophys. J.*, 14, 269, 1974. With permission.)

Recently, Gulotti et al.[43] investigated the picosecond fluorescence decay of the photosynthetic system of the green algae *Chlamydomonas* using single-photon counting methods. They determined the single step energy transfer time in the photosystem I to be 200 to 400 fs. They also showed that the excitation requires two to four excitations of the trapping site before it is finally effective. Photosystem II, however, must undergo many more excitations before the final trapping. This difference between the two photosystems is regarded as a decisive aspect of energy transfer regulation in higher plants.

Raman spectra of photosynthetic pigments have been studied, too. One example is RR spectra of chlorophyll a excited with various laser lines by Fujiwara and Tasumi.[51] Furthermore, with picosecond fluorescence and absorption measurements model systems in artificial photosynthesis were performed to understand and imitate the primary processes occurring in natural photosynthesis. This kind of study could also be important in the future for possible applications in the chemical storage of solar energy.[52]

2. Excitons and Energy Traps

As all pigment molecules in the antennae can absorb the light energy, there must be an efficient mechanism by which the energy is transferred nearly without degradation to the reaction center. In most cases, this can be described by a dipole-dipole interaction, the Förster energy transfer mechanism, which acts at a rate depending on the inverse sixth power of the distance between the donor and acceptor molecules, on their mutual orientation, and on the overlap between the absorption of the acceptor and the emission of the donor. This excitation is suggested to be hopping randomly between neighboring chlorophyll molecules ("random-walk approach") and is usually referred to as exciton migration.[44] The photosynthetic unit fulfills the conditions of closeness and orientation of the pigment molecules as well as a large spectral overlap for a rapid energy migration process. The detailed position of the pigments in the proteins may be important, but little is known about it at present. For a dipole-dipole interaction the maximal energy transfer is expected for a parallel orientation of the transition-dipole moments of the single molecules. On the other hand, the manner in which excitons interact *in vivo* should inform about the topology of the photosynthetic unit.

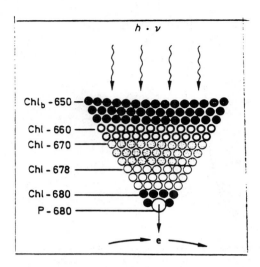

FIGURE 19.   Schematic presentation of the effective pigment array (trapping center) of photosystem II. Chl_b-650: chlorophyll b having an absorption maximum at 650 nm, etc. and P-680: reaction center (pigment 680: chlorophyll a dimer). (From Richter, G., *Stoffwechselphysiologie der Pflanzen*, Thieme Verlag, Stuttgart, 1981. With permission.)

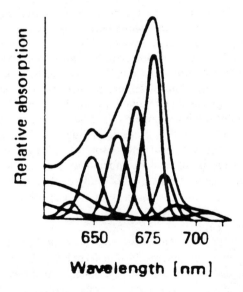

FIGURE 20.   Low-temperature absorption spectrum of spinach chloroplasts. (From Häder, D. P. and Tevini, M., *General Photobiology*, Pergamon Press, Elmsford, NY, 1987. With permission.)

The antennae pigments act *in vivo* like energy traps. The principle of such a trapping center is shown in Figure 19.[11,45] The excitons migrate from an excited singlet state of a donor molecule to an acceptor molecule in the ground state which undergoes a transition to its excited state and the donor molecule simultaneously to the ground state. In the trapping center this is realized by different types of pigment molecules having slightly different but overlapping absorption spectra (Figure 20) with components in photosystem II from maximum absorption at 660 to 700 nm (i.e., from higher to lower lying excited singlet states).

In this sequence they are arrayed towards the direction of the reaction center and the amounts of the single components decrease. Such an energy trap for absorbed quanta (which contains about 300 pigment molecules) allows an energy transfer to the reaction center with an effectivity of close to 100%. In the center of the trap a chlorophyll dimer is found which reduces an acceptor and oxidizes a donor.

Several experimental observations using single- or multiple-pulse laser excitation and streak camera detection were concerned with the exciton migration in photosynthetic systems. For example, a review was given by Campillo and Shapiro in 1978.[46] Singlet-singlet exciton annihilation occurring within the photosynthetic units of the green algae *Chlorella* was demonstrated. The dominant mechanism is singlet-singlet fusion, manifesting itself by a decrease in the measured lifetime and quantum efficiency of fluorescence for high-intensity single-pulse excitation. Differences in low-intensity (10^{13} to 10^{16} photons/cm^2) and high-intensity excitation (10^{17} photons/cm^2 and higher) and various quenching mechanisms are discussed.

A recent review by Voigt[44] about exciton states and interactions in biological systems, especially in chlorophyll *in vivo*, discusses the peculiarities of excitons in photosynthetic membranes starting from well-investigated exciton effects in solids. A comparison with recent results of laser spectroscopic measurements *in vivo* was done, for example, with fluorescence measurements performed with a nitrogen-pumped dye laser tunable from 400 to 650 nm.

The light-harvesting complex consisting of pigment molecules bound to proteins is quantum mechanically described by a disordered molecular system. Conclusions about the organization of chlorophyll *in vivo* in subunits could be drawn. For example in the emission attributed to photosystem II, three subbands could be resolved (exciton-split energy levels). From their preliminary results concerning time-shape analysis of absorption and emission lines of chlorophyll *in vivo* at different temperature, a dominating influence of inhomogenous line broadening mechanisms was concluded. These mechanisms result in a nearly complete overlapping of absorption and emission lines and, after the Förster theory, in a maximum hopping rate of excitons. Furthermore, from the investigation of molecular transitions of chlorophyll *in vivo*, using higher excitation energies, this is the first evidence of the "bound" state of two excited chlorophyll molecules in subunits of the light-absorbing pigments, similar to the case of the excitonic molecules in semiconductors.[44]

3. Reaction Centers

Information about processes in the reaction centers[50] were also obtained by Klevanik et al.[47] They investigated the dynamics of bacterial reaction centers with nanosecond and picosecond flash photolysis. They used a subpicosecond double-beam absorption spectrometer consisting of a passively mode-locked CW dye laser pumped by an argon laser and additionally an amplifier system. After amplification the beam was split into two parts, one of which was converted into a "white light" continuum by passing a cell containing H_2O and served as a probing pulse, while the second one was used as an exciting pulse. The wavelength region for the exciting pulse was extended by stimulated Raman scattering in cyclohexane. The pigments were excited at 718 nm; with the aid of delay lines the changes in the absorption spectra after excitation were followed. The transfer rate of an electron from the so-called bacteriochlorophyll P_{870} to the primary acceptor could be measured to be 7 ps. Furthermore, the distance between different chromophores (bacteriopheophytin and bacteriochlorophyll P_{870}) could be estimated to be 12 Å.

B. VISION

In the process of vision, light is absorbed by photoreceptor molecules like rhodopsin followed by a formation of several intermediate products (Figure 21). Many spectroscopic techniques have been applied to elucidate the primary photochemical process in rhodopsin.

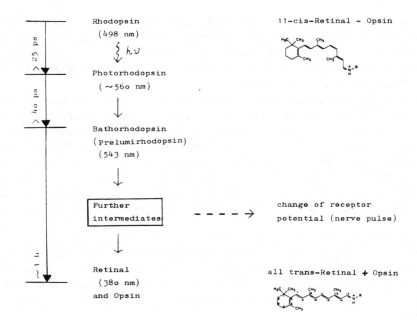

FIGURE 21.   Sequence of intermediates in the photolysis of rhodopsin (see text).

Because of the rapid formation of the first intermediate products, laser techniques are of special importance.

1. Structural Aspects

Earlier investigations of the first photochemical events have been made at low temperature. Under physiological conditions, these events have to be studied with picosecond spectroscopy. Investigations with pico- and nanosecond laser photolyses as well as low temperature spectrophotometry resulted in the detection of several intermediates in the photobleaching process of rhodopsin. After the discovery of lumirhodopsin in 1950, bathorhodopsin and hypsorhodopsin were detected by several authors during the next 20 years. Busch and co-workers[53] carried out the first application of a picosecond laser photolysis in the field of visual photobiology in 1972. Bathorhodopsin was detected at room temperatures by means of pico- and nanosecond laser photolyses.

In the same year a candidate of a precursor of bathorhodopsin, called hypsorhodopsin, was reported. The question whether hypsorhodopsin or bathorhodopsin is the first photoproduct arose and several experimental and theoretical results have been reported since then. The formation lifetime of bathorhodopsin, also called prelumirhodopsin at that time, was measured to be less than 6 ps at room temperature.[53] Another intermediate, photorhodopsin, was found by excitation of cattle, squid, and octopus rhodopsins with a 532-nm laser pulse.[54]

Rhodopsin consists of an 11-*cis* retinal chromophore and opsin, a protein. After the absorption of a photon the 11-*cis* bond of retinal is transformed into an 11-*trans* bond. From experiments with low temperature spectrophotometry, Yoshizawa et al.[58] concluded that bathorhodopsin has a highly constrained and distorted all-*trans* retinal as its chromophore and that the process from rhodopsin to bathorhodopsin should be due to *cis-trans* photoisomerization of the chromophore. Rentzepis[55] suggested a proton tunneling effect (e.g., in the Schiff base of retinal) for the mechanism of prelumirhodopsin (bathorhodopsin) formation, at least at low temperatures. The ultrafast rate observed for the formation of prelumirhodopsin could be interpreted as an activation without a barrier for the proton translocation, the distance through which the proton tunnels could be calculated to be 0.5 Å. Models for the proton

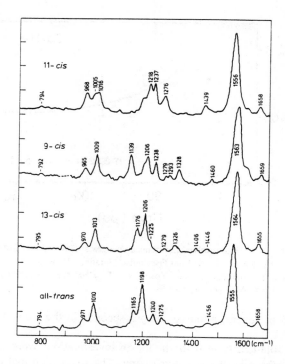

FIGURE 22.   Rapid-flow resonance Raman spectra of protonated Schiff
bases of retinals in ethanol (chloride salts of protonated *n*-butylamine
derivatives). (From Hillenkamp, F., Pratesi, R., and Sacchi, C. A., Eds.,
Lasers in Biology and Medicine, Plenum Press, New York, 1980. With
permission.)

translocation are also discussed by Applebury.[56] To reveal the sequence of conformational
changes occurring in the retinal chromophore group and in the protein, Raman spectra were
recorded.[57] RR spectra of the intermediates in the photolysis of rhodopsin are given in Figure
22. The spectra were measured with CW dye lasers using a rapid-flow technique to obtain
the spectra before a photolabile molecule is altered by light. The differences in the fingerprint
region (1100 to 1350 cm^{-1}) demonstrate the sensitivity of the vibrational spectrum to the
conformation.

The objections against a *cis-trans* photoisomerization mechanism as the formation process of bathorhodopsin came from the laser photolytic experiments in 1972 and 1977. Busch
et al.[53] had excited cattle rhodopsin with a 530-nm pulse (width: 6 ps) from a Nd^{3+}:glass
laser and monitored a transient increase in absorbance at 560 nm on a picosecond time scale.
A rapid increase in absorbance which reached a plateau within 6 ps was thought to be due
to the formation of bathorhodopsin because at that time bathorhodopsin was the only intermediate having an absorption spectrum in a wavelength region longer than that of rhodopsin.
It was concluded that the formation of bathorhodopsin within 6 ps would be too fast to be
due to such a conformational change as *cis-trans* isomerization of retinal.

In 1984, Yoshizawa et al.[58] described an examination of the proton translocation model
by a series of experiments. Based upon the experimental facts of five examples they conclude
that *cis-trans* isomerization of the chromophore is essential for formation of bathorhodopsin.
The opinion that bathorhodopsin has a distorted all-*trans* retinal as its chromophore is also
supported from a variety of data of RR spectroscopy which also contradicts the model
proposed earlier. Additionally, more recent experiments of picosecond laser photolysis have
revealed that bathorhodopsin is produced with a time constant of about 40 ps in the case of

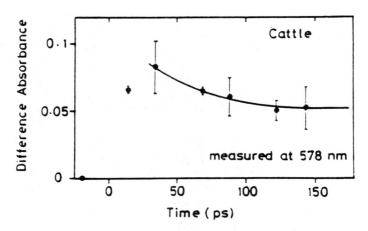

FIGURE 23.   Time-resolved absorbance changes at 578 nm after excitation
of cattle rhodopsin with 532 nm laser pulse (duration: 25 ps) at 18°C. (From
Shichida, Y., Matuoka, S., and Yoshizawa, T., *Photobiochem. Photophys.*,
7, 221, 1984. With permission.)

cattle rhodopsin and of several hundreds of picoseconds in the case of squid and octopus
rhodopsin systems.

In 1986 results obtained from bathorhodopsin investigations were summarized by Shichida.[59] He refers to the role of RR spectroscopy. The suggestion that a distorted all-*trans*
form is formed at liquid nitrogen temperature is regarded as confirmed by a direct measurement of distortion of the chromophore of bathorhodopsin by use of this technique. In
addition, RR spectroscopy proved the chromophore of bathorhodopsin to be a protonated
retinylidene Schiff base like that of rhodopsin. The motion of the retinylidene chromophore
during the process of the photoisomerization of rhodopsin is discussed on the basis of the
experimental data on both the excited state of rhodopsin and photorhodopsin. Unfortunately,
there are only few experimental data about the configuration and conformation of the chromophore of these states. Several picosecond experiments and CD measurements are discussed
which suggest that the movement of the chromophore may take place in the region near the
Schiff base. The determination of the Schiff base configuration in rhodopsin and bathorhodopsin as well as photorhodopsin is regarded as highly important for elucidation of the
primary photochemical reaction of rhodopsin.

2. The New Intermediate Photorhodopsin

Photorhodopsin, a precursor of bathorhodopsin, was first described by Shichida.[54] Transient absorption spectra of intermediates of rhodopsin were measured after excitation with
picosecond pulses from a frequency-doubled Nd:YAG laser at 532 nm. A photoproduct was
detected, of which the absorption maximum lay at a slightly longer wavelength (near 560
nm) than that of bathorhodopsin. It was produced with laser pulses of 25-ps duration and
it decayed to bathorhodopsin with a time constant of about 40 ps in the case of the cattle
rhodopsin system (Figure 23). The result is discussed in connection with former reports
about possible precursors of bathorhodopsin and it is concluded that their study clearly shows
that bathorhodopsin is produced from the early bathochromic photoproduct with a time
constant of about 40 ps after the excitation of cattle rhodopsin. This early bathochromic
photoproduct is called "photorhodopsin". Yoshizawa et al.[58] summarized the primary photochemical processes of rhodopsin studied by their group. They discussed the spectral properties of three intermediates and presented a new photobleaching process of rhodopsin at
room temperature. The presence of a precursor of bathorhodopsin or hypsorhodopsin had

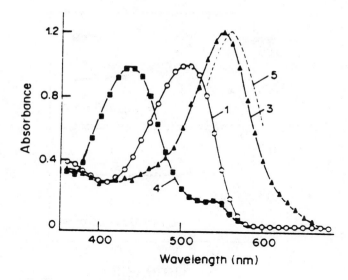

FIGURE 24.   Absorption spectra of (1) cattle rhodopsin, (3) bathorhodopsin, and (4) hypsorhodopsin measured by conventional spectrophotometry at liquid helium temperature. Curve 5 shows a spectrum of photorhodopsin roughly estimated from data obtained by a picosecond laser spectroscopy at room temperature. (From Yoshizawa, T., Shichida, Y., and Matuoka, S., *Vision Res.*, 24, 1455, 1984. With permission.)

been reported by several groups who regarded the observed precursors as an excited state of rhodopsin at that time. From later experiments it was followed that at least one intermediate and not an excited state prior to bathorhodopsin and hypsorhodopsin should be present in the primary photochemical process of rhodopsin. Yoshizawa et al.[58] concluded from the decay-time constants of 40 ps and several hundreds of picoseconds that the photoproduct is not an excited state of rhodopsin but an intermediate.

In Figure 24 the absorption spectra of three intermediates and of rhodopsin are shown. In 1986 it was stated again that the first photoproduct of rhodopsin is now thought to be the photorhodopsin which was recently detected.[59] The time constant of its formation is too small (less than 25 ps) to be estimated by the picosecond laser photolytic system (25-ps pulses) which was used in the described experiment. Shichida underlines the fact that photorhodopsin was not detected by means of conventional low temperature spectrophotometry even at liquid helium temperature. It is concluded that the conversion of rhodopsin to photorhodopsin is likely to be a *cis-trans* photoisomerization process of the chromophore probably without detectable change of conformation in the protein moiety.

3. Effects of Intense Laser Pulses

A new aspect which has to be considered is the use of very intense laser pulses. Shichida pointed out that little attention was paid, so far, to a multiphoton effect which could be included by excitation of rhodopsin with an intense laser pulse. Therefore, it is especially important to examine whether the results obtained include any multiphoton effect on rhodopsin and its intermediates when using picosecond laser photolysis. In Figure 25 an example for the dependence on photon density of the photochemical reaction of rhodopsin is given. When the excitation took place with a weak picosecond laser pulse, the absorbance at 570 nm decreased after an initial increase. Whereas when the excitation was performed with an intense laser pulse, however, a bathochromic photoproduct was produced which did not decay during the time of observation. Thus, the formation of photorhodopsin depends on the photon density of the picosecond laser pulse used for the excitation of rhodopsin.

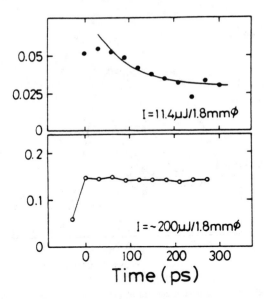

FIGURE 25.   Formation of photorhodopsin by excitation of cattle rhodopsin in 2% digitonin solution at 18°C with Nd^{3+}:YAG laser pulse (wavelength: 532 nm; pulse width: 25 ps). Difference absorbance vs. time. Time-
resolved absorbance changes at 578 nm after excitations with weak (●:
11.4 μJ/1.8 mm ø) and intense (○: ~200 μJ/1,8 mm ø) laser pulses. (From
Shichida, Y., *Photochem. Photobiophys.*, 13, 287, 1986. With permission.)

While investigating the initial photochemical process of rhodopsin by picosecond laser
photolyses, a photoproduct named hypsorhodopsin was detected. It was found with several
animal systems upon irradiation on the basis of low-temperature spectrophotometry and was
reported as a candidate of a precursor of bathorhodopsin. Contradictions and disagreements
among research groups arose about the formation of hypsorhodopsin at room temperature.
To answer the question whether hypsorhodopsin could be detected at room temperature and
whether hypsorhodopsin or bathorhodopsin is the first photoproduct was characterized by
Yoshizawa and Shichida[60] as the second era of the study by picosecond laser photolysis.
Recently, Matuoka et al.[61] examined in detail whether or not hypsorhodopsin is produced
by excitation with 532-nm laser pulses. The formation of hypsorhodopsin was observed on
excitation of squid and octopus rhodopsins at room temperature only, when these rhodopsins
were excited with a high energy pulse (Figure 26). There was no increase of the absorbance
observed at 420 nm, indicating that the formation of hypsorhodopsin did not take place.
The increase of the absorbance at 558 nm shows the formation of photorhodopsin and
bathorhodopsin. On the other hand, using intense laser pulses for excitation, the absorption
at 420 nm increases within 15 ps due to the formation of hypsorhodopsin. Matuoka et al.
did not observe the formation of cattle hypsorhodopsin even when exciting with high energy
pulses. However, the authors suppose that it may be possible to observe the formation of
hypsorhodopsin in this case if they used even more intense laser pulses; and they refer to
other investigations which were reported about the excitation of cattle rhodopsin at an energy
density of 60 J/m^2 which is at least a fourfold higher energy density compared to their own
experiments. An increase of absorbance near 460 nm was not assigned to hypsorhodopsin
by these authors but, according to Matuoka et al.,[61] it could be possible to assign it to
hypsorhodopsin taking other experiments into account. They came to the conclusion that
the dependency of formation on the excitation energy suggests that hypsorhodopsins of squid
and octopus are formed by a two-photon reaction.

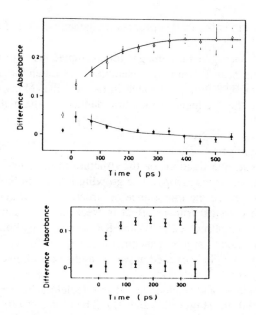

FIGURE 26. Absorbance changes at 558 (○) and at 420 nm (●) after excitation of squid rhodopsin with laser pulses of low and high intensity (excitation wavelength 532 nm; pulse width 25 ps). Upper: excitation energy approx. 120 μJ/1.8 mm ⌀ Bottom: excitation energy approx. 20 μJ/1.8 mm ⌀. (From Matuoka, S., Shichida, Y., and Yoshizawa, T., *Biochim. Biophys. Acta*, 765, 38, 1984. With permission.)

As the previously reported kinetic behavior of rhodopsin was observed by excitation with a very weak laser pulse, one has to consider that this did not induce multiphoton effects and, therefore, the kinetic behavior may differ.

In 1984 Yoshizawa et al.[58] stated that three kinds of photoproducts were observed in the process of rhodopsin photobleaching at room temperature: photorhodopsin, hypsorhodopsin, and bathorhodopsin. However, the experiments performed until that time were not sufficient to explain the role of hypsorhodopsin in the pathway. The candidate of the precursor of hypsorhodopsin was thought to be either the singlet excited state of rhodopsin or photorhodopsin.

The controversy about the primary photochemical event of rhodopsin can be possibly cleared up by regarding the energy of the laser pulse for the excitation of rhodopsin, as discussed by Yoshizawa and Shichida.[60] Single- and multiphoton reactions in the rhodopsin system may occur under special excitation conditions. Picosecond laser photolysis proved to be a powerful tool for detecting primary photochemical intermediates of rhodopsin at physiological conditions, but the contradictory results were often discussed without considering the energy density of the laser pulses used in the experiments.

Picosecond studies were also made of the closely related pigment bacteriorhodopsin, which is present in the purple membrane of the halophilic bacterium *Halobacterium halobium*. Although bacteriorhodopsin does not function as a visual transducer, its early-time photochemistry parallels that of animal rhodopsin.[62]

C. NUCLEIC ACIDS

Nucleic acids store and transmit genetic information and regulate metabolism in living cells. The genetic code is represented in the sequence of bases in the DNA helix (Figure 31). The translation of the genetic code is performed with the aid of other nucleic acids: transfer RNA and messenger RNA (ribonucleic acids). These molecules play an important role in protein synthesis in cells. In this process the genetic code is translated into a peptide sequence.[11]

UV radiation can cause genetic damage; for example, the action spectrum for killing bacteria has a maximum near 265 nm and resembles very much the absorption spectrum of DNA, indicating that the UV target is located in the genome. Many organisms have repair mechanisms to remove the products of radiation damage of the genetic material (see also Sections II and V).

1. Structural Investigations

It is well known that investigations on the structure of the nucleic acids and their bases have been done by Raman spectroscopy. The growing interest in molecular mechanisms of gene expression has increased the importance of structural investigations of nucleic acids.

Unfortunately, the classical Raman effect is weak and the Raman spectrum becomes crowded when studying large molecules. Using RR spectroscopy enhancements of the scattered light may be obtained. Large enhancements are seen (for vibrations associated with the chromophore) if the sample is excited at wavelengths near allowed electronic transitions of the molecules.

The bases, nucleosides, and nucleotides of the nucleic acids strongly absorb at wavelengths shorter than 280 nm (Figure 27 and 31). Therefore, efforts have been undertaken to obtain UV-excited Raman spectra which promise high sensitivity and selective enhancement.[28] Those RR enhancement spectra are expected to play an important role in elucidating the purine and pyrimidine electronic spectra, since the RR intensities are directly related to the nature of the excited states. High-quality spectra were obtained by Fodor et al.[63] for the deoxyribonucleotides of uracil, thymine, cytosine, guanine, and adenine (dUMP, etc.). They used a pulsed ND:YAG laser and a H_2 Raman shift cell operating at the first, second, and third anti-Stokes lines to convert the 266-nm output of the Nd:YAG laser to shorter wavelengths, Thus, four excitation wavelengths are available: 266, 240, 218, and 200 nm, opening the UV region down to 200 nm for RR studies. The spectra of Fodor et al. obtained at these variable wavelengths are rich in detail and reveal new vibrational features (Figure 27). The strongest enhancement for all of the nucleotides is given by the 266-nm excitation. All of the Raman bands are attributable to vibrational modes of the heterocyclic bases, whose electronic transitions are responsible for the resonance enhancement. The RR enhancements are examined in the light of the experimental electronic transition wavelengths, the intensities as well as other experiments. In the case of dUMP they observe only two RR enhancement patterns (Figure 27), consistent with there being only two resonant transitions at 265 and 215 nm. Enhancement intensities are at the highest with the excitation wavelengths nearest to the two transitions, while enhancement patterns may be similar.

The electronic spectrum of cytosine is more complicated than that of uracil. There appear to be four transitions of comparable intensity. Although the four excitation wavelengths used by the authors happen to be nearly coincident with each of these transitions, there are only three different enhancement patterns. From the fact that the spectra at 240 and 218 nm resemble each other quite closely and are within a factor of 2 to 3 in intensity, it is concluded that the two resonant transitions have similarly distorted excited states. The electronic spectrum of guanine shows additional complexities. However, in the case of this base as well as of adenosine the four laser wavelengths used lead to explainable RR spectra.

Summarizing, one can say that many features of the RR-enhancement pattern for the purines and pyrimidines can be understood on the basis of the currently available normal mode and Π-Π* transition moment analyses, even though some pieces of the puzzle do not yet fit.

The enhancement patterns of the RR bands are related to the normal mode compositions and to the excited-state bonding changes.[64] Normal coordinate calculations lead to an equation

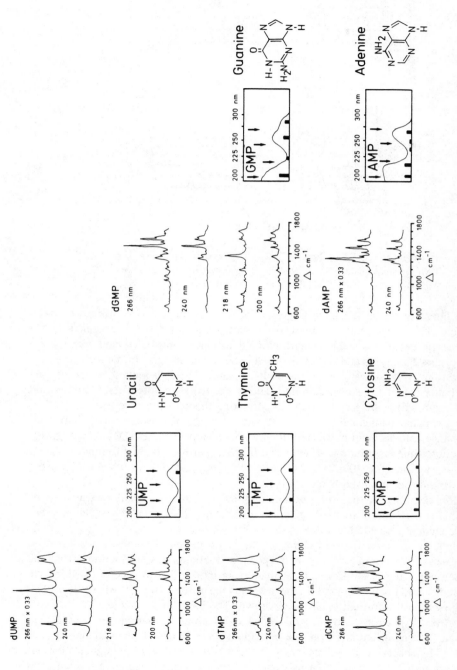

FIGURE 27. Nucleic acid bases: RR spectra, absorption spectra, and structure formula. Left column: RR spectra of aqueous deoxy-monophosphates (5 m*M*) of the five bases with 266-, 240-, 218-, and 200-nm excitation and 266 and 240 nm, respectively. The intensity of the 266-nm spectrum has been multiplied by 0.33 in the cases of dUMP, dTMP, and dAMP. Middle column: UV absorption spectra for the five mononucleotides. Estimated oscillator strengths are given graphically by height of blocks. Vertical arrows indicate the laser frequencies utilized in this study. Right column: structures of the five bases in aqueous solution at pH ~7. (From Fodor, S. P. A., Rava, P. R., Hays, T. R., and Spiro, T. G., *J. Am. Chem. Soc.*, 107, 1520, 1985. With permission.)

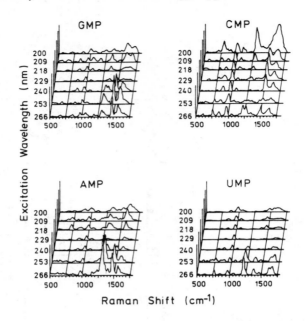

FIGURE 28.   RR spectra of the nucleotide bases GMP, AMP, CMP, and
UMP. (From Kubasek, W. L., Hudson, B., and Peticolas, W. L., *Proc.
Natl. Acad. Sci., U.S.A.*, 82, 2369, 1985. With permission.)

in which the intensity of the RR band is proportional to the bond-order changes between
the ground and the excited states. Peticolas has shown that some bands obtained with a
dUMP experiment are in reasonable accord with RR enhancements calculated with the aid
of normal coordinate analyses and CNDO-derived bond-order changes for the longest wavelength excited state of dUMP. However, major discrepancies were also noted, which suggest
the need for better calculations of the ground- and/or excited-state potentials. Here we find
an important example for the role of lasers in correcting theoretically obtained results.

Other experiments with nucleic acid bases were done by Kubasek et al.[65] They constructed resonance enhancement profiles for the four 5'-monophosphates UMP, CMP, AMP,
and GMP in dilute aqueous solutions (Figure 28). In this experiment eight different excitation
wavelengths were used: 200, 209, 218, 229, 240, 253, 266, and 299 nm, obtained by using
the higher harmonics of a pulsed Nd:YAG laser.

The authors demonstrate the utility of obtaining RR spectra at a variety of wavelengths
when looking for particular structural or ionization effects. The variable wavelengths enhancement dependency on the pH value is shown for some cases.

The application of Raman spectroscopy for the investigation of nucleic acids has been
limited because of the complexity of their spectra. Overlapping contributions from the
different bases, e.g., the close proximity of many ring modes, led to spectra in which the
individual bands could not be resolved. In the last years the utility of far-UV RR spectroscopy
for obtaining structural information has been shown. The bases of the nucleic acids have a
complex electronic spectrum with several broad overlapping transitions in the 280- to
200-nm region. The contributions from the individual bases to nucleic acid Raman spectra
may be resolved by observing the enhancement patterns using variable-wavelength excitation.
In the last years additional information was obtained by observing changes in Raman band
intensities as a function of wavelength. It became increasingly more evident that information
about DNA conformation can be obtained from the enhanced vibrations of the chromophoric
aromatic portions of nucleic acids. By applying variable-wavelength excitation on nucleic
acids, the enhancement pattern can be altered dramatically. This application of RR spectroscopy increases the specificity and the amount of vibrational information for a sample.

It has been known for about 30 years that upon duplex formation of nucleic acids a loss in strength for the ~260-nm absorption bands is observed (e.g., see Reference 66). This effect is called hypochromism. Some workers later noted losses in purine and pyrimidine Raman band intensity upon duplex formation and attributed this effect to a preresonance enhancement mechanism. The availability of far UV-excitation lines allows this effect to be investigated directly.

Fodor and Spiro[67] compared data which were obtained from measurements with synthetic polydeoxyribonucleotide poly(dG-dC) and with equimolar dGMP and dCMP at different UV-excitation wavelengths. Ratios of Raman band intensities and ratios from the absorption spectra were compared considering theoretical aspects. It was found that the ratio of the absorptivities at 240 nm is in good agreement with a corresponding value from the 240-nm Raman excitation. It was concluded that the data are in accord with a simple model in which the Raman bands are wholly or partially resonant with the ~260-nm transitions and contain varying contributions from more energetic, less hypochromic transitions.

Furthermore, it is concluded by Fodor and Spiro that the Raman intensity gain associated with unstacking of the bases may be useful in monitoring single-stranded DNA. They think it is possible to estimate both the extent and composition of single-stranded regions because bands associated with individual bases can be isolated by adjusting the excitation wavelength.

Raman spectra of the purine and pyrimidine bases exhibit characteristic frequency shifts in their exocyclic modes upon duplex formation.[68] The carbonyl and amine groups on the purine and pyrimidine rings are directly involved in H-bonding between the bases when the duplex is formed. Electronic changes due to base pairing and stacking and vibrational coupling across the H-bonds of the base pair occurs. Therefore, it was expected that the vibrational frequencies associated with these groups are perturbed upon duplex formation and that frequency shifts occur. For G-C pairs, in contrast to A-T pairs, vibrational coupling is particularly important because they possess two symmetrically disposed $-C = O$ . . . H-NH linkages. In the cases of poly(dG-dC) and polyG.polyC the specific frequencies were found to be quite similar which is consistent with vibrational coupling being more important than electronic effects in determining the duplex frequencies. The base pairs in poly(dA-dT).poly(dA-dT) have only one $-C=O$ . . . H-NH- interaction. In this case, the Raman spectra exhibited much smaller effects on the exocyclic modes upon duplex formation, which emphasizes the importance of vibrational coupling in G-C pairs.

Conformationally sensitive bands: conversion of the normal B DNA conformation to the left-handed Z form is obtained for poly(dG-dC) in high concentrations of salt. The stacking arrangement of the bases is quite different in the two forms, the orientation of the guanine bases changes, and the conformation of the sugar attached to G changes. Raman spectral differences between the B and Z forms have been noted by several authors. Fodor and Spiro[67] compared RR spectra with 240- and 218-nm excitation of poly(dG-dC) in B-form and Z-form. First they reported a cytidine band shift associated with Z-DNA. This 1526- to 1517-cm^{-1} band is prominent with 218-nm laser excitation and is expected to be useful as a B-Z marker for natural DNA.

2. Excited States

The fluorescence of nucleic acids and of nucleic acid components (e.g., of the bases) is difficult to be measured at room temperature and at neutral pH, as the fluorescence quantum yields lie in the range of 10^{-5} to 10^{-4}. For this reason many investigations have been carried out at low temperatures because under these conditions the quantum yield of fluorescence increases to about 10^{-3} to 10^{-2} due to the decrease of radiationless processes (see also Figure 3). The small quantum yield and the short lifetime of the first excited singlet state (about several picoseconds) makes it difficult to investigate the properties of excited singlet states in liquid aqueous solutions by traditional luminescence methods.

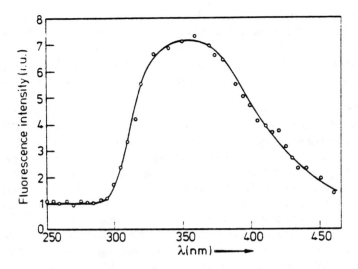

FIGURE 29.   Fluorescence spectrum of calf thymus DNA at room temperature in neutral aqueous solution. Excitation at 266 nm. (From Anders, A., *Chem. Phys. Lett.*, 81, 270, 1981. With permission.)

Laser methods have been introduced to examine the excited states of molecules of nucleic acids and nucleic acid bases.[7,34,70] On one hand the high spectral intensity is of importance to measure such low-fluorescence quantum yields and on the other hand the short laser pulses are necessary to investigate the short-living excited singlet states, the deactivation of the electronic states, and the very fast energy transfer processes. UV lasers have to be used for electronic excitation (Figures 2 and 6).

The fluorescence spectrum of calf thymus DNA in neutral aqueous solution was measured by Anders[69] after excitation with a frequency-doubled pulsed dye laser at 266 nm (Figure 29); the quantum yield at room temperature was evaluated to about 2×10^{-5}. An increase of the quantum yield by a factor of 2 to 3 was observed when the excitation wavelength was shifted from the first absorption maximum (about 266 nm) of DNA to its 0-0 energy, the long wavelength absorption edge. This increase can be explained by a decrease of the competitive radiationless processes in big molecules when they are excited at the 0-0 energy (energy between the lowest vibrational level of the ground state and of the first excited singlet state, respectively).

Picosecond single-photon counting fluorescence spectroscopy of nucleic acids was performed by Rigler et al.[70] They used a krypton-pumped dye laser and the excitation wavelength was 290 nm. Lifetimes of excited states of nucleosides and synthetic polynucleotides were determined at room temperature. One example shall be mentioned: the lifetime of poly d(A-T): 94 ps.

Angelov et al.[71] investigated the properties of excited states of nucleic acid bases, DNA, and RNA by the method of picosecond UV laser photolysis, as well as with the aid of picosecond kinetic spectroscopy.[72] The fourth harmonic wavelength of a mode-locked Nd:YAG laser at 266 nm near the maximum of the first electronic absorption band of nucleic acids was used; the irradiation intensity amounted to about 1 GW/cm^2. Photodecomposition of DNA bases in a liquid aqueous solution by high power UV laser pulses was caused by a two-step excitation of the nucleic acid bases through an intermediate state. This leads to dissociation and/or ionization of the bases. The powerful UV irradiation simultaneously produces two-photon dissociation or ionization of the water. Free radicals of water are produced followed by chemical reactions and the formation of stable photoproducts. On the

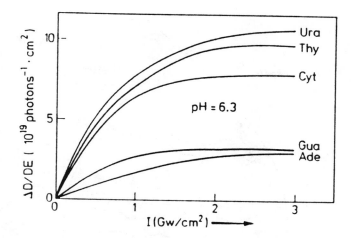

FIGURE 30.   Dependences of photoproduct yield vs. irradiation intensity for all five DNA bases in neutral aqueous solution (D: optical density).[71,72]

contrary, under low-intensity UV irradiation the probability of absorption of two quanta by the same molecule is negligible. Figure 30 shows the dependence of the photodecomposition efficiencies on the irradiation intensity. The photoproduct yield was determined by the relative change in optical density. The molecular character of action was found to be different for each type of base showing a selective action of high power picosecond UV-laser pulses in the process of producing photoproducts of the bases. Information about the formation of hydrated electrons and single-strand breaks in DNA were obtained with picosecond flash photolysis at 266 nm. On the probing site of the apparatus the wavelength 627 nm was used to detect the hydrated electron which absorbs in the red region. This wavelength was generated by stimulated Raman scattering of the Nd:YAG laser in cyclohexane.

Comparisons between the experimental dependencies and the theory yielded a number of characteristic data of the excited states of the bases, e.g., the lifetime of the first excited singlet state and the absorption cross section of the excited singlet state. The lifetime of the first excited singlet state in polynucleotides (poly U) was estimated to be >6.6 ps and the absorption cross section 1.7×10^{-16} cm^2.[72] Information was also obtained about energy transfer processes. From the results and theoretical considerations the distance of electronic energy transfer along the DNA polynucleotide chain was estimated to be less than 170 nucleotide pairs in double-stranded DNA of a bacteriophage and less than 50 pairs in double-stranded segments of tobacco mosaic virus RNA.

In the nucleus, DNA is bound to proteins. These DNA-protein-complexes which play an important role in genetics. To investigate the interaction of DNA binding proteins with nucleic acids by fluorescence methods, time-resolved spectroscopy is necessary. Such measurements of DNA-binding protein and its complex with DNA were done by Greulich et al.[73] The DNA-binding protein from a filamentous bacteriophage (Pf1) was used; this protein exhibits fluorescence from a single tryptophan residue (lifetime: 7,8 ms). The fluorescence lifetime was measured using excitation with a frequency-doubled dye laser at 298 nm (pulse width: 10 ps). Information about changes in the fluorescence behavior on DNA binding were obtained, e.g., the native nucleoprotein complex exhibits a similar fluorescence lifetime (6.5 ns) and an approximately equal fluorescence yield, indicating the absence of tryptophan-DNA stacking. With fluorescence depolarization measurements the rotational mobility of the tryptophan in the nucleoprotein complex could be examined. Also distances between amino acids and DNA bases could be estimated with fluorescence investigations and the use of the Förster law for energy transfer. For example, a distance of 7 Å was estimated between the tyrosin-40 component of the coat protein and the bases in the phage Pf 1.[74]

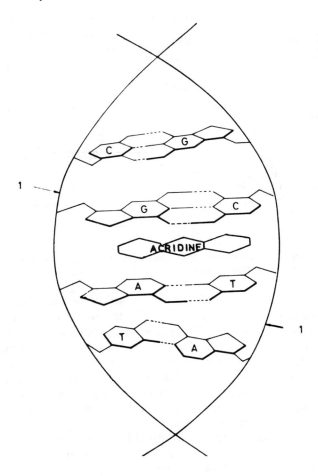

FIGURE 31.   Section from the DNA-helix. DNA mainly consists of four
components: the two purine bases (A) adenine and (G) guanine and the
two pyrimidine bases (T) thymine and (C) cytosin. Adenine is connected
with thymine and cytosine with guanine via hydrogen bonds. Each of the
four bases is also connected with a sugar molecule (desoxyribose) and
phosphoric acid. The latter two components form the "backbone" (1) of
the DNA. All three components (base, sugar, and phosphoric acid) together
are called nucleotide. The two components (base and sugar) are called
nucleoside. In this figure, additionally an intercalated molecule (acridine)
is drawn between two base pairs (see text).

3. Energy Transfer

Energy transfer in nucleic acids, like an exciton transfer along the DNA chain, plays
an important role for UV-produced defects in the DNA of cells as well as for the interaction
between DNA and small molecules. In addition to the above mentioned experiments data
about energy migration in nucleic acids can also be obtained from investigations of DNA-
dye complexes (see Figure 31).

Time-resolved observations of the energy transfer in DNA-acridine-orange-complexes
were performed by Shapiro et al.[75] They excited the complexes in solution with 10-ps pulses
from a mode-locked Nd:YAG laser at 265 nm. The measurement of the dye fluorescence
rise time provides the time interval over which the exciton transfer along the DNA takes
place. Experiments with frequency-doubled pulsed dye lasers (218 to 310 nm) were done
using dye complexes with DNA of varying base contents, isolated from different organisms,

and with synthetic polynucleotides of a fixed-base sequence.[76] The energy transfer from DNA to the dye in complexes, where the different mean spacings of the dye molecules along the DNA chain varied, rendered the range of the exciton transfer along the DNA (see below). A dependence of the exciton transfer on the excitation wavelength and on the base composition of the DNA was found. The transfer increases near the 0-0 energy of DNA; this may be explained by a decrease of radiationless processes. The exciton transfer drops to 1/e, along 80 base pairs in *Escherichia coli* DNA (49.8% GC) and along 60 base pairs in GC polynucleotides (100% GC). Applying the Förster theory to the excitation transfer between the bases in a DNA chain, a mean distance of the excitation transfer of about 55 base pairs was calculated.

The interaction of biologically active molecules such as dyes with nucleic acids is of interest because such molecules are used as drugs in chemotherapy (e.g., antibiotics, antitumor substances). On the contrary other ones can induce mutations and tumors. They form intermolecular complexes with nucleic acids. A special class are dyes interacting with DNA by intercalation. These dye molecules are inserted between two base pairs (Figure 31). A spectroscopic pecularity of such complexes are base-sequence-dependent effects concerning the fluorescence behavior due to the energy transfer inside the complex. These processes can depend on the type of base pairs between which the dye molecules are inserted. Base-sequence-dependent energy transfer processes from DNA-bases to intercalated dye molecules in solution were measured by Anders[77] using a frequency-doubled pulsed dye laser. To cover the absorption range of the DNA in the UV frequency doubling with KDP, LFM, and KB5 crystals yielded radiation from 218 to 320 nm. The quantum yield of the dye fluorescence after excitation in the visible and the UV range (where not only the dye but also the DNA absorbs) render the relative amount of energy transfer from the bases to the intercalated dye molecules. This energy transfer to the dyes (e.g., proflavine, acridine orange, ethidium bromide) depends on the excitation wavelength and on the kind of bases between which the dye molecules are inserted.[77] Furthermore, in DNA-dye-complexes a selective action of narrow band laser excitation was proved. Andreoni et al.[78] investigated a selective photo-damage of dye molecules bound to polynucleotides. The experiments were based on a time-delayed two-step photoionization (Figures 8 and 12). The selectivity arises from the fact that the lifetime of the first excited electronic state of the dye may be sensitive to the binding site (Section II.C). The preferential excitation of distinct DNA bases via a one-step electronic excitation (Figure 6) was performed with a dye laser and was tested by the interaction of these bases with intercalated dye molecules.[79] In order to analyze selective effects, the above-mentioned energy transfer in the three possible base pair units after the preferential excitation of one kind of base was compared in different complexes. Two tendencies were found: (1) the energy transfer in one kind of base pair unit is larger if mainly these units are excited, e.g., with ethidium bromide intercalated between AT-AT units, the transfer is stronger in that wavelength region in which more AT than GC pairs are excited and (2) a higher transfer appears in some units if the neighbored ones are selectively excited; for example, proflavine inserted in AT-AT units shows a greater transfer if GC pairs are excited and the energy is transferred along the DNA to the AT pairs.

4. Conformation and Motion

Information about conformation and dynamic properties of nucleic acids were obtained by Rigler et al.[70] and Claesens and Rigler[80] using pulsed fluorescence and time-resolved fluorescence spectroscopy. For example, rotational motion of molecules can be investigated by exciting their fluorescence with short pulses of polarized light and observing the time dependence of the polarized light emission. For a symmetric body the anisotropy of the emission (Figure 32) is characterized by three rotational relaxation times, each of which is a function of the rotational diffusion constants around the main axes of rotation. The cor-

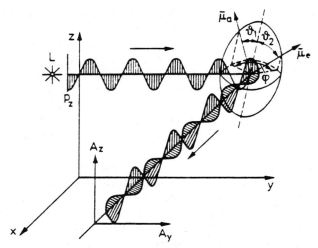

Time dependence of anisotropy r
for a symmetric rotor

FIGURE 32.   Rotational diffusion of molecules as detected in pulsed
fluorescence experiments. r(t): anisotrophy emitted by a symmetric rotor
after pulsed excitation with polarized light L. μ_a and μ_c: absorption and
emission vectors with coordinates 1, 2, and 0 (molecular frame). P_z: polarizer, A_z and A_y analyzers in z and y directions of the laboratory frame.[4]

responding amplitudes a_i depend on the position of the absorption vector μ_a and the emission
vector μ_c within the coordinates of the rotating unit.

For investigating structural information of nucleic acids fluorescent labels are often
necessary. The dye ethidium bromide or the natural hypermodified nucleotide wybutine
served as local structural probes for investigating t-RNA molecules.[80] t-RNA plays, as
mentioned above, an important role in protein synthesis by translating the genetic code into
a peptide sequence by codon-anticodon interaction at the ribosome particles. Several models
for describing this mechanism and the structure of the anticodon loop were proposed from
mainly X-ray crystallographic studies and nuclear resonance measurements.

Fluorescence lifetimes and anisotropies were measured in solution using time-correlated
single-photon counting and a mode-locked synchronously pumped and frequency-doubled
dye laser as excitation source. From the analysis of lifetimes (τ) and rotational relaxation
times (τ_R) the various structural states of wybutine — situated in the anticodon loop of yeast
t-RNA — could be concluded: one stacked conformation where the base has no free mobility
and only rotational motion reflects the mobility of the whole t-RNA molecule ($\tau = 6$ ns
and $\tau_R = 19$ ns) and an unstacked conformation where the base can freely rotate ($\tau = 100$
ps and $\tau_R = 370$ ps) and an intermediate state ($\tau = 2$ ns and $\tau_R = 1.6$ ns).[80]

Upon complex formation with the complementary codon (*E. coli* t-RNA) a change in
the environment of wybutine takes place which leads to a completely different distribution
of the fluorescence lifetimes in which a conformation of wybutine specified by a very short
lifetime of 100 ps is the main feature (see Figure 33). Furthermore, anisotropy curves were
calculated from the experimental unpolarized and parallel polarized fluorescence decays and
led to information about the motion of the wybutine itself and the rotation of the whole t-
RNA molecule. Concluding, it was shown that the fluorescence properties of wybutine
showed the existence of different structures in the anticodon loop and from the results it is
evident that the anticodon loop undergoes a conformational change on binding a complementary codon triplet.[80]

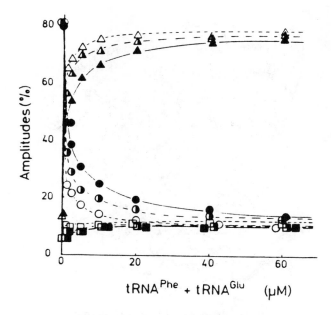

FIGURE 33. Distribution of the relative amplitudes a_1 (▲, △, △), a_2 (■, ◨, □), and a_3 (●, ◐, ○) of the fluorescence lifetimes of wybutine in yeast t-RNA-*E. coli* t-RNA complex as a function of t-RNA concentration. The curves are shown for 20°C (▲, ■, ●), 12°C (△, ◨, ◐), and 4°C (△, □, ○). (From Claesens, F. and Rigler, R., *Eur. Biophys. J.*, 13, 331, 1986. With permission.)

D. HEME PROTEINS

Ultrafast processes which require ultrashort laser pulse methods for studying are ubiquitous in biology. A special type of laser application in biology is the investigation of heme proteins. The biological system is disturbed from equilibrium by means of light pulses and the subsequent recovery of the system is followed by laser spectroscopy.[9,81]

Heme proteins are molecules of highly complex structure (Figure 34). They contain iron porphyrins which absorb visible light. Hb, the oxygen carrier in red blood cells, and Mb, a muscle protein which stores the oxygen, belong to this group. In both proteins molecular oxygen is reversibly bound to the iron atom in the heme group of porphyrin structure. The binding and release of O_2 (or CO) is critical to the biological function. The dynamics of these motions are most productively studied by time-resolved spectroscopic methods. Hb and oxyhemoglobin (HbO_2) are known to have different equilibrium structures which are considered to have significance for the mechanism of cooperativity. A laser pulse rapidly breaks the bond between the ligand and the heme iron atom. The heme iron atom is left in a nonequilibrium, yet biologically important state.

1. Fast Recovery Processes

Examples of performing photodissociation experiments are found in the nanosecond, picosecond, and recently femtosecond range. Spectra obtained with 300-fs pulses from a passively mode-locked CW dye laser and femtosecond pump-probe experiments have shown that the dissociation of HbCO and MbCO occurs with a time constant of 350 fs regardless of the species (referred to in Reference 9). Hb and Mb, respectively, appear in 350 fs and, therefore, this time is assumed to be that one for the structural change in the Fe region to occur. Until 1984 only the response of the heme region of the molecule to photolyzing pulses was studied; however, it was also assumed that effects of deligation at the amino acid side chains or the polypeptide backbone may be measured without difficulties using current

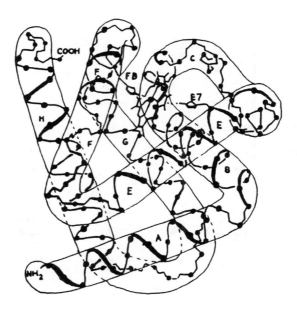

FIGURE 34.   Mb protein chain with Hb in upper center.

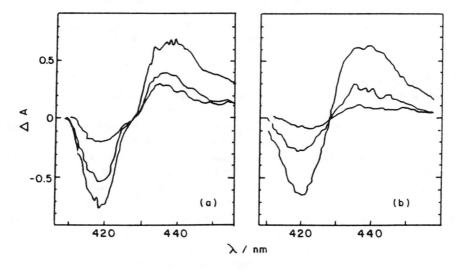

FIGURE 35.   Transient absorption spectra. Absorption in myoglobin of (a) Whale. Spectra at 4, 10, and 35 ps and (b) Elephant. Spectra at 4, 20, and 28 ps. (From Madge, D., Martin, J.-L., Wilson, K. R., Dupuy, C., Zhu, J. G., Campbell, B. F., Mitchell, M., and Marsters, J., *SPIE*, 533, 2, 1985. With permission.)

pulsed-laser technology. In 1985 Madge et al.[82] published early results of studying distal side effects in myoglobin and model compounds with the help of femtosecond laser pulses. A comparison of heme proteins from different species, which vary in specific ways, was performed. A mode-locked ring laser with a four-stage Nd:YAG-pumped dye amplifier generates femtosecond pulses at 630 nm with a repetition rate of 10 Hz. Frequency doubling to 315 nm provides photolysis pulses, while continuum generation offers probe wavelengths for transient spectroscopy throughout the visible. An example for the comparison of whale and elephant is given in Figure 35.

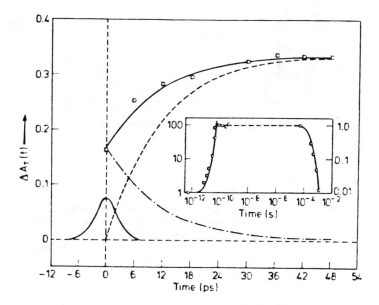

FIGURE 36. Absorbance changes in HbCO at 440 nm from t = 0 to t = 48 ps after single-pulse excitation at 530 nm in solution; □ experimental $A_T(t)$; —— calculated $A_T(t)$. The excitation pulse profile is indicated at the origin. Inset: a log-pot plot covering 10 orders of magnitude in time, shows the exponential kinetics of both the dissociation and the recombination initiated by a single 530-nm pulse. The left ordinate, A_{max}/I, $A_{max} - A_t(t)I$, corresponds to the photodissociation and the right ordinate, $A(t)/A_{max}$, corresponds to the recombination, —— calculated line; 0 experimental points. (From Noe, L. J., Eisert, W. G., and Rentzepis, P. M., *Proc. Natl. Acad. Sci. U.S.A.*, 75, 573, 1978. With permission.)

Examples for photodissociation experiments in the picosecond range include transient absorption spectra as well as Raman spectra. The spectra of the first type are presented as difference curves. One example shows the change in absorbance between excitation and no excitation vs. time (Figure 36). The absorbance change with HbCO is demonstrated by Noe et al.[83] In this experiment a mode-locked Nd:glass laser at 530 nm with 6-ps pulses was applied as an excitation source. A variety of new phenomena was discovered using nonlinearly generated continuum techniques. In Figure 37 the transient difference spectra of MbCO after photolysis by a 353-nm laser pulse are shown.

Picosecond Raman spectroscopy can be carried out in a pump-probe configuration. The first experiments were carried out with one pulse acting both as pump and probe. More recently, studies of Mb were performed by Hochstrasser and Johnson[9] in collaboration with Spiro and co-workers applying pump-probe Raman experiments. At 30 ps certain Raman transitions were significantly shifted from those observed on the 10-ns time scale (Figure 38), indicating that a significant relaxation process occurs between those times.[9] As this relaxation was not found in Hb it could be followed that a difference exists between Hb and Mb dynamics after photolysis. Thus, picosecond RR spectroscopy has discovered a difference not known before from any other investigations. This pump-probe Raman experiment was carried out with a Nd:YAG laser-pumped dye laser and a gas Raman shifter both together serving as sources for pump and probe light.

For taking nanosecond absorption spectra two synchronized Nd:YAG lasers were used. One laser pumps the sample at 530 nm, while the other one pumps a fluorescent dye to generate a broadband nanosecond continuum. Difference spectra could be plotted at times between 1 ns and 100 ms after photolysis. On photolysis of HbCO, for example, a nascent

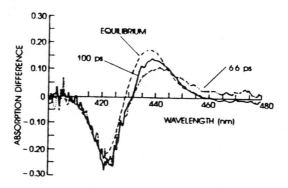

FIGURE 37.   Transient difference spectra of carboxymyoglobin after photolysis by a 353-nm laser pulse. (From Prince, M. R., Deutsch, T. F.,
Shapiro, A. H., Margolis, R. J., Oseroff, A. R., Fallon, J. T., Parrish,
J. A., and Anderson, R. R., *Proc. Natl. Acad. Sci., U.S.A.*, 83, 7064,
1986. With permission.)

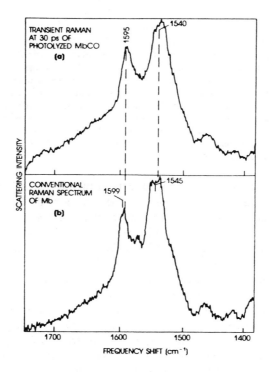

FIGURE 38.   Raman spectra showing time-dependent shifts occurring on
a picosecond time scale in myoglobin: (a) transient Raman spectrum of
photolyzed carboxymyoglobin within 30 ps of photolysis and (b) Raman
spectrum of stable deliganded myoglobin. (From Hochstrasser, R. M. and
Johnson, C. K., *Laser Focus/Electro-Optics*, 100, 1985. With permission.)

Hb structure and CO is generated. The nascent Hb structure approaches the so-called equilibrium form of Hb only after some microseconds and the bleaching process could be followed
by computer analysis.

Earlier, Friedman and Lyons[84] reported information which was obtained about changes
between quarternary structures in the protein matrix of Hb. The authors recorded RR spectra

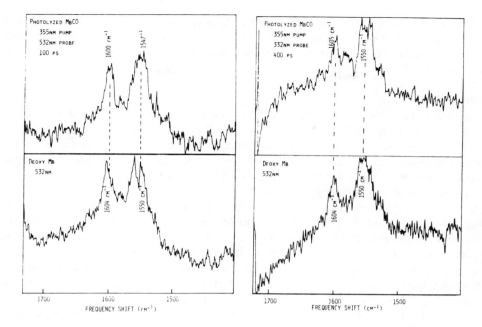

FIGURE 39.   Picosecond Raman spectrum of carboxymyoglobin 100 and 400 ps after photolysis (top) and Raman spectrum of stable Myoglobin (bottom). (From Prince, M. R., Deutsch, T. F., Shapiro, A. H., Margolis, R. J., Oseroff, A. R., Fallon, J. T., Parrish, J. A., and Anderson, R. R., *Proc. Natl. Acad. Sci. U.S.A.,* 83, 7064, 1986. With permission.)

after photolysis of HbCO. They used a 10-ns Nd:YAG laser pulse at 532 nm for photolysis and monitored the Raman spectra of the transient species (occurring with 10 ns after photolysis) with a N_2-pumped dye laser.

This is an example for the combination of photodissociation and Raman laser techniques in the nanosecond range. The Raman spectrum reflects the structure of heme during the first 10 ns of the evolution of the protein from the liganded to the deliganded conformation. It was observed that the vibration corresponding to the iron-histidine stretching motion undergoes significant shifts corresponding first to local changes, later to changes in the overall configuration of a protein subunit and the rebinding of the CO to the Fe. Effects of interrelations among subunits were observable at later times through the vibrational spectrum.

In the field of application of lasers to heme protein investigations other examples may be mentioned. Recently, picosecond time-scale motions for the tryptophans in Mb were discovered using fluorescence polarization. In experiments on tuna Mb it was confirmed that the relevant part of the structure is not stationary on the time scale of 10 ps. Johnson and Hochstrasser[81] demonstrated how biologically relevant interpretations could be reached only by application of lasers and not by other experimental techniques. They recorded pump-probe picosecond Raman spectra of photolyzed HbCO. By the results it could be verified that the observation of a shift in HbCO Raman spectrum in an earlier single-pulse experiment was not due to heating of the sample by the nanosecond laser pulse as was discussed before.

In 1986 Johnson and Hochstrasser[81] published results showing the picosecond Raman spectra of MbCO 100 and 400 ps after photolysis (Figure 39). The differences between the spectrum at 7 ns and the spectrum at equilibrium are clearly presented by the curves. These spectra are compared with those of stable Mb.

V. PHOTOMEDICINE

The use of lasers in medical practice has gained growing importance.[6,85] Many of these applications in surgery and coagulation rely on the high intensity and the small focus. Another class of laser application in medicine represents the use of low-intensity radiation. In non-thermal use, the laser properties playing a major part are high spectral intensity and tunability of the wavelength.

A. SURGERY

In clinical medicine — using thermal effects of laser radiation — the laser systems of choice are argon, CO_2, and Nd:YAG laser.[86] The biological effects of heating are, of course, nonspecific, although the choice of the wavelength can change the penetration depth of the light into different kinds of tissue and as a consequence, the region in which heat is deposited. Numerous laser applications exist in the following fields: ophthalmology, plastic and general surgery, dermatology, gynecology, gastroenterology, urology, and neurosurgery. In many applications, e.g., in dermatology, the argon laser is very useful because the red pigment of the blood renders blood vessels very susceptible to absorption of green light and resulting destruction. In gasteroenterology, bleeding can be stopped by means of photocoagulation with lasers using endoscopes and fiberoptic techniques. In neurosurgery the laser is applied, for example, for tumor removal. An advantage is the "no touch" removal technique, and with a simultaneous coagulation of the surrounding blood vessels a prevention of tumor cell spreading is possible.

Excimer laser radiation at 193 and 248 nm has been used to produce well-defined cuts in tissue by ablation, e.g., applied to corneal surgery or the removal of atherosclerotic lesions.[87] At these wavelengths the absorption length of many polymeric molecules is in the order of 0.01 to 1.0 μm, so that the energy is absorbed in a very thin layer of the surface. Furthermore, the fragmentation of gallstones and kidney stones using pulsed visible laser radiation was recently demonstrated *in vitro* and in humans with pulsed lasers (laser litho-tropsy).[88] The fragmentation mechanism is not completely understood. The temporal and spectral behavior of the light flash accompanying fragmentation of gallstones was studied with dye laser pulses. Time-resolved emission spectra showed a continuum upon which line spectra are superimposed. This indicates the initiation of a plasma, impulsively expanding and generating intense acoustic transients which fracture the stones.

B. PHOTOCHEMOTHERAPY OF TUMORS

A new, promising nonthermal laser therapy of cancer is based on selective tumor destruction combining light and an administered drug, a dye, called photochemotherapy or photoradiation therapy (PRT) of tumors[40] (Figure 15). Some dyes, like hematoporphyrin or acridine orange, have a tendency to concentrate more in malignant, rather than normal, tissue. Such dyes can become very toxic by exposure to light; thus, tumors can be selectively destroyed in comparison with the surrounding normal tissue. The origin of the differences in concentration *in vivo* is still rather speculative; the initial uptake of hematoporphyrin in cells, e.g., depends on lipid solubility or aggregation states in the cells. With time-resolved fluorescence spectroscopy using picosecond pulses from a dye laser, differences in the fluorescence spectra from human gastric cancer compared to those of the surrounding tissue were found, showing that molecular dimers play an important role in the selective accumulation properties.[89]

In clinical treatment hematoporphyrin is intravenously injected and for irradiation (absorption maximum at 630 nm) mainly lasers (dye lasers or gold vapor lasers), not lamps, are used because they can be employed in conjunction with endoscopes and optical fibers.

Dyes attached to monoclonal antibodies which localize strongly and selectively on specific structures of the tumor cells are a new, interesting direction in research.

C. LOW-INTENSITY THERAPY

Low-intensity laser radiation is supposed to affect photobiological phenomena such as the growth of bacteria, synthesis of DNA, stimulation of ATP synthesis, or cell proliferation.[90] He-Ne lasers (633 nm) or diode lasers (904 nm) are applied in low-power therapy of, for example, wound healing or pain reduction. The "biostimulation effect" seems to occur only within a narrow range of parameters (dose, wavelength, pulse frequency). Contradictive results are found and further research is necessary to explain the mystery of laser biostimulation with well-known photobiological phenomena and clarify the fundamental mechanisms of this effect.

D. DIAGNOSIS

The use of lasers for diagnostic purposes is well established in some areas. It includes various techniques like cytofluorometry[36] (e.g., cell sorting and counting), Doppler techniques (e.g., measurement of blood flow), various kinds of spectroscopy, and holography.[91]

Flow cytometry and sorting is a technique for rapidly measuring optical properties of cells and to selectively separate special groups of cells. One example of the apparatus is presented in Figure 14. Now, flow cytometers and sorters are standard features in many research laboratories and they found their way into clinical applications. There is a variety of instruments with which special techniques can be carried out, such as fluorescence polarization measurements, using different fluorescent labels. A wide range of methods of data analysis to extract the maximum amount of biologically relevant information is integrated. There are also many examples of applications, such as the determination of the DNA or protein content of cells, drug effects or the presence of virus in infected cells with fluorescent antibody techniques, the classification of human blood cells (also leukemias) or chromosomes and the recognition of diagnostically important precancerous cervical or bladder cells.

There are many examples of laser spectroscopic applications in medicine, such as a laser nephelometer, to determine different protein concentrations in the blood by means of Mie scattering techniques, Doppler measurements of blood flow, or the CARS microscope to image the microscopic spatial distribution of biological compounds with two scanning laser beams.[92]

E. PHOTOMEDICAL RESEARCH

Laser microbeam systems (Figure 13) are used for microsurgery or stimulated microfluorescence.[6] Spots of about 0.01 μm diameter can be produced. Damages even smaller than the diffraction limits of the wavelengths can be reached, because of the "Gaussian hot spot" in the center of the laser beam. By selectively altering or destroying of small regions inside living cells or chromosomes and studying the cell afterwards, it is possible to evaluate the control of basic cellular processes and cell division ("selective genetic engineering"). Laser beams can be applied to induce fluorescence within small regions of cells, resulting also in the study of cell membranes because special fluorescent molecules can be attached on the cell surface and then excited with the laser beam. Another example is the laser-induced fusion of cells.[38] An excimer laser-pumped dye laser beam at 340 nm was suitable for inducing fusion of mammalian lymphocytes with myelomid cells or plant protoplasts. Laser fusion of cells can be — besides other methods — of interest for the production of therapeutic antibodies, e.g., immune therapy of cancer.[38]

An improved knowledge about photoinduced processes in man is of great interest as a basis for the nonthermal medical therapy with light and photosensitizing drugs.[85] Figure 40 shows the optical properties of the skin. Incident radiation is partly reflected, partly absorbed or scattered in the different layers of skin. Thus, the action of light on skin is a very complex process which is difficult to be resolved into its various basic mechanisms. Spectroscopic properties of isolated human epidermis were measured with dye lasers. Structures in transmission or absorption spectra could be resolved, which were not found with usual spectro-

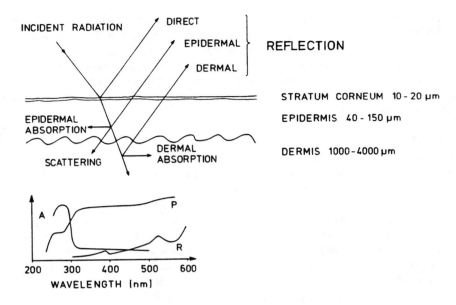

FIGURE 40.   Left: spectral distribution of absorption (A) and reflection (R) of radiation in skin, and depth of penetration (P) of light into the skin, measured with spectrophotometers. Right: absorption, reflection, and scattering of optical radiation in skin (schematic).

photometers.[93] The optical behavior of isolated skin treated with photosensitizing dyes was also investigated with tunable lasers and changes in transmission after irradiation with different laser wavelengths were detected.[93] Such dyes are used in photochemotherapy of dermatosis, e.g., psoriasis and tumors. Photosensitizing dyes work through different mechanisms in the cell. For example, psoralens react under long-wavelength UV irradiation with the thymine in DNA by cyclobutane addition forming mono- and diadducts. In this way the synthesis of nucleic acids is blocked. Pathological cell growth, as in the skin disease psoriasis, is thus prevented.

In such kinds of skin diseases a therapy with UV light is applied in the UVB and UVA (see Figure 2). Also for cosmetic tanning high doses of UVA light are applied. These longer UV wavelengths penetrate into the lower layers of the skin (Figure 40). Therefore, the different action spectra direct on probands should be known with high accuracy to avoid unwanted effects. Such action spectra of medical interest are the erythema curve (the action spectrum of sunburn), the pigmentation curve, both spectra after oral or local administration of photosensitizers, and action spectra of the photochemotherapy of special dermatosis (optimum range of healing).

Dye lasers with their high spectral intensity and tunability are ideal light sources to get an improved knowledge about such UV-induced effects in skin. As an example of an action spectrum the erythema effectiveness curve investigated with frequency-doubled pulsed dye lasers in the UV by Anders et al.[94] is shown in Figure 41. Small test areas of probands were irradiated with different wavelengths and energies, and the skin reactions were determined. Figure 41 shows the results in comparison with those which were obtained with lamps and monochromators (Freeman and Berger) and Hg spectral lines (Hausser). By using lamps and monochromators of relatively broad spectral bandwidths (about 10 nm) the minimum of the curve could not be resolved. The use of narrow band irradiation sources, like Hg spectral lines or tunable lasers, gives similar results; the smaller erythema bandwidth of the latter can partly be explained by the larger number of measured points, which is possible with a tunable laser.[94]

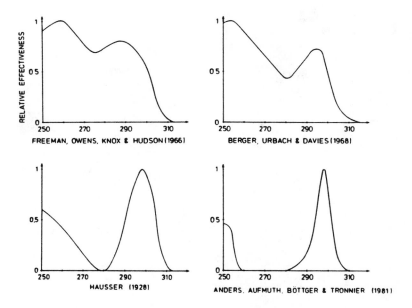

FIGURE 41.   Comparison of erythema effectiveness curves obtained 24 h after irradiation (see text and Figure 2). (From Anders, A., Aufmuth, P., Böttger, E.-M., and Tronnier, H., *Dermatosen*, 32, 153, 1984. With permission.)

Another example for laser spectroscopic investigations of tissue is the laser induced fluorescence of normal and native cancerous tissue *in vitro*.[95] The fluorescence was excited with a CW argon ion laser operating at 488 nm; the beam was focused on the front surface of the tissue. The fluorescence profiles from the cancerous and normal tissue, respectively, are substantially different, showing characteristic maxima (an example is displayed in Figure 42). The peaks are assigned to fluorochroms, such as flavins and porphyrins, in different environments in the two types of cells. Taking different kinds of tissue, the prominent maxima of the cancerous tissue spectra are blue shifted and located around 521 nm, whereas the typical maxima of the normal one are located at about 531 nm. Furthermore, time-resolved fluorescence kinetics and fluorescence polarization spectroscopy excited by pico-second pulses from a mode-locked Nd-laser at 353 nm yielded differences for both kinds of tissues. The degree of polarization and the emission lifetimes from normal and malignant rat kidney tissues are different.[95]

VI. CONCLUSION

Photosynthesis, vision or heme proteins, and nucleic acids are very important examples indeed, but other biological processes and systems have been investigated with laser spectroscopy as well. Molecular motion of biomolecules occurring on time scales less than 1 ps must be considered as one of the processes of fundamental importance in biological systems. A complete understanding of such motions could explain the function of many biological molecules including nucleic acids, proteins, amino acids, enzymes, antibodies, or membrane molecules.[9,34]

Typical experiments on molecular motion involve fluorescence polarization or transient absorption of excited states and ground state absorption polarization spectroscopy. Rapid dynamics of biomolecules were evaluated with lasers in heme proteins and nucleic acids (see above). Other examples include fluorescence probe studies of proteins and membranes.[22] Different fluorochromes when incorporated into membranes or proteins exhibit different

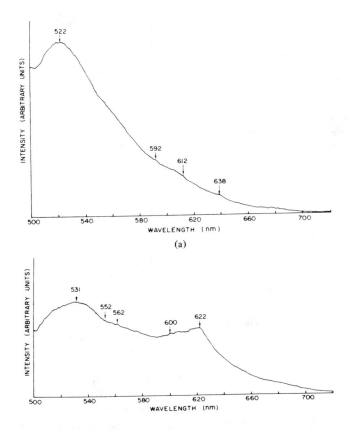

FIGURE 42. (a) Fluorescence spectrum of rat kidney tumor and (b) Fluorescence spectrum of normal rat kidney. (From Tata, D. B., Foresti, M., Cordero, J., Tomashefsky, P., Alfano, M. A., and Alfano, R. R., *Biophys. J.*, 50, 463, 1986. With permission.)

types of interactions. The decay of the anisotropy yields information about the environments. Studies of membrane dynamics and lipid-protein interactions were described by Hudson et al.[22]

Application of laser Raman spectroscopy in biology is, of course, a widely applied technique. Among the biomolecules and biological structures examined by laser Raman spectroscopy are[25,26] sugars, amino acids, proteins, polynucleotides, chromosomes, lipids, enzymes, hormones, toxins, photoreceptors, cell membranes, nerve fibers, viruses, and tissue. Enzyme-substrate reactions are other objects of investigations with laser Raman spectroscopy.[96,97] For a full understanding of an enzyme-catalyzed reaction it is important to know the position of the atoms and electrons of the substrate and the enzyme for each species along the reaction pathway. Vibrational spectroscopy has several advantages when studying enzyme interactions between the "static" structures and the dynamic events.

Further information can be obtained from laser light scattering.[98] A common experiment of biological interest is the analysis of the angular dependence on the intensity of scattered laser radiaton from macromolecules or molecular aggregates in solution. Quasielastic laser light scattering is an accurate method to measure the translation diffusion coefficient of macromolecules. The diffusion coefficient is a parameter that depends on the size and shape of the macromolecules and on the thermodynamic and hydrodynamic interaction between the macromolecules. Radiation which is scattered from a moving particle is shifted in frequency by the Doppler effect; the magnitude of the Doppler shift allows an exact determination of the velocity of the object.

Laser spectroscopic methods are established in biomedical investigations, especially because of their sensitivity to structural parameters and their short-pulse capabilities. There are examples where lasers are advantageous but often laser spectroscopy is necessary. Laser spectroscopy can shed new knowledge and suggest new mechanisms not only for processes of small molecules but also for complicated biological phenomena. The field of laser medicine is continuing to improve therapy and diagnosis in both new and established areas.

Strong efforts were made to improve technological development of lasers where the generation of short pulses tunable from the vacuum UV to the IR is of special importance. Laser methods are valuable supplements to other methods such as conventional optical spectroscopy, electron spin resonance, or nuclear magnetic resonance spectroscopy and X-ray scattering.[99]

The life sciences photobiology and photomedicine are interdisciplinary sciences which have undergone dramatic development in the last years. This led to a collaboration of scientists from all sides and gives hope for the future to solve questions of major biomedical interest.

ACKNOWLEDGMENT

We gratefully acknowledge Prof. Dr. E.-G. Niemann for his helpful discussions.

REFERENCES

1. **Mollenhauer, L. F. and White, J. C.,** Eds., *Tunable Lasers,* Springer-Verlag, Heidelberg, 1987.
2. **Hänsch, T. W. and Shen, Y. R.,** Eds., *Laser Spectroscopy VII,* Springer-Verlag, Heidelberg, 1985.
3. **El-Sayed, M.,** Ed., Laser Applications in Chemistry and Biophysics, Vol. 620, Proc. of SPIE, The Int. Soc. Opt. Eng., Washington, 1986.
4. **Hillenkamp, F., Pratesi, R., and Sacchi, C. A.,** Eds., *Lasers in Biology and Medicine,* Plenum Press, New York, 1980.
5. **Pratesi, R. and Sacchi, C. A.,** Eds., *Lasers in Photomedicine and Photobiology,* Springer-Verlag, Heidelberg, 1980.
6. **Berns, M. W.,** Lasers in biomedicine, *Laser Focus,* 6, 19, 1983.
7. **Anders, A.,** Laser spectroscopy of biomolecules, in *Topics in Current Chemistry,* Boschke, F. L., Ed., Springer-Verlag, Heidelberg, 1984, 23.
8. **Kleinermanns, K. and Wolfrum, J.,** Laser chemistry — what is its current status?, *Angew. Chem.,* 26(1), 38, 1987.
9. **Hochstrasser, R. M. and Johnson, C. K.,** Lasers in Biology, *Laser Focus/Electro-Optics,* 100, 1985.
10. **Regan, J. D. and Parrish, J. A.,** Eds., *The Science of Photomedicine,* Plenum Press, New York, 1982.
11. **Häder, D. P. and Tevini, M.,** *General Photobiology,* Pergamon Press, Elmsford, NY, 1987.
12. **Andrews, D. L.,** *Lasers in Chemistry,* Springer-Verlag, Heidelberg, 1986.
13. **Cantrell, C. D.,** Ed., *Multiple-Photon Excitation and Dissociation of Polyatomic Molecules,* Springer-Verlag, Heidelberg, 1986.
14. **Letokhov, V. S.,** Ed., *Laser Picosecond Spectroscopy and Photochemistry of Biomolecules,* Adam Hilger, London, 1987.
15. **Andreoni, A., Cubeddu, R., De Silvestri, S., Laporta, P., and Svelto, O.,** Time-delayed two-step selective laser photodamage of dye-biomolecule complexes, *Phys. Rev. Lett.,* 45, 431, 1980.
16. **Mourou, G. A. et al.,** Eds., *Picosecond Electronics and Optoelectronics,* Springer-Verlag, Heidelberg, 1985.
17. **Holtom, G. R.,** Laser research conducted at the regional laser and biotechnology laboratories — an overview of the techniques and applications, *SPIE J.,* 482, 2, 1984.
18. **Stwally, C. W. and Lapp, M.,** Eds., Advances in laser science. I, American Institute of Physics Conference Proc., no. 146, New York, 1986.
19. **Kobayashi, S., Yamashita, M., Sato, T., and Muramatsu, S.,** A single-photon sensitive synchroscan streak camera for room temperature picosecond emission dynamics of adenine and poly adenylic acid, *IEEE J. Quantum Electron.,* 12, 1383, 1984.

20. **Pratesi, R.,** Diode lasers in photomedicine, *IEEE J. Quantum Electron.,* 12, 1433, 1984.
21. **Meech, R. S., Stubbs, C. D., and Phillips, D.,** The application decay measurements in studies of biological systems, *IEEE J. Quantum Electron.,* 12, 1343, 1984.
22. **Hudson, B., Harris, D. L., Ludescher, R. D., Ruggiero, A., Cooney-Freed, A., and Cavalier, S. A.,** Fluorescence probe studies of proteins and membranes, in *Application of Fluorescence in the Biomedical Sciences,* Taylor, D. L., Waggoner, A. S., Murphy, R. F., Laani, F., and Birge, R. R., Eds., Alan R. Liss, New York, 1986, 159.
23. **Holten, D. and Windsor, M. W.,** Picosecond flash photolysis in biology and biophysics, *Ann. Rev.,* 7, 189, 1978.
24. **Greene, B. J., Hochstrasser, R. M., and Weisman, R. R.,** Picosecond transient spectroscopy of molecules in solution, *J. Chem. Phys.,* 70, 1247, 1979.
25. **Gaber, B. P.,** Biological applications of laser Raman spectroscopy, *Am. Lab.,* 3, 15, 1977.
26. **Carey, P.,** Raman studies of biological molecules, *Opt. Pura Apli.,* 17, 219, 1984.
27. **Parker, F. S.,** *Applications of Infrared, Raman, and Resonance Raman Spectroscopy in Biochemistry,* Plenum Press, New York, 1983.
28. **Hudson, B.,** *Ultraviolet resonance Raman spectroscopy of biopolymer components,* AIP Conf. Proc., 146, 690, 1986.
29. **Chu, B.,** Dynamics of macromolecular solutions, *Phys. Scr.,* 4, 458, 1979.
30. **Earnshaw, J. C. and Steer, M. W., Eds.,** *The Application of Laser Light Scattering to the Study of Biological Motion,* Plenum Press, New York, 1983.
31. **Kam, Z. and Rigler, R.,** Cross-correlation laser scattering, *Biophys. J.,* 39, 7, 1982.
32. **Shank, C. V.,** Investigation of ultrafast phenomena in the femtosecond time domain, *Science,* 233, 1276, 1986.
33. **Hilinski, E. F. and Rentzepis, P. M.,** Biological applications of picosecond spectroscopy, *Nature,* 302, 481, 1983.
34. **Rigler, R., Claesens, F., and Kristensen, O.,** Picosecond fluorescence spectroscopy in the analysis of structure and motion of biopolymers, *Anal. Instrum.,* 14, 525, 1985.
35. **Siegmann, A. E. and Fleming, G. R., Eds.,** *Ultrafast Phenomena V,* Springer-Verlag, Heidelberg, 1986.
36. **van Dilla, M. A., Dean, P. N., Laerum, O. D., and Melamed, M. R., Eds.,** *Flow Cytometry: Instrumentation and Data Analysis,* Academic Press, London, 1985.
37. **Fodor, S. P. A., Rava, P. R., Hays, T. R., and Spiro, T. G.,** Ultraviolet resonance Raman spectroscopy of the nucleotides with 266-, 240-, 218-, and 200-nm pulsed laser excitation, *J. Am. Chem. Soc.,* 107, 1520, 1985.
38. **Wiegand, R., Weber, G., Zimmermann, K., Monajembashi, S., Wolfrum, J., and Greulich, K.-O.,** Laser induced fusion of mammalian cells and plant protoplasts, *J. Cell Sci.,* 88, 145, 1987.
39. **Doiron, D. R. and Profio, A. E.,** Laser fluorescence bronchoscopy for early lung cancer, in *Lasers in Photomedicine and Photobiology,* Springer-Verlag, Heidelberg, 1980.
40. **Kessel, D. and Dougherty, T. J., Eds.,** *Porphyrin Photosensitization,* Plenum Press, New York, 1983.
41. **Seibert, M. and Alfano, R. R.,** Probing photosynthesis on a picosecond time scale (evidence for photosystem I and photosystem II fluorescence in chloroplasts), *Biophys. J.,* 14, 269, 1974.
42. **Binnie, N. E., Haley, L. V., Mattioli, T. A., and Koningstein, J. A.,** Energy transfer between dimeric chlorophyll a and all *trans* β carotene *in vitro* as resolved by fluorescence photoquenching, *J. Lumin.,* 35, 25, 1986.
43. **Gulotty, R. J., Mets, L., Alberte, R. S., and Fleming, G. R.,** The origin of chloroplasts fluorescence decay kinetics: picosecond fluorescence of mutants and sub-chloroplast particles of *Chlamydomonas reinhardii* and barley, *Photochem. Photobiol.,* 41, 487, 1985.
44. **Voigt, J.,** Exciton states and exciton interaction processes in chlorophyll *in vivo, Photobiochem. Photobiophys.,* 12, 21, 1986.
45. **Richter, G.,** *Stoffwechselphysiologie der Pflanzen,* Thieme Verlag, Stuttgart, 1981.
46. **Campillo, A. J. and Shapiro, S. L.,** Picosecond fluorescence studies of exciton migration and annihilation in photosynthetic systems. A review, *Photochem. Photobiol.,* 28, 975, 1978.
47. **Klevanik, A. V., Kryukov, P. G., Matveets, Y. A., Semchishen, V. A., and Shuvalov, V. A.,** Measurement of electron and energy transfer rates in physical stages of photosynthesis with subpicosecond time resolution, *JETP Lett.,* 32, 97, 1980.
48. **Alfano, R. R., Ed.,** *Biological Events Probed by Ultrafast Laser Spectroscopy,* Academic Press, New York, 1982.
49. **Geacintov, N. E. and Breton, J.,** Application of pulsed lasers to the study of energy transfer and fluorescence phenomena in photosynthetic systems, in *Advances in Laser Spectroscopy,* Gareto, B. A. and Lombardi, J. R., Eds., Heyden, London, 1982, 213.
50. **Netzel, T. L., Bucks, R. R., Boxer, S. G., and Fujita, J.,** A report on picosecond studies of electron transfer in photosynthetic models, in *Picosecond Phenomena II,* Hochstrasser, R. M., Kaiser, W., and Shank, C. V., Eds., Springer-Verlag, Heidelberg, 1980, 322.

51. **Fujiwara, M. and Tasumi, M.**, Resonance Raman spectra of chlorophyll a excited with various laser lines near the Soret and Q bands, in Proc. 9th Int. Conf. on Raman Spectroscopy, Tokyo, 1984, 142.

52. **Doizi, D. and Mialocq, J.-C.**, Picosecond laser pulses: a tool for the study of model systems in artificial photosynthesis, in Opto 84, Fourth Optoelectronic Meeting, ESI Publications, Paris, 1984, 94.

53. **Busch, G. E., Applebury, M. L., Lamola, A. A., and Rentzepis, P. M.**, Formation and decay of prelumirhodopsin at room temperatures, *Proc. Natl. Acad. Sci. U.S.A.*, 69, 2802, 1972.

54. **Shichida, Y., Matuoka, S., and Yoshizawa, T.**, Formation of photorhodopsin, a precursor of bathorhodopsin, detected by picosecond laser photolysis at room temperature, *Photobiochem. Photophy.*, 7, 221, 1984.

55. **Rentzepis, P. M.**, Probing ultrafast biological processes by picosecond spectroscopy, *Biophys. J.*, 24, 272, 1978.

56. **Applebury, M. L.**, The primary processes of vision: a view from the experimental side, *Photochem. Photobiol.*, 32, 425, 1980.

57. **Aton, B., Doukas, A. G., Narva, D., Callender, R. H., Dinur, U., and Honig, B.**, Resonance Raman studies of the primary photochemical event in visual pigments, *Biophys. J.*, 29, 79, 1980.

58. **Yoshizawa, T., Shichida, Y., and Matuoka, S.**, Primary intermediates of rhodopsin studies by low temperature spectrophotometry and laser photolysis, *Vision Res.*, 24, 1455, 1984.

59. **Shichida, Y.**, Primary intermediates of photobleaching of rhodopsin, *Photochem. Photobiophys.*, 13, 287, 1986.

60. **Yoshizawa, T. and Shichida, Y.**, Picosecond laser analysis of rhodopsin, in *Biomolecules*, Nagata, C., Ed., Japan Scientific Societies Press, Elsevier, 1985, 111.

61. **Matuoka, S., Shichida, Y., and Yoshizawa, T.**, Formation of hypsorhodopsin at room temperature by picosecond green pulse, *Biochim. Biophys. Acta*, 765, 38, 1984.

62. **Matveetz, Y. A., Chekalin, S. V., and Sharkov, A. V.**, Molecular dynamics of primary photoprocesses in bacteriorhodopsin: subpicosecond study of absorption and luminescence kinetics, *J. Opt. Soc. Am.*, B, 2, 634, 1985.

63. **Fodor, S. P. A., Rava, P. R., Hays, T. R., and Spiro, T. G.**, Ultraviolet resonance Raman spectroscopy of the nucleotides with 266-, 240-, 218-, and 200-nm pulsed laser excitation, *J. Am. Chem. Soc.*, 107, 1520, 1985.

64. **Peticolas, W. L., Lagant, P., and Vergoten, G.**, Ultraviolet resonance Raman spectroscopy of nucleic acid chromophores, *Proc. SPIE in Soc. Opt. Eng.*, 620, 89, 1986.

65. **Kubasek, W. L., Hudson, B., and Peticolas, W. L.**, Ultraviolet resonance Raman excitation profiles of nucleic acid bases with excitation from 200 to 300 nm, *Proc. Natl. Acad. Sci., U.S.A.*, 82, 2369, 1985.

66. **Tinoco, I.**, Hypochromism in polynucleotides, *J. Am. Chem. Soc.*, 82, 4785, 1960.

67. **Fodor, S. P. A. and Spiro, T. G.**, Ultraviolet resonance Raman spectroscopy of DNA with 200—266 nm laser excitation, *J. Am. Chem. Soc.*, 108, 3198, 1986.

68. **Howard, F. B., Frazier, J., and Miles, H. T.**, Interbase vibrational coupling in G:C polynucleotide helices, *Proc. Natl. Acad. Sci. U.S.A.*, 64, 451, 1969.

69. **Anders, A.**, DNA fluorescence at room temperature excited by means of a dye laser, *Chem. Phys. Lett.*, 81, 270, 1981.

70. **Rigler, R., Claesens, F., and Lomakka, G.**, Picosecond single photon counting fluorescence spectroscopy of nucleic acids, in *Ultrafast Phenomena, Part IV*, Auston, D. H. and Eisenthal, K. B., Eds., Springer-Verlag, Heidelberg, 1984, 472.

71. **Angelov, D. A., Kryukov, P. G., Letokhov, V. S., Nikogosyan, D. N., and Oraevsky, A. A.**, Selective action on nucleic acids components by picosecond light pulses, *Appl. Phys.*, 21, 391, 1980.

72. **Nikogosyan, D. N., Oraevsky, A. A., and Letokhov, V. S.**, Two-step picosecond UV-excitation of polynucleotides and energy transfer, *Chem. Phys.*, 97, 31, 1985.

73. **Greulich, K. O., Wijnaendts van Resandt, R. W., and Kneale, G. G.**, Time resolved fluorescence of bacteriophage Pf1 DNA binding protein and its complex with DNA, *Eur. Biophys. J.*, 11, 195, 1985.

74. **Greulich, K. O. and Wijnaendts van Resandt, R.**, Estimation of tyrosine-40-DNA distance in the filamentous phage Pf1 by analysis of its intrinsic fluorescence properties, *Biochim. Biophys. Acta*, 782, 446, 1984.

75. **Shapiro, S. L., Campillo, A. J., Kolman, V. H., and Goad, W. B.**, Exciton transfer in DNA, *Opt. Commun.*, 15, 308, 1975.

76. **Anders, A.**, Energy transfer in nucleic acid-dye complexes, *Opt. Commun.*, 26, 339, 1978.

77. **Anders, A.**, Models of DNA-dye-complexes: energy transfer and molecular structures as evaluated by laser excitation, *Appl. Phys.*, 18, 333, 1979.

78. **Andreoni, A., Cubeddu, R., De Silvestri, and Svelto, O.**, Laser selective photobiology: dye-biomolecule complexes, in *Lasers in Photomedicine and Photobiology*, Springer-Verlag, Heidelberg, 1980.

79. **Anders, A.**, Selective laser excitation of bases in nucleic acids, *Appl. Phys.*, 20, 257, 1979.

80. **Claesens, F. and Rigler, R.**, Conformational dynamics of the anticodon loop in yeast tRNAPhe as sensed by the fluorescence of wybutine, *Eur. Biophys. J.*, 13(6), 331, 1986.

81. **Johnson, C. K. and Hochstrasser, R. M.,** Studies of fast processes in biological systems, *AIP Conf. Proc.,* 146, 686, 1986.
82. **Madge, D., Martin, J-L., Wilson, K. R., Dupuy, C., Zhu, J. G., Campbell, B. F., Mitchell, M., and Marsters, J.,** Femtosecond laser studies of distal side effects in myoglobin and model compounds, *SPIE,* 533, 2, 1985.
83. **Noe, L. J., Eisert, W. G., and Rentzepis, P. M.,** Picosecond photodissociation and subsequent recombination processes in carbon monoxide hemoglobin, *Proc. Natl. Acad. Sci. U.S.A.,* 75, 573, 1978.
84. **Friedman, J. M. and Lyons, K. B.,** Time resolved resonance Raman scattering as a probe of hemoglobin dynamics, in *Lasers in Photomedicine and Photobiology,* Springer-Verlag, Heidelberg, 1980.
85. **Parrish, J. A. and Deutsch, T. F.,** Laser photomedicine, *IEEE J. Quantum Electron.,* QE-20(12), 1386, 1984.
86. **Apfelberg, D. B., Ed.,** *Evaluation and Installation of Surgical Laser Systems,* Springer-Verlag, Heidelberg, 1987.
87. **Prince, M. R., Deutsch, T. F., Shapiro, A. H., Margolis, R. J., Oseroff, A. R., Fallon, J. T., Parrish, J. A., and Anderson, R. R.,** Selective ablation of athomeras using a flashlamp-excited dye laser at 465 nm, *Proc. Natl. Acad. Sci. U.S.A.,* 83, 7064, 1986.
88. **Teng, P., Nishioka, N. S., Anderson, R. R., and Deutsch, T. F.,** Optical studies of pulsed-laser fragmentation of biliary calculi, *Appl. Phys. B,* 42, 73, 1987.
89. **Yamashita, M., Nomura, M., Kobayashi, S., Sato, T., and Aizawa, K.,** Picosecond time-resolved fluorescence spectroscopy of hematoporphyrin derivative, *IEEE J. Quantum Electron.,* QE-20(12), 1363, 1984.
90. **Karu, T. I.,** Photobiological fundamentals of low-power laser therapy, *IEEE J. Quantum Electron.,* QE-23(10), 1703, 1987.
91. **v Bally, G., Ed.,** *Holography in Medicine and Biology,* Springer-Verlag, Heidelberg, 1979.
92. **Duncan, M. D., Reintjes, J., and Manuccia, T. J.,** Imaging biological compounds using the CARS microscope, *SPIE,* 482, 46, 1984.
93. **Anders, A., Aufmuth, P., Böttger, E.-M., and Tronnier, H.,** Spectroscopic properties of human skin and photodestruction of tumors, in *Lasers in Photomedicine and Photobiology,* Springer-Verlag, Heidelberg, 1980.
94. **Anders, A., Aufmuth, P., Böttger, E.-M., and Tronnier, H.,** Investigation of the erythema effectiveness curve with tunable lasers, *Dermatosen,* 32, 153, 1984.
95. **Tata, D. B., Foresti, M., Cordero, J., Tomashefsky, P., Alfano, M. A., and Alfano, R. R.,** Fluorescence polarization spectroscopy and time-resolved fluorescence kinetics of native cancerous and normal rat kidney tissues, *Biophys. J.,* 50, 463, 1986.
96. **Carey, P. R. and Storer, A. C.,** Characterization of transient enzyme-substrate bonds by resonance Raman spectroscopy, *Ann. Rev. Biophys. Bioeng.,* 13, 25, 1984.
97. **Peticolas, W. L., Bajdor, K., Patapoff, T. W., and Wilson, K. J.,** New methods of studying enzyme-substrate interactions using ultraviolet resonance Raman and microscopic Raman difference techniques, in *Laser Scattering Spectroscopy of Biological Objects,* Stepanek, J., Anzenbacher, P., and Sedlacek, B., Eds., Elsevier, Amsterdam, 1987, 249.
98. **Stepanek, J., Anzenbacher, P., and Sedlacek, B., Eds.,** *Laser Scattering Spectroscopy of Biological Objects,* Elsevier, Amsterdam, 1987.
99. **Ehrenberg, A., Rigler, R., Gräslund, A., and Nilsson, B., Eds.,** *Structure, Dynamics and Function of Biomolecules,* Springer-Verlag, Heidelberg, 1987.

Chapter 5

EMISSION AND LASER RAMAN SPECTROSCOPY OF NUCLEIC ACID COMPLEXES

Robert F. Steiner, Gary Holtom, and Yukio Kubota

TABLE OF CONTENTS

I. EXPERIMENTAL ASPECTS OF TIME-RESOLVED LASER SPECTROSCOPY

A. INTRODUCTION

Fluorescence measurements are one of the fundamental spectroscopic methods and have widespread application in biophysics. Extending fluorescence techniques to provide information about the evolution in time of the excited state adds to the information that may be extracted, which is particularly helpful for complex molecular systems. The fundamentals of time-correlated single-photon counting (TCSPC) are well covered in the literature[1-3] and the reader is expected to be familiar with the basic concepts. Special advantages (and problems) of lasers and the methods needed to obtain the best possible data will be the focus of this section.

Flashlamps with ultrafast discharge circuits are traditional and still useful light sources, and it is likely that more flash sources than lasers will be in use as long as lasers alone cost more than an entire lamp-based TCSPC apparatus. The only other contender for short-pulse sample excitation is synchrotron radiation, which can be a subnanosecond white light source but which also has serious drawbacks of accessibility and limited choices in repetition rate.[4] A comparison of lamps vs. lasers for excitation is given in Table 1.

Modulation methods have been improved and are a viable alternative to TCSPC for fast fluorescence measurements.[5-6] Lasers or conventional xenon lamps may be used, and it is possible that in the absence of a laser there may be an advantage for modulation methods. Both methods have their proponents and the only substantial drawback of the modulation method is the inability to directly observe the time evolution of the excited state.

It is worth noting that state-of-the-art research with modulation apparatus involves attention to many of the problems important to TCSPC, especially with respect to laser sources, detection optics, and sample polarizing optics.

B. INSTRUMENTAL DESIGN CONSIDERATIONS
1. Obtaining Reliable Data

The single most important requirement is to develop techniques for verifying that data is free of artifacts. Appropriate performance checks and means for obtaining repeatable data are essential. A general problem with the TCSPC method is that subtle changes (perhaps

TABLE 1
Experimental Limits of Pulsed Light Sources

	Lamp	Laser
Pulse width (half maximum)	0.7 ns	<<.02 ns
Repetition rate	50 KHz	<=4 MHz
Photons/pulse imaged on sample	10^7-10^{9a}	10^9 (UV)
		2×10^{10} (Vis)

^a Polychromatic — a filter decreases this. The higher outputs are
for wider pulses at a reduced repetition rate.

hiding within the width of the instrument function) can produce data that will not fit expected models, or worse, will appear to fit an incorrect model.[1,7] Timing shifts are particularly difficult, and measures to identify and eliminate them will be described.

Conventional photomultiplier tubes (PMTs) have two artifacts in both end- and side-window configurations: imaging problems, in which the photoelectron transit time to the dynodes of the tube are dependent on position of the image on the photocathode;[8] and color shifts caused by different kinetic energy of the photoelectrons as the energy of the photon changes.[9] Both of these affect the transit time of the electron as it moves to the anode. In favorable conditions the transit-time spread (TTS) of the photoelectrons is the dominant contribution to the instrument function and the TTS may be much narrower than the anode pulse width.

The proximity-focused microchannel PMT is essentially free of the above problems and is worth using despite its considerable expense in order to produce confidence in the data acquired. Due to the compact geometry, the TTS of these tubes is very small and instrument functions below 60 ps full-width at half-maximum (FWHM) are reported regularly for tubes with 12-μm channels[10,11] and 20 ps FWHM has been reported for 6-μm tubes.[12] A major part of this section deals with reducing all other sources of jitter or TTS to insignificance compared to the PMT contribution.

Only very recently has the time resolution of the TCSPC method approached that of streak cameras. While streak cameras have ultimate time resolution of 1 ps, this is available only at considerable expense and with trade-offs in sensitivity, sweep linearity, dynamic range, and noise level. The number of research applications accessible to convenient study has increased greatly with improved time resolution resulting from the advanced microchannel plate PMT.

The TCSPC method is limited by the photoelectron TTS, while modulation methods are limited by anode frequency response. The loss of modulation depth of the anode signal occurs at a frequency less than the transform of the TTS or instrument function width.[13]

2. Dye Lasers, Dumping, and Tunability

A cavity-dumped mode-locked dye laser is the most versatile light source for TCSPC.[3] Figure 1 shows the usual arrangement. A continuous wave (CW) laser is mode locked, usually with an acoustooptic modulator in the cavity. This produces a train of pulses about 100 ps wide with a spacing of about 12 ns, depending on the cavity length of the laser used. Both argon ion lasers and Nd:YAG lasers are widely used for this purpose. The YAG laser must be frequency doubled to produce green light at 532 nm while the argon laser is usually operated at 514 nm, the most powerful line and the one most easily mode locked. The new material Nd:YLF is a replacement for ND:YAG and is useful for applications requiring shorter pulses.

While it is possible to use the green beam or its second harmonic in the UV directly, a serious problem is the very high repetition rate. This affects the electronics used in timing

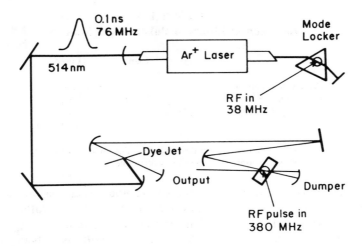

FIGURE 1.   Laser setup. The argon laser is mode-locked and pumps a cavity-dumped dye laser which has the same cavity length.

TABLE 2
Approximate Tuning Ranges for Dyes

Dye[a]	Visible (nm)	Frequency doubled (nm)
R560	540—600	270—300
R575	556—620	278—310
R590	568—640	284—320
DCM	606—710	303—355
LDS-698	660—770	330—385
LDS-751	720—840	360—840
LDS-820	800—940	400—470

[a]   Exciton Chemical Co. number.

and also limits the usefulness with samples having fluorescence lifetimes greater than about 2 ns. The dye laser performs these very important functions:

1.   Decreases the repetition rate as desired
2.   Increases the energy per pulse to maintain good average power even at longer pulse spacing
3.   Reduces the pulse width substantially
4.   Provides tunability

Increasing the pulse energy means more peak power which improves the conversion efficiency of the dye laser into the UV.

Typical tuning ranges for some common dyes are given in Table 2. Note that the dye R560 cannot be excited with a Nd:YAG laser but provides the bluest available light with an argon ion laser. As the dye wavelength becomes redder than 700 nm, the fluorescence becomes increasingly faint. As a consequence, rather few dye laser systems are operated in the blue region (375 to 475 nm) due to the inconvenience of operating in the fundamental at 750 to 950 nm. Operation of a dye laser from 475 to 540 nm in the fundamental requires a UV or blue mode-locked laser source of perhaps 0.5 W which is recently avaiable by frequency tripling Nd lasers. This makes the blue region accessible and also should result in significant far-UV light. This creates the possibility of exciting phenylalanine or DNA bases not accessible with green-pumped dye lasers.

3. Anisotropy and Sample Compartment Considerations

For experiments in which fluorescence lifetime (decay of the excited state) is the only important observable, it is sufficient to use polarized light to excite the sample and to use a magic angle polarizer for emission detection. This removes the effects of molecular reorientation, as is well documented. However, in many cases information about molecular motion is the most important part of the time-dependent measurement,[14-16] and appropriate control of excitation and emission polarization must be built into the apparatus.

The fluorescence anisotropy A(t) is defined as

$$A(t) = \frac{I_\parallel(t) - gI_\perp(t)}{I_\parallel(t) + 2gI_\perp(t)} \tag{1}$$

where $I_\parallel(t)$ is the vertical excitation, vertical emission decay curve and $I_\perp(t)$ is the vertical excitation, horizontal emission decay curve. The factor g is an experimentally determined correction factor for sensitivity of the detection system since monochromator throughput is sensitive to polarization. This factor might also include fluctuations in excitation power if the data $I_\parallel(t)$ and $I_\perp(t)$ are not collected simultaneously. Certainly provision must be made for determining the g factor, which is done conveniently by rotating the excitation polarization to horizontal and then performing the vertical and horizontal emission experiment (for the case of 90° geometry in detecting emission). Since the emission dipole of the molecule must rotate by 90° in either case, the observed fluorescence is identical and any changes are due to a nonunity g factor in the detection system. Means for changing polarization must be built into the sample optics and it may be useful to employ computer control so that signal averaging may be used to minimize the effects of long-term power fluctuations.

Temperature regulation may be important in some experiments and it may be necessary to provide a thermal jacket. It is also convenient to incorporate a rotating holder to quickly change between samples and to obtain instrument functions from a scattering solution without having to swap cells.

Samples which are photochemically unstable require special attention. It may be sufficient to use the most sensitive detection possible (wide bandpass filter) and attenuate the excitation as much as possible. However, in extreme cases a flow system must be provided to provide a continuously fresh sample.

4. General Utility Measures

A well-designed TCSPC apparatus can perform experiments quickly and is likely to be used for a variety of measurements. Changing excitation and emission wavelength conveniently deserves attention. In particular, having more than one dye circulator is useful since some time is required to clean up and change dyes. At this laboratory six dye circulators, three mirror sets for the dye laser, and three doubling crystals are used routinely and are changed often, even daily or in the middle of an experiment.

A small monochromator is also convenient in that a large selection of interference or bandpass filters are not required. Often the monochromator is used with wide slits to perform as a tunable filter. Scattered light is a problem with some samples, and a single monochromator may not be sufficient to avoid artifacts due to scattered excitation light. Additional red-pass filters or a double monochromator may be helpful. Ordinary interference filters are usually worse in rejection than a single monochromator.

Monitoring the pulse width of the dye laser is important as the instrument function becomes less than 100 ps. Small cavity length misadjustments can produce afterpulses 10 to 50 ps behind the main laser pulse whose intensity changes with time. Not only can these degrade the instrument function, but they can produce a bad fit to even very simple data due to a tiny change in the profile of the instrument function. An on-line autocorrelator is most useful for checking performance and is essential for producing pulses of 5 ps which are needed for the very fastest detectors. It is worth noting that dyes differ in their pulse width. For example, DCM at 620 nm tends to have a pulse with a long "tail" where R590 produces a very clean pulse at the same wavelength.

A topic of considerable interest now is deciding which laser is better: argon ion or Nd:YAG. Initial costs are very similar. The Nd:YAG laser is less expensive to maintain since it does not have an expensive plasma tube and it requires much less electricity and water. However, the makers of argon lasers have invested considerable effort in improving reliability and extended warranties are available, which puts their costs on a more equal basis. The argon laser does not require frequency doubling and no hazardous high-power IR beams are produced. Pointing stability is good. It is worth noting that the expensive KTP crystals that make efficient green light generation possible with YAG lasers are vulnerable to misalignment. The pulse width of the YAG is somewhat shorter which makes the dye laser slightly more efficient, but the ability to reach down to 540 nm with R560 is lost, which may eliminate this advantage. Finally, for those who live dangerously and do not mind risking damage to the KTP crystal, more power is available from a Nd:YAG laser than an argon ion laser. However, controlling the pulse width of the dye laser, particularly while it is being cavity dumped, becomes a real problem and in most cases high power is not a major requirement since the TCSPC method is inherently sensitive.

The krypton ion laser is similar to the argon laser and has the advantage of a strong red line. It may be mode locked easily on this or on a weak green line, but is generally more difficult to maintain than the argon ion laser. The deep red styril dyes which can be pumped with a green laser remove the only real advantage of the krypton ion laser.

C. ACHIEVING EXPERIMENTAL LIMITS, ESPECIALLY FAST RESPONSE TIMES

1. Adding Multiple Sources of Jitter

Usually the PMT is the limiting factor in instrumental function width and the object is to make other contributions less than the PMT TTS. The observed FWHM is usually considered to be a sum of all contributions to jitter as follows:

$$\text{total jitter} = [\Sigma \ (\text{each contribution})^2]^{1/2} \qquad (2)$$

Experimentally this means that the most attention must be paid to jitter sources which are a substantial fraction of the TTS of the PMT, since even a number of sources of small jitters do not have much effect on the observed instrumental FWHM.

2. Dye Laser Requirements

An ordinary dye laser with a three-plate tuning device and a single-gain jet has a width of about 15 ps when operated at high power, as measured by the FWHM of the autocorrelation trace. This is no problem for a 12-μm PMT but will severely limit performance of a 6-μm device. It may be necessary to add a saturable absorber or to use a less dispersive tuning filter to reduce pulse width of the dye laser, and both of these make the dye laser less convenient to operate and more vulnerable to misadjustments.

Performance of the cavity dumper is also important, particularly with visible excitation. Rejection of adjacent pulses in the dye laser cavity about 12 ns away from the dumped pulse is important for long lifetime samples. In principle, a convolution with the instrument function will eliminate the effects due to adjacent pulses but their amplitude tends to be unstable,

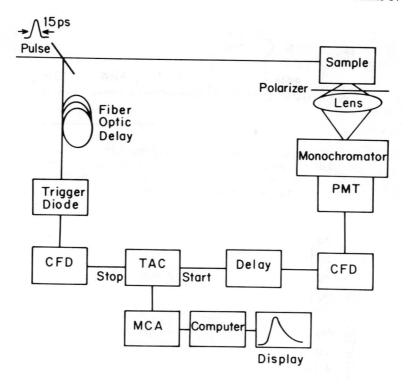

FIGURE 2.   Timing electronics.

making it important to have good rejection. Typically the pulses are down by a factor of 10^2 to 10^3 in the visible and are insignificant in the UV.

Usually a repetition rate of about 4 MHz is appropriate for the dye laser. A 1% stop rate corresponds to 40 Khz, which is near the limit for data acquisition with ordinary electronics. Typically, the pulse spacing is chosen to be at least 8 to 10 fluorescence lifetimes. If the sample lifetime exceeds 25 ns it is necessary to decrease the dye laser dump rate. It is possible or even likely that the pulse width of the dye laser will depend on the dump rate, making it necessary to collect a new instrument function for each dump rate.

Working with the fundamental beam has another pitfall: substantial broadband fluorescence from the dye laser gain medium may overlap fluorescence from the sample and produce a spurious time-dependent component. This may be eliminated with a cleanup filter such as a straight-through dispersion prism in the dye laser beam followed by a slit.

3. Time Zero Trigger

A timing pulse marking the arrival time of the light pulse in the sample is necessary for the TCSPC method. While several schemes have been used, picking off part of the visible dumped beam toward a photodiode is the simplest. It is possible to match the rise time of the PMT fairly closely which means that the two timing channels in the constant fraction discriminator (CFD) may have the same delay (Figure 2).

Given the possibility for a dye laser to have long- and short-term intensity fluctuations or drift, it is useful for the timing circuit to be insensitive to amplitude, which is of course the function of the CFD. Some photodiodes have noticeably better performance than others and it is important to check their "walk", the shift in time as the intensity varies. The RCA avalanche photodiode works well and has the advantage of some optical gain, making it possible to generate trigger pulses of volts with about 1 mW of average power from the dye laser operating at 4 MHz. A simple circuit is given in Figure 3. Providing a separate monitor

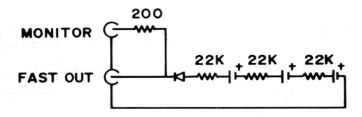

FIGURE 3.   Diagram of photodiode assembly. The RCS avalanche diode C30921E is mounted as close as possible to the SMA fast out connector. Bias is provided by three $67\frac{1}{2}$ V batteries. This is about 40 V below breakdown voltage and provides a gain of about 10.

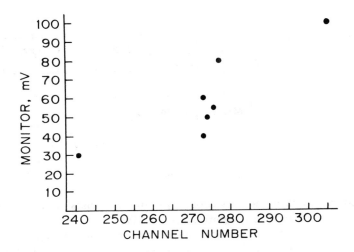

FIGURE 4.   Center channel position as a function of pulse amplitude at 0.67 ps/channel. The monitor oscilloscope has a bandwidth of 300 MHz and sees $\frac{1}{5}$ of the fast signal. Assuming a fast signal rise time of 300 ps, that signal will have 20 times the peak amplitude of the monitor channel.

channel for pulse height measurement is useful since even adding a T connector to the cable connecting the photodiode and CFD causes a noticeable degradation in walk. No special construction measures are necessary except to make all connections to the diode and to the SMA connector as short as possible. The junction capacitance of the photodiode is sufficient to produce about 5 V into a 50-Ω cable. Adding extra capacitance produces the risk of blowing out the input stage of the CFD.

The performance is simple to check. With the dye laser well adjusted (about 3% intensity fluctuation), two photodiodes are used to provide the start and stop pulses for the time-to-amplitude converter (TAC). It should be possible to produce an instrument function of 10 ps FWHM that falls sharply to zero on both sides of the peak. One diode provides a constant signal while the other one receives an adjustable light level from a rotating polarizer or a wedge filter. Adding neutral density filters provides a timing shift and should be avoided. With a calibration of 0.67 ps per channel, a measurement of walk is given in Figure 4. With almost a factor of 2 change in intensity the walk is minimal, which is useful in maintaining good timing stability for extended periods.

Previously, HP 4220 photodiodes were used in two identical circuits.[3] One had significantly less walk than the other but neither was as good as the RCA avalanche diode.

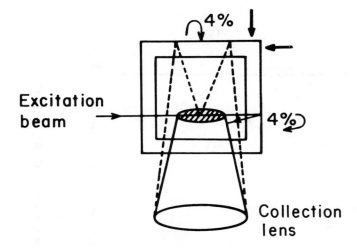

FIGURE 5.   Reflections from walls of a 1-cm cuvette. The two sides with arrows could be made of black glass. Assuming a refractive index of 1.4, the round trip time is about 8 ps/mm path distance for wall reflections. Inner surfaces are essentially index matched to the solvent and make a much smaller contribution to the problem.

4. Sample Optics

Most experimental scientists are more concerned with sample preparation than they are with the immediate environment of the sample cell. Laser light transmitted through an optically thin sample can scatter from the cell holder and produce a substantial signal at a short delay. This is a particular problem for collecting instrument functions and can be minimized by using a Raman-shifted frequency. A solvent such as pentane has a strong Raman band at 3100 cm^{-1} shift, but must be of exceptional purity to avoid trace fluorescence. This is a convenient way to collect the instrument function since scattered light is not detected.

Another problem which is not usually addressed is the reflection from cell walls. Typically 1-cm fluorescence cuvettes are used, and reflections have a substantial time delay which produce artifacts. The reflected excitation pulse and the fluorescence bouncing from the glass-air interfaces at cell walls each produce an amplitude 4% of the main peak with a time delay of 40 to 80 ps, depending on geometry. A reasonable possible solution for liquid cells is to use special cuvettes with two adjacent black glass walls as shown in Figure 5.

Imaging is also a problem. If a sample is excited perpendicular to observing its fluorescence, a significant time spread elapses as the beam crosses the sample cell. If an entire 1-cm cuvette is imaged onto a detector, the spread is about 40 ps. The situation is worse with front face excitation. Limiting the image size is an effective control for this time spread. Focusing the laser gently serves to eliminate spreading across the depth of the cuvette.

5. Monochromator Dispersion

The obvious problem of scattered light in the filter or monochromator has been described above, but a much more subtle problem has been reported recently.[17] A single monochromator has an optical path length which depends on which edge of the grating is illuminated. This problem is easily identified by comparing an instrument function obtained with illumination of the whole grating to that obtained with a mask blocking all but the center of the grating (Figure 6). To make the matter worse, this broadening increases with wavelength, making the effective instrument function wavelength dependent. For example, a calculation of TTS in a 0.2-*M* f/4 monochromator with a 1800-groove per millimeter grating shows a half-width of 55 ps at 300 nm to over 100 ps at 600 nm, assuming uniform grating illumination.

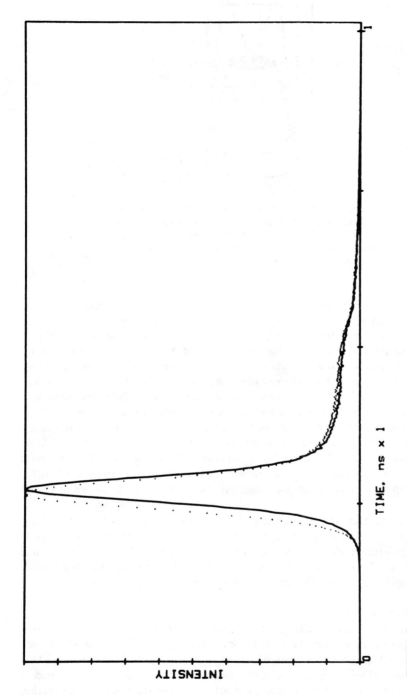

FIGURE 6. Instrument function with (solid line) and without (dots) a grating mask using a single monochromator, recorded at 325 nm. The FWHM increases from 46 to 56 ps due to TTS from the grating.

A general solution to the scattering and the TTS problems is to use a subtractive dispersion monochromator. At the present time, it is necessary to make compromises between cost, size, throughput, and bandwidth of commercially available monochromators, which are less than ideal for this application.

6. Electronics

The rise times of the diode trigger and of the PMT anode pulse are in the vicinity of 200 ps. Preserving the rise time and pulse amplitude is a problem, particularly with small diameter coaxial cable and BNC connectors. Semirigid coax and SMA screw connectors offer better performance at high frequencies, and the length should be kept as short as possible, preferably 1 m or less.[18]

A key component is the CFD, which produces a logic pulse that is very reproducible from analog pulses of varying amplitude. The stock Tennelec model 455 is adequate for a TTS of 100 ps or so, and modifications can be made to improve its performance for the ultrafast PMTs. Protective diodes can be removed, and off-the-board components like the fraction modules can be replaced with very small resistors with short leads and no sockets. However, the most important need is to use a delay cable shorter than possible with an external loop. A 4-cm length of RG-174 coax soldered directly to the board was selected after several trials with other delays.

Delay modules with switch-selectable lengths of cable are convenient for locating the proper relative arrival times of the pulses to the TAC. High quality coax with good connectors should be used; RG-58 seems to be sufficient for up to 100 ns but probably RG-174 will cause excessive loss of rise time.

Two-plate microchannel plate dynode assemblies with 12-μm channels are used in the Hamamatsu R-1564U-07 PMT. Operation at the rated maximum voltage produces pulses barely over the threshold of the CFD, indicating the need for a pulse amplifier. The Minicircuits ZHL-2000 was found satisfactory, with a 1-db feed-through attenuator on the input side to provide a DC leakage path. Many RF amplifiers do not specify phase (inverting or not), and the wider bandwidth ZFL-4200 does invert. This is a nuisance since a broadband pulse inverter must be used with it.

Selection of the amplifier requires some care, since an important requirement for low walk is providing substantial signals (volts) to the CFD, without clipping or saturation which would affect the pulse shape and thus the timing. An output of 17 dba is from the ZHL-2000 sufficient (about 2 V peak into 50 Ω), but doubling that would be useful. This is particularly important since PMTs produce a wide range of pulse amplitudes, with the lower amplitude pulses producing the greatest contribution to the walk. An amplifier with a greater output should be an improvement. A bandwidth of 2 GHz is sufficient for a FWHM of 50 ps, but for the 6-μm tube with a shorter anode pulse width a 3- to 4-GHz amplifier would be appropriate.

At this point it is appropriate to describe an important performance check. Totally uncorrelated pulses are provided to the start and stop of the TAC to check for systematic errors. A convenient method is to use the laser or a pulse generator produce stop pulses at several megahertz and start pulses from a random source such as room lights hitting a piece of white paper in the sample chamber. A totally uniform distribution of counts per channel should be seen on the multichannel analyzer (MCA), with the deviation or noise level decreasing as the square root of the mean number of counts. At 10^4 counts per channel the deviations should be about $\pm 1\%$ and totally random. Performing an autocorrelation is useful to expose systematic effects. This checks for linearity of the time scan in the TAC, conversion accuracy or differential linearity of the MCA, and electronic crosstalk in the CFD or other components.

In this manner a very serious artifact in the CFD was discovered. When one of the four channels in the Tennelec 455 is triggered, other channels are either moved slightly forward or back in time to produce a very distinct ripple of several percent in the intensity (Figure 7). This can be moved by adjusting the relative appearance of photoelectron pulses and timing diode signals. Following the suggestion of Anfinrud,[19] 20 m of large core optical fiber was used to delay the arrival of the trigger pulse until long after the sample fluorescence. Note that a long length of coaxial cable before the CFD would seriously degrade the pulse rise time, and any delay after the CFD would not eliminate the crosstalk.

D. COMPUTERS AND DATA HANDLING
1. Capabilities Needed

Fitting data requires storing and viewing a substantial amount of data. A microcomputer with floppy discs is sufficient for storing data. Bit-mapped graphics capability is suitable for quickly viewing thousands of points which might be required to decide whether a data set is good or not. Huge memories are not needed and the speed of processing data is essentially limited by the floating-point multiplication and addition times. A small computer with a fast math coprocessor and 100% availability to a single user competes very favorably with a much larger time-shared system.

Traditionally TCSPC apparatus have used a stand-alone MCA for collection of the timing pulses. An off-line computer captures a data set and is used for curve fitting. Within the past few years MCAs which can be built into microcomputers have become available. This allows examination of the data and curve fitting to take place conveniently in near real time.

A number of criteria for estimating the accuracy of fitting experimental data to a given model has been described in the literature. The residuals between the raw data and the convolution of the instrument function with the model function should be examined as a function of time. Any nonrandom deviations coinciding with the instrument function suggest that there is an artifact, although of course a very short component could be missing from the model. Other tests, such as the autocorrelation function of the residuals[7,20] or the runs test,[9] check for subtle systematic deviations which are obscured by the point-to-point noise in the deviations and which may not be obvious from the chi-squared test. It should be easy to make these statistical checks to determine the success of an experimental run before hours or days are invested in collecting many data sets.

A special problem with fast detectors is that with an ordinary calibration factor, perhaps 512 points covering a total of 5 ns, is that the instrument function is covered by only a few channels. The exponential fast convolution routines implicitly assume a continuous function[21] and the method fails as the coverage becomes more and more discrete. Several methods are appropriate for dealing with this:

1. Using a nonlinear mapping in which the calibration factor or point spacing is very small immediately under the instrument function and becomes larger at longer time delays.
2. Global methods in which data is collected with more than one sweep speed in order to represent long and short components adequately.[22]
3. Special interpolation methods to help overcome the effects of discontinuous or sparse sampling. The method described by Wahl[23] has very little cost in computational effort and does work better for short lifetime components than the usual "trapezoidal integration".[21] In addition, the method of Wahl can be extended to quadratic interpolation.

The third method is a limited solution but is the most convenient and may be employed along with the other two methods. For covering a wide range of lifetimes global fitting may be necessary.

Nonexponential curves for lifetimes and convolution has been suggested for modeling special cases. These include energy transfer in large molecules with multiple linked chromo-

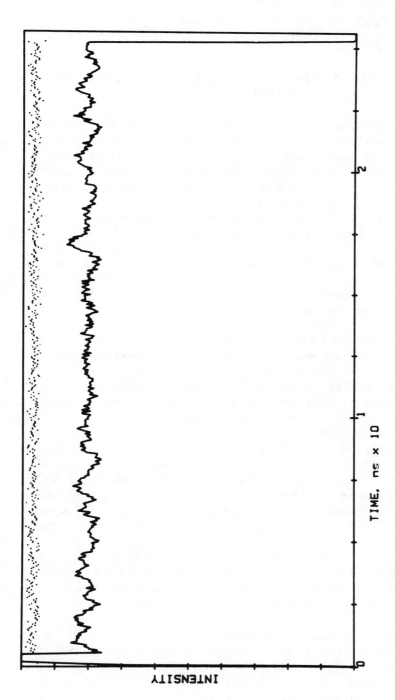

FIGURE 7.   Crosstalk appears as a nonrandom deviation in the solid curve and is eliminated by moving the trigger pulse to a later time as shown in the dotted curve. Conditions are identical except that the delay the stop channel is done electrically following nearly coincident CFD pulses in the solid curve and is performed optically before the CFD in the dotted curve. A reversed timing mode is used, in which the PMT provides the start pulse.

phores,[24] Förster energy transfer,[25] and rotational correlations with hindered motions.[15,26,27] The trivial convolution methods for generating curves and their derivatives are no longer applicable and computational effort increases by orders of magnitude. Proper handling of complex models is an important next step in solving interesting problems and work is under way to meet this challenge. Choice of a fast computer with a good environment for program development must be done carefully.

II. EMISSION SPECTROSCOPY OF NUCLEIC ACIDS AND NUCLEIC ACID-DYE COMPLEXES

During the past 10 years, laser emission techniques have become indispensable in biochemistry and biophysics because of their great advantages over techniques using conventional light sources. High intensity and monochromacity of laser sources are ideal for spectroscopic investigations, especially for fluorescence measurements with low quantum yields such as nucleic acids. Short pulses allow one to measure fluorescence lifetimes of biomolecules down to the picosecond time domain.

Emission techniques whose major parameters are spectrum, quantum yield, lifetime, and polarization (anisotropy) can provide important information about the structure and dynamics of biomolecules. Emission can be observed either of biomolecules themselves or suitable dye molecules bound to special biopolymers. This section describes applications of emission spectroscopy to nucleic acids and nucleic acid-dye complexes. While the stress would be upon results obtained with laser sources, background information obtained with conventional light sources would also be included.

A. NUCLEIC ACIDS AND NUCLEIC ACID BASES

In order to understand the mechanisms of photophysical and photochemical processes occurring in nucleic acids, it is necessary to have detailed knowledge of the excited-state properties of nucleic acids and their components. Fluorescence spectroscopy is important in probing the nature and dynamics of the excited singlet states. Fluorescence measurements, however, have been difficult to carry out at room temperature because of the very low fluorescence quantum yield ($\phi_f \simeq 10^{-4}$). Most work has been carried out at 77 K in glasses where the large fluorescence enhancement is observed ($\phi_f \simeq 10^{-1} \sim 10^{-2}$)[28] or at extreme pH values where some nucleic acid bases are known to exhibit fluorescence, which enables a normal recording of emission spectra. In recent years, the steady-state fluorescence behavior at room temperature has been reported by using the photon-counting method which allows an improvement of the signal-to-noise ratio by increasing the counting time. The weak fluorescence from nucleic acid bases is now well characterized and their fluorescence quantum yields are found to be of the order of 10^{-4} to 10^{-5}.[29-34]

For more quantitative elucidation of the excited-state properties of nucleic acids and their components, direct measurements of the excited-state lifetimes at room temperature are very desirable. Judging from the low quantum yields and the radiative lifetimes ($\sim$5 ns) for DNA bases, their fluorescence lifetimes are expected to be of the order of $\sim$1 ps.[33,34]

1. Nucleic Acid Bases and Synthetic Polynucleotides

The first direct measurements of room-temperature fluorescence decays of DNA components, four nucleosides, have been made by Ballini et al.[35] using synchrotron nanosecond exciting light pulses (FWHM of 1.76 ns). The fluorescence decays of thymidine, cytosine, and guanosine were too fast for their single-photon counting fluorometer to resolve, suggesting that the fluorescence decay times must be below 100 ps. On the other hand, the fluorescence decay of adenosine showed significant deviation from the flash lamp profile, suggesting the existence of a longer lifetime component.

The same measurements have been applied to polynucleotides [poly(A) and poly(dA-dT)·poly(dA-dT)· and DNA.[35,36] The fluorescence decay of poly(A) is found to be dependent on the emission wavelength, confirming the multicomponent nature of the emission. The emission from poly(dA-dT)·poly(dA-dT) and DNA is shown to consist of at least two components, the major one decaying very fast and the minor one decaying more slowly than the exciting pulse. Quantitative analysis indicates that a postpulse exponentially decaying emission has a lifetime of 2.9 ns for calf thymus DNA and 3.0 ns for poly(dA-dT)·poly(dA-dT); it is suggested that this may be due to an excimer fluorescence. However, the questions still remain unsolved due to the instrumental restriction of the time resolution.

New developments in laser techniques and in signal detection have now made it possible for the direct observation of the generation and decay of the excited state of nucleic acids and their components on the picosecond time scale.

Picosecond single photon counting fluorescence spectroscopy has been applied to synthetic polynucleotides to obtain fluorescence lifetimes on the picosecond time scale.[37,38] A large-frame Kr-ion laser with an acoustooptic mode locker was used for generation of picosecond laser pulses and the multichannel plate detector with short transient time jitter was used for single photon detection. The decay analysis shows two main lifetime components with values ($\tau_1 = 872$ ps and $\tau_2 = 94$ ps for poly(dA-dT)·poly(dA-dT) and $\tau_1 = 2926$ ps and $\tau_2 = 255$ ps for poly(dG-dC)·poly(dG-dC)), which are similar to those obtained from synchrotron experiments.[36] From quantum yields and radiative lifetimes, fluorescence lifetimes of few picoseconds are predicted.[33,34] In practice, the criterion of the reduced chi-square (χ_R^2)[39] indicates additional lifetime components. The lowest picosecond range, however, has not been penetrated in sufficient detail because of the restriction of the detector system (FWHM ~ 43 ps).[37,38]

Using a highly sensitive synchroscan streak camera which incorporates a microchannel plate and synchronizes with UV picosecond pulses generated inside the cavity of a mode-locked CW ring dye laser, Yamashita and co-workers[40] have reported the first direct measurement of picosecond time- and wavelength-resolved emissions from adenine and poly(A) in aqueous solution at room temperature. The time-resolved emission spectra for adenine show that the 340-nm emission disappears within 42 ps after the excitation, while a red-shifted emission peaking at 390 nm is predominant over several hundred picoseconds. The deconvolution of the decay curve indicates that the lifetime of the 340-nm emission is 6 ps, whereas that of the 390-nm emission is 790 ps; the former is attributable to the monomer fluorescence and the latter to some aggregation resulting in excimer fluorescence.[35] The measured value of 6 ps is comparable to the calculated lifetime of ~ 1 ps.[33,34]

On the other hand, the time-resolved emission spectra for poly(A), unlike adenine, consists of not only the 340-nm emission but also the 390-nm emission in the shortest time range. The wavelength-resolved fluorescence decay curves of poly(A) are displayed in Figure 8. The analysis shows that the shortest component of the decay curve has a lifetime of 8 ps which is dominant at the 340-nm emission, while the longer component consists of two decay times of 510 and 1300 ps. The quenching results of poly(A) emission by metal ions indicates that the longer wavelength emission of poly(A) (390 nm) is quenched by not only Co^{2+} and Ni^{2+} but also Mg^{2+}, suggesting that this emission cannot be assigned to phosphorescence.[41] Therefore, two decay times observed at the longer wavelength emission may be due to excimer fluorescence in different conformations.[35,36]

Yamashita et al.[40] found that 9-methyl adenine exhibits a red-shifted emission at 380 nm in addition to the monomer fluorescence as well as adenine and poly(A). Deconvolution analysis shows that there are two distinct decay components: a fast component (lifetime of 5 ps) which is dominant in the short wavelength region of 300 to 360 nm and a slow component (lifetime of 330 ps) which is dominant in the long wavelength region of 360 to 450 nm. The fast component is ascribable to the monomer fluorescence as in the case of

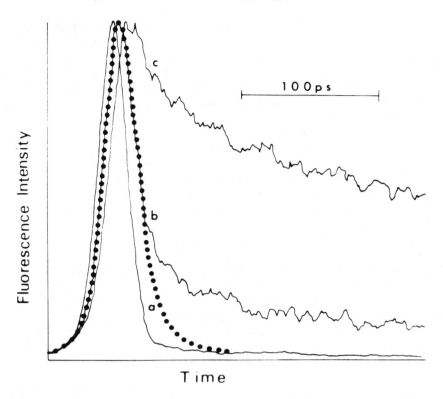

FIGURE 8.   Wavelength-dependent fluorescence decay curves measured for poly(A) in aqueous solution at room temperature; the decay was observed at (a) 320 nm (Raman scattering profile), (b) 340 nm, and (c) 380 nm. The excitation wavelength was 291 nm. Solid circles represent the theoretical decay curve with the lifetime of 8 ps. (From Kobayashi, S., Yamashita, M., Sato, T., and Muramatsu, S., *IEEE J. Quantum Electron.*, QE-20, 1383, 1984. With permission.)

adenine.[40] Since the addition of paramagnetic metal ions does not affect the emission from 9-methyl adenine, the slow decaying emission band is not attributed to phosphorescence but to excimer emission. The decay profile of 9-methyl adenine does not exhibit the behavior characteristic of a diffusion-controlled excimer formation, i.e., a negative-amplitude exponential component and a considerable time delay relative to the exciting pulse.[43] Judging from the 5-ps excited-state lifetime of the monomer, it appears unlikely to form the intermolecular excimer by diffusion-controlled processes. Further, it is well known that nucleic acid bases in water easily form weakly coupled stacking dimers.[44] On the basis of the above arguments, Yamashita et al.[42] concluded that the excimer is formed through preformed sandwich-like pairs in the ground state. This situation seems similar to the formation of intramolecular excimer observed with dinucleoside monophosphates such as CpC and ApA.[45-47]

The emission behavior of dinucleotides and polynucleotides is complex and is marked by the occurrence of red-shifted excimer emissions in addition to the monomer emission which may originate in distinct stacked conformations.[45-47] Different lifetimes observed for synthetic polynucleotides confirm the multicomponent nature of the emission. It should be noted that the decay behavior depends upon solvent, concentration, temperature, pH, and so on. In order to assign different lifetimes and to elucidate the dynamics of the excimer formation, further investigation would be necessary under various experimental conditions, especially using dinucleotides.

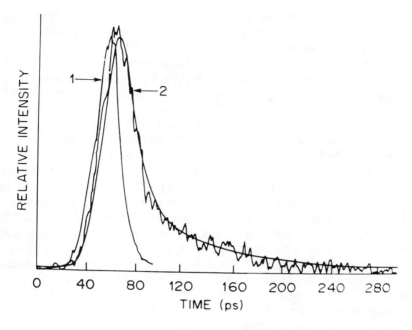

FIGURE 9.   Fluorescence decay profile of calf thymus DNA at room temperature (2) with excitation pulse (1) and double exponential fit to the data. The sample was excited at 266 nm with a pulse width of 25 ps. The emission was observed through a WG305 Schott filter which transmits practically the whole emission band. The buffer employed was 0.5 *M* sodium cacodylate, 0.01 *M* NaCl, pH 6.6. (From Georghiou, S., Nordlund, T. M., and Saim, A. M., *Photochem. Photobiol.*, 41, 209, 1985. With permission.

2. DNA and Methylated DNA

The fluorescence of DNA is not well understood. The question of whether the DNA emission originates from an excited-state molecular complex (exciplex or excimer), formed from the interaction between an excited base and a neighboring one in the ground state, has often been asked.[28,29,31-33,48,49] The first emission spectrum at room temperature showed considerable similarity to that found at 77 K where the fluorescence apparently was entirely of an exciplex nature (fluorescence maximum at 350 nm, $\phi_f \simeq 0.01$).[28] This spectrum was confirmed in both peak position and half-width by using narrow band laser excitation.[48] Aoki and Callis,[49] however, reported somewhat different results using conventional excitation. The DNA spectrum closely resembled that from a mixture of mononucleotides between 300 and 360 nm but showed a broad, low level shoulder around 450 nm, which might be of an exciplex nature. On the other hand, Morgan and Daniels[45,46] have suggested that similar red-shifted emissions from synthetic polynucleotides might be phosphorescence. Direct lifetime measurements would help to choose between these extreme possibilities.

Georghiou et al.[50] first observed the fluorescence decay of DNA at room temperature on the picosecond time scale; the samples were excited with single, 25 ps, 266-nm laser pulses from a frequency-quadrupled Nd:YAG laser and the fluorescence was detected with a streak camera-optical MCA system that has a time jitter of about 2 ps. Two decay components were detected for calf thymus DNA (GC 42%) (Figure 9); a major one has a 10-ps decay time, while a minor one has a corresponding decay time of 65 ps and makes a contribution of 0 to 10%, depending on the transmission characteristics of the emission filter employed. The decay profile in the 310- to 390-nm spectral region could be best described by a single exponential with a decay time of 10 ps. Therefore, a long decay time may correspond to the shoulder around 450 nm observed by the steady-state emission measurements.[48,49] For *Escherichia coli* DNA (GC 50%), two decay components of 6- and 40-ps decay times were found. These results imply that the base composition of DNA may play a role in the decay behavior.

The decay profile of DNA did not exhibit the behavior characteristic of a diffusion-controlled exciplex (or excimer) formation (Figure 9).[43] Therefore, the long decay time of 65 ps may originate from the preformed stacked dimers in the ground state which are favorable for exciplex (or excimer) formation. From the fluorescence decay time $\tau = 10 \pm 3$ ps of the major component and the average value of the fluorescence quantum yield ($\phi_f \simeq 4 \times 10^{-5}$),[48,49] the following ranges for the radiative (k_f) and the sum of the nonradiative (Σk_i) rate constants are calculated on the basis of equations of $\tau = 1/(k_f + \Sigma k_i)$ and $\phi_f = k_f/(k_f + \Sigma k_i)$: $k_f = (3 - 6) \times 10^6$ s^{-1} and $\Sigma k_i = (8 - 10) \times 10^{10}$ s^{-1}. The identity of the nucleic acid residues that are responsible for emission is still unclear.

The room-temperature picosecond decay measurements have been extended to methylated DNA[50,51] in which guanine residues are methylated at the N-7 position. Methylated DNA elicits interest because it has one major fluorophore whose fluorescence intensity is much enhanced compared to nonmethylated DNA and it is a major product of mutagenic and carcinogenic methylating agents.[52] The fluorescence decay profile for methylated DNA was found to comprise two decay components, as in the case of nonmethylated DNA; a major one with a decay time of 20 ps and a minor one with a decay time of 80 ps. In contrast, the decay profile for 7-methyl GMP was approximately single exponential with a decay time of 180 to 210 ps depending on the emission filter.

In contrast to 7-methyl GMP, the fluorescence spectrum of methylated DNA depends strongly on the excitation wavelength, shifting to the blue and becoming narrower as the wavelength increases. The fluorescence quantum yield exhibits a pronounced drop at long excitation wavelengths relative to that for excitation at 265 nm. Georghiou and Saim[51] interpreted these observations in terms of a combination of a heterogeneous environment of the methylated guanine residues, which results from sequence-dependent stacking interactions, and transfer of excitation energy from the other residues to the fluorescing methylated guanine residues. From the fluorescence quantum yields ($\phi_f = 0.012$ for 7-methyl GMP and $\phi_f = 0.0031$ for methylated DNA) and the decay times ($\tau = 180$ ps for 7-methyl GMP and $\tau = 20$ ps for methylated DNA), the following values are obtained for the radiative (k_f) and the sum of the nonradiative (Σk_i) rate constants: $k_f = 7 \times 10^7$ s^{-1}, $\Sigma k_i = 6 \times 10^9$ s^{-1} for 7-methyl GMP and $k_f = 1.6 \times 10^8$ s^{-1}, $\Sigma k_i = 5 \times 10^{10}$ s^{-1} for methylated DNA. The Σk_i value of methylated DNA is by about an order of magnitude larger than that of 7-methyl GMP. This enhancement of the nonradiative transitions appears to stem from stacking interactions. In agreement with this interpretation, it is concluded that DNA has stronger stacking interactions than methylated DNA, since the apparent value of $\Sigma k_i \simeq 1 \times 10^{11}$ s^{-1} for DNA is larger.

Further picosecond decay studies would provide a better clue to the elucidation for processes such as base-base interactions, modes of deexcitation of excited states, time-dependent conformational changes, and excitation energy transfer along the helix.

3. Dinucleotides and Polynucleotides Containing 1,N⁶-Ethenoadenosine

In view of the high fluorescence yield of 1,N⁶-ethenoadenosine (ϵA) produced by the chloroacetaldehyde modification reaction, this reaction has an obvious applicability to physical studies of nucleic acids since it offers a means of introducing a fluorescent reporter group.[53] Because of the high fluorescence quantum yields ($\phi_f \simeq 0.5$) and long fluorescence lifetimes ($\tau \simeq 25$ ns) of ϵA and its nucleotides,[54,55] a conventional single photon counting technique using a nanosecond discharge lamp has been applicable for fluorescence decay studies.

The first application was to ϵApϵA by Baker et al.[56] who observed a temperature-dependent fluorescence decay which could be resolved into two exponential terms, $\tau_1 \simeq 4$ ns and $\tau_2 \simeq 30$ ns at 25°C. Kubota et al.[57] obtained somewhat different results using five dinucleoside monophosphates containing ϵA, ϵApϵA, ϵApG, GpϵA, ϵApU, and UpϵA. Fluorescence quantum yields of these dimers are greatly reduced compared to that of ϵAMP, suggesting that intramolecular base-base interactions may be responsible for fluorescence quenching. It is found that the fluorescence decay kinetics for all dimers does not obey a simple decay law but that the decay data can be well described as a sum of three exponentials. This implies that these dimers cannot be characterized as a two-state system, an equilibrium between stacked and unstacked conformations, but can be described as systems consisting of three or more conformational states. Nuclear magnetic resonance (NMR)[58] and fluorescence-detected circular dichroism[59] of dinucleotides containing ϵA suggest that a discrete multistate hypothesis is most plausible for conformations of these dimers. The fluorescence quenching and decay parameters of GpϵA and UpϵA indicate a higher degree of base-base interaction than in their ϵApG and ϵApU counterparts.[57]

The same measurements have been applied to poly(A) containing ϵA [poly(ϵA)] and poly(ϵA)-poly(U) complexes.[60,61] The fluorescence quantum yield of poly(ϵA) decreases with increasing degree of ϵA substitution and is much smaller than that for ϵAMP even for low degrees of ϵA substitution. This result suggests that the nearest-neighbor interactions, such as ϵ-adenine-ϵ-adenine and adenine-ϵ-adenine, may play a significant role in fluorescence quenching. It is also found that the fluorescence decay kinetics obeys a three-exponential decay law as well as dinucleoside monophosphates containing ϵA (ϵ-dinucleotides), suggesting that there exist at least three different stacked conformational states. The decay behavior of poly(ϵA) is somewhat different from that of ϵ-dinucleotides. The proportions of the long lifetime τ_{long} for poly(ϵA) and τ_{long} are considerably larger than the corresponding values for ϵ-dinucleotides. Due to the limitations of sugar puckering and backbone rotations in polynucleotide chains, it seems likely that although ϵA groups in poly(ϵA) are immersed in the hydrophobic region, all of them cannot orient with their neighbors in favorable geometries to produce efficient quenching. The fluorescence decay behavior for poly(ϵA)-poly(U) complexes is very similar to that for poly(ϵA).[61] The fluorescence decay parameters and quantum yields indicate that the base-stacking interactions in complexes play an important role.

These studies show that the fluorescence decay measurements are helpful in understanding the multistate behavior in polynucleotide systems.

4. Intrinsic Fluorescent Base, Wyebutine Base

Among the modified fluorescent nucleosides isolated from nucleic acids, the highly substituted guanosine derivative called wyebutine (Wye) base is the only indigenous nucleoside that has shown utility as a fluorescent probe of tRNA tertiary structure. tRNA, because of its importance in gene translation, is widely studied.[62] Recent attempts have been made to obtain information about the solution structure of tRNA, especially about the anticodon loop of yeast tRNA^{Phe}. The existence of multiple conformations for tRNA has been proposed to explain experimental observations and biological function. As studied by fluorescence spectroscopy, the most extensive data for tRNA have been obtained using either extrinsic labels such as ethidium bromide or the intrinsic label Wye at position 37 in the anticodon loop of yeast tRNA^{Phe}.[62,63]   Very recently two research groups[64,65] have independently investigated conformational and dynamic properties of the anticodon loop of yeast tRNA^{Phe} by using fluorescence spectroscopy. Claesen and Rigler[64] have analyzed the picosecond time-resolved fluorescence of the Wye base serving a local structure probe adjacent to the anticodon G_mAA on its 3' side. Fluorescence lifetimes and anisotropies were measured by using TCSPC and a mode-locked synchronously pumped and frequency-doubled dye laser as exciting source. From the analysis of fluorescence lifetimes (τ) and rotational relaxation times (τ_R), it is concluded that the Wye base occurs in various structural states: (1) one stacked conformation where the base has no free mobility and the only rotational motion reflects the mobility of the whole tRNA molecule ($\tau = 6$ ns, $\tau_R = 19$ ns), (2) an unstacked conformation where the base can freely rotate ($\tau = 100$ ps, $\tau_R = 370$ ps), and (3) an intermediate state ($\tau = 2$ ns, $\tau_R = 1.6$ ns). Under biological conditions, i.e., in the presence of Mg^{2+} and neutral salts, the Wye base is found to be in a stacked and immobile state which is consistent with the crystallographic picture.[66]

Wells and Lakowicz[65] used the technique of variable frequency phase-modulation fluorometry (frequency-domain fluorometry) to examine the effects of magnesium ion on the fluorescence intensity and anisotropy decays of the Wye base in yeast tRNA[Phe]. In the frequency domain one measures the phase angle and modulation of emission relative to the intensity-modulated incident light. This technique also uses a laser as an exciting source and allows the resolution of multiexponential fluorescence decays.[67-69] Because of the ease, rapidity, and accuracy of the measurements, frequency-domain fluorometry will be now widely useful in biochemical and biophysical research.

Wells and Lakowicz[65] obtained somewhat different results compared to those of Claesen and Rigler.[64] They found that the fluorescence decay of the Wye base obeys at least a double-exponential decay law in the presence and absence of Mg^{2+}, but the multiexponential character of the decay is more pronounced in the absence of Mg^{2+}.[65] An explanation consistent with this result is that the anticodon loop is quite flexible at low concentration of Mg^{2+}; (1) the bases are freely twisting and tilting, allowing greater exposure to the solvent and (2) upon addition of Mg^{2+} or at very high ionic strength, the loop is constrained. This two-conformer model is consistent with the fluorescence anisotropy decays, which show two components due to overall tRNA rotational diffusion and to local torsional motions and the amplitude for Wye base torsional motion is decreased twofold in the presence of Mg^{2+}. Based on the above results, it is concluded that two conformers as exemplified by the anticodon loop are a loose flexible loop and a constrained, taut loop.

The fluorescence of the Wye base from tRNA[Phe] has been found to display multiple decay times, suggesting the existence of multiple conformers.[64,65] The biological relevance of different tRNA conformations is still unknown but the conformational dynamics of tRNA may be of importance for understanding phenomena such as codon recognition.

B. NUCLEIC ACID-DYE COMPLEXES

The interaction of fluorescent dyes with nucleic acids is of special interest because of the biological and physicochemical aspects of both the process of small-molecule binding to macromolecules and the process of energy transfer or charge transfer within nucleic acids.[70-73] By far the most intensively studied of fluorescent dyes are acridine derivatives and ethidium bromide, of which the most important are shown in Figure 10. Many of the acridine derivatives have important biological activities, acting as mutagens, carcinogens, and bacteriostatic agents. Ethidium bromide also induces several biological effects such as the inhibition of nucleic acid synthesis *in vivo*. These fluorescent dyes have been found to interact with nucleic acids through intercalation between adjacent base pairs in the double helix.[74,75]

Emission spectroscopy is one of the most sensitive techniques available for studying the binding interaction of small molecules with macromolecules and for studying the structure and dynamics of macromolecules. Time-resolved emission spectroscopy on the nanosecond or picosecond time scale can provide valuable information regarding the microenvironments at binding sites, the specific interactions between dyes and binding sites, the intramolecular orientation and the local motion within nucleic acids, and the energy transfer processes within nucleic acids.[73,76]

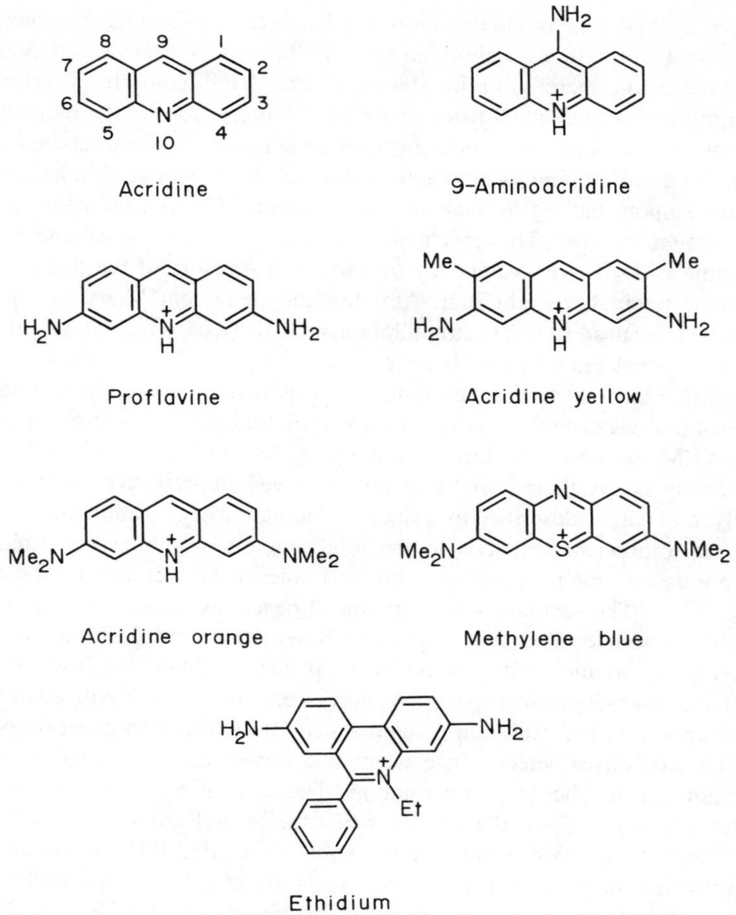

FIGURE 10. The structures of some important intercalating dyes.

1. Fluorescence Decay and Lifetime

Kubota and collaborators[77,78] have systematically investigated the binding interaction of acridine dyes with DNAs of various base composition by a combination of steady-state and nanosecond pulse fluorometry. It is found that the fluorescence decay behavior of the DNA-aminoacridine complexes is not so simple, but rather complicated. This is, of course, expected from the content of DNA bases and the existence of different binding modes. Nanosecond pulse fluorometry reveals that the fluorescence decays of aminoacridines such as 9-aminoacridine and acridine yellow (3,6-diamino-2,7-dimethylacridine) bound to poly(dA-dT)·poly(dA-dT), which contains only one type of binding site, are single exponential but those of these aminoacridines upon binding to DNA are complicated. The deconvolution analysis of fluorescence decay curves indicates that the decay kinetics of 9-aminoacridines upon binding to DNA obey a three-exponential decay law,[77] whereas that of 3,6-diaminoacridines (such as acridine yellow) is consistent with a double-exponential decay law.[78] This decay behavior has its origin in the heterogeneity of emitting sites. There is strong evidence that GC base pairs quench the fluorescence of bound aminoacridines.[73] The difference between the decay behavior of 9-aminoacridines and 3,6-diaminoacridines is ascribed to the different quenching ability of GC pairs against both groups of aminoacridines; GC pairs have stronger quenching ability for 9-aminoacridines than for 3,6-diaminoacridines.

The fluorescence lifetime of ethidium bromide has been shown to be dramatically changed by its environment, in particular, upon intercalation between successive DNA base pairs;[79] there is about a 3.5-fold increase in the lifetime of free dye in going from water to D_2O. In order to interpret this behavior, Olmsted and Kearns[79] proposed that the major pathway for deactivation of free ethidium bromide in aqueous solution involves excited-state proton transfer from the ethidium amino groups to water and the pronounced enhancement of the dye fluorescence upon binding to nucleic acids is attributable to a reduction in the rate of excited-state proton transfer. This mechanism suggests that the enhancement of the fluorescence lifetime in D_2O over water may be used as a measure of the degree of exposure to solvent of the intercalated ethidium. Atherton and Beaumont[80] have demonstrated that this technique can be used to probe structural changes of DNA which allow greater solvent accessibility to intercalated ethidium bromide.

Sommer et al.[81] have measured the fluorescence spectrum of ethidium bromide dissolved in both glycerol and water and intercalated into DNA following excitation by a laser pulse with 30-ps FWHM, in which the time resolution of the streak camera based spectrometer was approximately 20 ps. It is found that time-resolved fluorescence spectra of ethidium bromide in glycerol can be described by a linear combination of two components. The slowly decaying major component responds to the total amount of fluorescent ethidium in the solution, whose decay time is consistent with the fluorescence lifetime of ethidium bromide in glycerol (5.9 ns). The second minor component describes a fast ($\sim$100 ps) red shift of the ethidium fluorescence. This time-dependent fluorescence red shift can be attributed in part to phenyl group rotation with respect to the phenanthridium ring in the excited state. On the other hand, time-dependent spectral changes were not observed with ethidium bromide in aqueous solution or when ethidium is intercalated into DNA. Any spectroscopic shift for water could be too fast to detect, since water is a low-viscosity solvent which offers no significant hindrance to phenyl group rotation. The lack of an observable relaxation for ethidium intercalated into DNA demonstrates that the phenyl group is in an unconstrained local environment. This is consistent with the known structural information on intercalated ethidium,[82] in which the phenyl group appears to protrude unhindered into the solvent.

Other interesting fluorescent dyes are the bisbenzimidazole dye Hoechst 33258 (2'-(4-hydroxyphenyl)-5-(4-methyl-1-piperazinyl)-2,5'-bi-1H-benzimidazole) and 4',6-diamidine-2-phenylindole (DAPI), which have been used as a probe of nucleic acid structure as well as in cytological applications.[73,83] The structures of these dyes are depicted in Figure 11.

It is well recognized that Hoechst 33258 does not bind to DNA by intercalation, but rather binds by an attachment to the groove of the DNA double helix.[84-87] Kubota et al.[88] have studied the interaction of DNA with Hoechst 33258 at a high molar ratio of DNA phosphate to dye (P/D > 200) by means of steady-state and nanosecond time-resolved fluorescence spectroscopy. It is found that there are two types of bound species; one type (type I) predominates at high ionic strength, whereas the other (type II) occurs at low ionic strength. The fluorescence properties observed are summarized in the following; for type I, the fluorescence peak excited at 335 nm (λ_{max}^{f}) is 460 nm and the fluorescence lifetime (τ) is 2.0 to 2.5 ns and for type II, $\lambda_{max}^{f} = 470$ nm and $\tau = 4$ to 5 ns. This behavior is interpreted in terms of solvent-solute relaxation; type I corresponds to the less hydrated bound species which is buried deeply within the minor groove of the helix, while type II corresponds to the more hydrated bound species, which is bound in the major groove and is exposed to a considerable extent to the surrounding solvent. This is consistent with the recent crystallographic analysis[89] on the complex between Hoechst and a synthetic B-DNA dodecamer which demonstrates that the dye can bind strongly within the minor groove, by displacing the spine of hydration, with the deepest values of electrostatic potential. The fluorescence quantum yield of bound Hoechst decreases with increasing GC content of DNA, suggesting that there occurs a specific interaction between a GC pair and the bound Hoechst 33258 as well as some aminoacridines.[77,78]

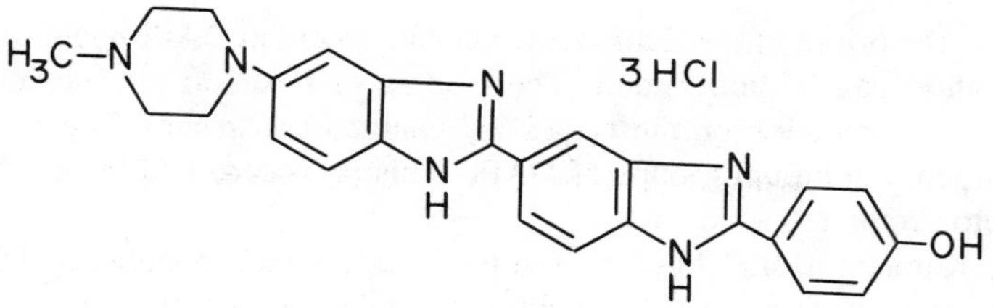

Hoechst 33258

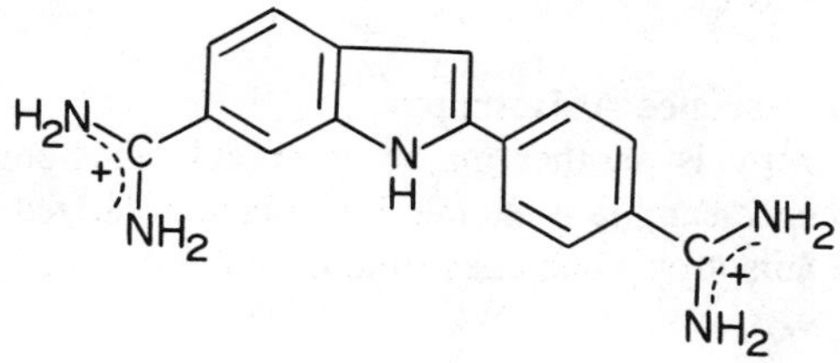

DAPI

FIGURE 11.   The structures of Hoechst 33258 and DAPI.

It is well known that DAPI shows a very marked increase in fluorescence quantum yield when bound to natural DNAs and synthetic polynucleotides containing AT, AU, and IC base pairs.[91] However, it is not fully understood whether DAPI intercalates between base pairs in DNA, binds with electrostatic interactions to the phosphate backbone of DNA, or interacts by another form of binding.

In order to obtain further information, Szabo and collaborators[92,93] have examined the fluorescence decays of DAPI when free or bound to calf thymus DNA, poly(dA-dT)·poly(dA-dT), and poly(dG-dC)·poly(dG-dC). The time-resolved fluorescence measurements were made using TCSPC and a mode-locked synchronously pumped and cavity dumped dye-laser system which gives excitation pulses with a pulse width of 20 ps. It is found that the fluorescence decay of DAPI complexed with polynucleotides at a high P/D ratio (P/D = 150) obeys a three-exponential decay law with lifetimes of 3.86 to 3.98 ns, 0.87 to 1.79 ns, and 110 to 130 ps; the medium lifetime component (1.79 ns for calf thymus DNA, 1.20 ns for poly(dA-dT)·poly(dA-dT), and 0.87 ns for poly(dG-dC)·poly(dG-dC)) seems dependent upon the base composition.[92] The lifetime results suggest that there occur at least three binding complexes.

On the other hand, the fluorescence of free DAPI in solutions at pH 3 to 9 has been shown to decay with double exponential kinetics with decay times of 2.86 ns and 144 ps.[91] The fluorescence spectra could be resolved into two components, a 2.86-ns component having a spectral maximum at 450 nm and a 144-ps component having a spectral maximum at 490 nm. The results are interpreted in terms of the existence of two different conformations of DAPI, one of which undergoes a rapid protonation of the indole ring by proton transfer from the 6-amidinium group in the excited state. The 144-ps component is ascribed to the fluorescence from the excited state of the protonated indole ring, while the 2.86-ns component is attributed to the conformation of DAPI in which rapid intramolecular protonation of the indole ring does not occur. The rapid intramolecular proton transfer in the excited state is responsible for the low fluorescence quantum yield of DAPI in aqueous solution. In view of these results, the origin of the fluorescence enhancement in DAPI-nucleic acid complexes at high P/D values can be understood. The binding of DAPI to nucleic acids involves an electrostatic interaction between the negatively charged phosphate of nucleic acids and the positively charged amidinium groups of DAPI. This process would inhibit the above intramolecular proton transfer.

Recently, Kubista et al.[94] have reexamined the interaction between DAPI and DNA using flow linear dichroism, circular dichroism, and fluorescence techniques. The results lead to the conclusion that DAPI has two binding sites on DNA, which correspond to the binding of DAPI in the minor and major grooves. The stronger binding most probably occurs in the minor groove since the interior of the major groove is less negative.[90]

In order to obtain further information on the nature of the interaction of Hoechst 33258 or DAPI with DNA, more detailed experiments including wavelength-dependent fluorescence decay measurements would be necessary.

2. Time-Resolved Fluorescence Anisotropy

Fluorescence anisotropy is another major observable phenomenon in emission spectroscopy. If the fluorescence sample is excited by a plane-polarized and infinitely short light pulse, the fluorescence anisotropy becomes time-dependent, its decay being defined as

$$A(t) = \frac{I_{\parallel}(t) - I_{\perp}(t)}{I_{\parallel}(t) + 2I_{\perp}(t)} = \frac{d(t)}{s(t)} \tag{3}$$

where $I(t)_{\parallel}$ and $I_{\perp}(t)$ are the fluorescence intensities observed through a polarizer whose transmission axis is aligned parallel and perpendicular, respectively, to the direction of polarization of the exciting light. The function $s(t)$ is proportional to the total fluorescence intensity.

Time-resolved fluorescence anisotropy is a powerful technique for studying internal and overall motions of nucleic acids and has been widely applied on both nanosecond and picosecond time scales. The intercalated ethidium cation has generally been used as a fluorescence probe since it has a relatively long fluorescence lifetime and high fluorescence quantum yield.[73] When intercalated dye molecules are excited with a picosecond laser pulse, the emission will be initially polarized and then the polarization will decay as the excited molecules undergo rotatory Brownian motion. Since intercalated dye molecules are closely attached to DNA base pairs, the anisotropy decay will reflect the torsional motions of the DNA double helix.

Subnanosecond time-resolved anisotropy techniques with a 100-ps laser pulse excitation have been used to study fast dynamic processes (0.5 ~ 100 ns) in nucleic acids by monitoring the reorientation of ethidium bromide intercalated in nucleic acids.[95,96] It is found that at least three different processes contribute to the anisotropy in high molecular weight DNA. Rapid initial decay has been attributed to the wobbling of the dye in the intercalation site; this will be discussed later. Further decay of the fluorescence anisotropy in the 1- to 50-ns region results primarily from torsional motions of DNA, and finally, the long time decay arises from the bending of the DNA double helix. Theoretical models have been presented to interpret the depolarization caused by internal rotatory Brownian motion in the DNA helix: an elastic model of semiflexible chain macromolecules[97] and a model which consists of a series of identical rigid rods connected by torsional springs.[98,99] Both models predict that the anisotropy decay of an intercalated dye should be complex. If DNA behaves as an uniformly elastic rod, the anisotropy decay $A(t)$ is mainly due to the torsional motions of the DNA chain and is expected to obey the following decay law:[97]

$$A(t) = A_0[1/4 + 3/4e^{-\Gamma(t)}] \tag{4}$$

where A_0 is the limiting anisotropy at $t = 0$ and $\Gamma(t)$ is the torsional decay function. Here DNA is considered as a semiflexible cylindrical rod of length $2L$ and radius b immersed in a liquid medium of viscosity η. The torsional decay function can be approximated at intermediate times by[97]

$$\Gamma(t) = (2k_B T/\pi)\sqrt{t/(b^2 \eta C)} \qquad (5)$$

where C is the torsional rigidity, k_B the Boltzmann constant, and T the absolute temperature. Millar et al.[95] analyzed their data in terms of a simplified elastic model and obtained a torsional rigidity $C = 1.30 \times 10^{-19}$ erg cm for calf thymus DNA. Thomas et al.[100] also measured time-dependent anisotropy of ethidium intercalated in viral ϕ29 DNA and found that only the intermediate decay zone of the torsion-spring model, which is equivalent to the elastic model, can best fit their data with a torsional rigidity $C = 1.29 \times 10^{-19}$ erg cm. Both values are in excellent agreement with the values estimated from supercoiling data (1.1×10^{-19} erg cm) and persistence length data (1.75×10^{-19} erg cm).[95]

Millar et al.[101] have extended their earlier work[95,96] and analyzed anisotropy data with a more complete elastic model, including both torsional and bending motions and dye wobbling. Figure 12 shows a typical set of experimental data for the calf thymus-ethidium bromide system in which a satisfactory agreement between the experimental data and the analysis can be seen. The results of the fluorescence anisotropy experiments are summarized in Table 3. We note from Table 3 that (1) helical structure plays a major role in the torsional rigidity; the order of torsional rigidities is triplet strand > double strand > single strand, (2) supercoiled DNA is more rigid than linear DNA, (3) the polyribonucleotide has a significantly larger torsional rigidity than its deoxy counterpart, and (4) the dye-wobble angle is relatively large for all polydeoxyribonucleotides but very small for the polyribonucleotide poly(rA)·poly(rU).

As is seen in Table 3, all the polynucleotides exhibit a nonzero wobbling angle $\Xi^{1/2}$, suggesting that there is an initial loss of fluorescence anisotropy. Magde et al.[102] have extended anisotropy decay measurements down to the picosecond regime by using a short laser pulse (30 ps) and a jitter-free streak camera system in which the time resolution is improved by a factor of ten or more over earlier studies.[95,96,100,101] There is a fast decay component of amplitude 0.025 and characteristic time of 100 ps for ethidium bromide intercalated in DNA, which is affected very weakly or not at all by solution viscosity. This fast relaxation has been attributed to wobbling of the dye within an intercalation site.

It should be noted that the Barkley-Zimm theory[97] is an approximation which holds rigorously only in the limit that the contour length of a helix is much larger than its persistence length. Schurr and co-workers[103,104] have tested theoretical models by using anisotropy data obtained with short DNA fragments. Wu et al.[105] have measured the time-resolved fluorescence anisotropy of ethidium intercalated in short DNA fragments, and determined the friction factor for rotation of DNA about its symmetry axis, using the model of an elastic filament with mean local cylindrical symmetry and a fixed contour length.[99,104] The value of the hydrodynamic radius, 12.0 ± 0.6 Å, is significantly larger than that suggested by the cross-sectional area perpendicular to the symmetry axis. This implies that a significant fraction of water in the grooves is rotating as though it is more or less rigidly attached to the DNA.

Time-dependent fluorescence anisotropy techniques are now applicable to more complex systems. Ashikawa et al.[106] have studied the internal motions of DNA in a nucleosome core particle by monitoring the nanosecond fluorescence anisotropy of the intercalated ethidium bromide. The results suggest that DNA in a nucleosome core particle has a torsional rigidity similar to that of DNA in solution, while the dynamics of linker DNA in chromatin reflect the overall structural state of the chromatin. Ashikawa et al.[107] have also reported that the rigidity of Z-DNA is much smaller than that of B-DNA, suggesting that the torsional stress

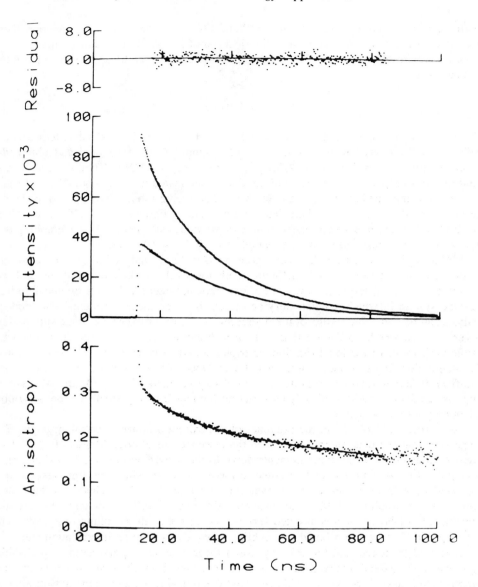

FIGURE 12.   Experimental $I_{\parallel}(t)$ and $I\perp(t)$ curves (center graph) and anisotropy r(t) (lower graph) for the calf thymus-ethidium complex in aqueous Tris buffer, 0.01 M NaCl, pH 7.7, 23°C. The points are experimental data and the smooth lines are the best fit of the data on the basis of τ = 23.3 ± 0.05 ns, C = 1.59 ± 0.11 × 10^{-19} erg cm, $\Xi^{1/_2}$ = 16.3° ± 1.2°, and X_R^2 = 1.65. Also shown (upper graph) are the weighed residuals for d(t) = $I\|(t)$ − $I\perp(t)$. (From Millar, D. P., Robbins, R. J., and Zewail, A. H., *J. Chem. Phys.*, 76, 2080, 1982. With permission.)

in the DNA may be largely absorbed by the Z-DNA if Z-DNA exists in negatively supercoiled DNA. Langowski et al.[108] have applied anisotropy techniques to investigate the internal dynamics of pBR322 supercoiled DNA upon binding of protein. They found that the torsional rigidity for the DNA-protein complex is significantly reduced compared to the free DNA. The observation is consistent with the notion that the binding of protein is accompanied by a gradual loss of supercoils and saturates when the supercoil twist is largely removed.

Since the fluorescence lifetime of the intercalated ethidium bromide is about 23 ns, all

TABLE 3
A Summary of Torsional Rigidities[a]

Sample	τ_n/s	$C \times 10^{19}$/erg cm	$\Xi^{1/2}$/deg[b]
Calf thymus DNA	23.0 ± 0.5	1.43 ± 0.11	17 ± 2
Poly(dG-dC)·poly(dG-dC)	23.5 ± 0.2	1.50 ± 0.15	19 ± 2
Poly(dA-dT)·poly(dA-dT)	25.4 ± 0.4	0.90 ± 0.22	12 ± 2
Poly(dG)·poly(dC)	22.2 ± 0.4	1.19 ± 0.10	10 ± 2
Poly(dA)·poly(dT)	19.9 ± 0.2	1.27 ± 0.17	12 ± 7
Poly(dA)·poly(dU)	22.4 ± 0.9	0.83 ± 0.12	15 ± 2
Poly(rA)·poly(rU)	27.6 ± 0.7	1.53 ± 0.13	6 ± 4
Poly(dA)·poly(rU)·poly(rU)	23.9 ± 0.5	2.65 ± 0.33	15 ± 5
PBR 322[c]	23.2 ± 0.2	1.95 ± 0.18	20 ± 2
Denatured calf thymus DNA	22.6[d]	0.90 ± 0.10	0 ± 2

[a]   In 0.1 M Tris·HCl, pH 7.7/0.15 M NaCl; P/D ≃120. From Reference 101.
[b] Ξ is the wobble parameter and the square root of Ξ measures the width of the distribution of dye orientations after wobbling.
[c]   Supercoiled pBR 322 plasmid DNA.
[d]   From Reference 95.

slower molecular motions such as end-over-end tumbling and rotation of the entire DNA will not be observed within the excited state lifetime. In order to probe these slower motions, phosphorescence anisotropy techniques have been developed; this will be discussed later.

3. Steady-State Fluorescence Anisotropy

Time-resolved fluorescence anisotropy has been verified to be a sensitive probe of rotational dynamics of macromolecules. However, the instrumentation and software necessary to perform time-resolved measurements are expensive to acquire and maintain. Because of this, steady-state anisotropy measurements recently have been widely used to investigate the local internal motions in DNA by monitoring the fluorescence of intercalated dyes.[109-114] The advantages of steady-state measurements are that they are rapid, accurate, and easily conducted on an ordinary spectrophotofluorometer.

Recently Härd and Kearns[110-112] have demonstrated that the steady-state fluorescence anisotropy of intercalated dyes can be used to distinguish between anisotropic twisting and bending motions in nucleic acids. The method is based on the fact that anisotropic motions have different effects on the anisotropy of emitted light that results from excitation of different electronic transitions.[115] Calculated effective end-over-end tumbling and twisting times for short DNA fragments (6-118 base pairs) are found to be shorter than predicted for overall motions of DNA, indicating that internal motions and/or dye wobbling contribute to the depolarization of fluorescence. The data can be interpreted in terms of a model[116] where DNA fragments are considered to be rigid against bending but torsionally flexible, and where dye can undergo some wobbling within the intercalated site without effecting the overall hydrodynamic behavior of the fragment.

From the analysis of the anisotropy data on DNA, poly(dA-dT)·poly(dA-dT), and poly(dG-dC)·poly(dG-dC) complexed with intercalating dyes, Härd and Kearns[112] have shown that twisting motions are responsible for most of the depolarization and that bending (or out-of-plane) motions are slow compared to the fluorescence lifetimes. The combined linear dichroism and time-resolved anisotropy data of the DNA-ethidium complex are consistent with significant out-of-plane mobility of ethidium, in addition to the DNA torsional motions and wobbling of the dye.[113] Furthermore, 9-aminoacridine is found to be more mobile than ethidium bromide, indicating that 9-aminoacridine is able to undergo extensive wobbling motions within the intercalating site.[112]

Some limitations of the two-wavelength technique used by Härd and Kearns[110-112] have been pointed out.[114] The use of a second long-lived emitting state has been suggested as a means of obtaining more reliable information about the slow tumbling motions from steady-state anisotropy measurements.[114]

4. Energy Transfer

When dyes such as proflavine, acridine orange, and ethidium bromide complexed with DNA are excited by UV light below 310 nm, sensitized fluorescence of the dyes can be observed.[73] This phenomenon has been attributed to energy transfer from the first excited singlet state of DNA bases to intercalated dye molecules. Anders[117,118] has studied sensitized fluorescence in DNA-dye complexes after excitation by a tunable dye laser and found that energy transfer depends on the excitation wavelength and the base composition of DNA. The energy transfer between the nucleic acid bases and the dye is helpful in understanding photochemical reactions of nucleic acids and photobiological actions of dyes. For example, the decrease of quantum yield of thymine dimerization in DNA-dye complexes is related to the trapping of energy by the dye.[73]

Andreoni et al.[119] have reported interesting results concerning time-delayed two-step selective laser photodamage of proflavine-poly(dA)·poly(dT) and proflavine-poly(dG)·poly(dC) systems. The dye-polynucleotide system is first pumped to the singlet excited state by 430-nm laser beam with a 250-ps duration and then is photoionized by time-delayed second laser excitation (337-nm and 550-ps duration). The result shows that the proflavine-poly(dA)·poly(dT) complex is more easily damaged than the proflavine-poly(dG)·poly(dC) complex and the probability of the photodamage increases to a much higher degree for a delay of 8 ns than for a delay of 5 ns. Most of the excited state population is expected to decay at 8 ns delay, because the fluorescence lifetime of proflavine bound to poly(dA)·poly(dT) is 6.7 ns. Therefore, it is concluded that the photodamage arises from two-step photoionization starting from the triplet state of the bound dye. The nonfluorescent exciplex formation may be responsible for the low damage probability for the proflavine-poly(dG)·poly(dC) system.

5. Fluorescence Quenching

Collisional quenching of fluorescence is of considerable interest for physical chemistry and biochemistry. Collisional quenching requires contact between the fluorophore and quencher during the excited-state lifetime. Consequently, the quenching process yields information about diffusional motions in solution and the accessibility of fluorophore in macromolecules to externally added quenchers.

It is well established that some metal ions are essential for the accurate and efficient replication of DNA and, indeed, play significant roles in many biological processes. Accordingly, it is of considerable interest to study the quenching of fluorescence from DNA-intercalated ethidium excited states by the transition metal ions.

Atherton and Beaumont[80,120,121] have reported that metal ions such as Cu^{2+}, Ni^{2+}, and Co^{2+} lead to double-exponential fluorescence decay. The faster component of fluorescence (lifetime of about 5 ns) is considerably slower than that of free ethidium in water (1.8 ns) and arises from a population of ethidium bromide excited states strongly quenched by metal ions, while the slower component arises from a population less strongly perturbed. There is evidence that the majority of metal ions are associated with DNA, and there is an equilibrium between metal bound at the phosphate groups and metal mobile around the helix. Therefore, it is concluded that the fast component results from ethidium bromide intercalated at a site which has a metal ion bound to a nearby phosphate group and the slower component results from ethidium bromide similarly intercalated but with no nearby metal ion. Quenching of this slower component is by metal ions that diffuse into the quenching sphere from other locations. A quantitative treatment of data shows that the metal ions can be arranged in order of increasing mobility as $Cu^{2+} < Ni^{2+} < Co^{2+}$. The mobility of Co^{2+} is a factor of about 20 less than in water. Quenching has been suggested to occur via electron transfer from excited ethidium to the metal ion.[120]

In the quenching experiments by Baguley and Le Bret, who studied fluorescence quenching of DNA-intercalated ethidium bromide by the antitumor drugs, 9-anilinoacridine derivatives, the double-exponential decay behavior was also observed but the slower component of fluorescence decayed at the same rate as the fluorescence in the absence of quencher. There is evidence that these quenchers bind to DNA by intercalation as well as ethidium bromide and thus are totally immobile on the time scale of ethidium fluorescence. Consequently, the double-exponential decay behavior is attributed to a proportion of the ethidium bromide molecules being highly quenched, with the proportion varying for different drug binding sites. Quenching has been suggested to occur as a result of reversible formation of electron-transfer complexes between the antitumor agent and ethidium excited state.[122] It is interesting that the antitumor activity of 9-anilinoacridine derivatives is significantly correlated with the degree of quenching.

It has been shown that methylviologen, one of the reduction-oxidation indicators, also quenches the fluorescence of ethidium bromide intercalated in DNA.[123] Laser flash photolysis[123] reveals that the quenching reaction occurs via charge transfer from the excited ethidium bromide to methylviologen to yield reduced viologen.

6. Phosphorescence and Delayed Fluorescence

In addition to fluorescence, delayed emissions (phosphorescence and delayed fluorescence) are another important emission phenomena. Fluorescence techniques are available for studying dynamic processes in macromolecules on time scales ranging from picoseconds to nanoseconds, while methods based on delayed emissions make the microsecond and millisecond time ranges accessible.

Berkoff et al.[124] have studied the dynamic quenching of the triplet state of methylene blue by O_2 to observe changes in the rate at which O_2 can penetrate the DNA helix. They showed that the rates of triplet quenching are a strong function of DNA base composition. The results indicate that the interior of poly(dA)·poly(dT) is more accessible than poly(dG)·poly(dC) to small molecules such as O_2. This is expected from the rigidity of the helix as revealed from the phosphorescence anisotropy decay studies.[125]

Laser flash spectroscopy[126] confirms that the triplet decay times of methylene blue are increased upon binding to DNA and synthetic polynucleotides, compared to the free dye, and show a decreased sensitivity of oxygen quenching. The results are interpreted in terms of the reduced accessibility of oxygen to the bound dye.

Milton and Galley[127] have shown that the mobility of solvent associated with native DNA can be monitored from the temperature-dependent red shift in the phosphorescence spectra of acridines bound to DNA. They found that 9-aminoacridine is less exposed to solvent than acridine orange and proflavine, DNA-associated solvent is considerably less mobile than bulk solvent, and there is a heterogeneity in the mobility of DNA-associated solvent. The solvent rigidities associated with polynucleotides are shown to be in the order of poly(dG)·poly(dC) > poly(dG-dC)·poly(dG-dC)~DNA > poly(dA-dT)·poly(dA-dT).[127] This can be compared with the order of the helix rigidities, poly(dG)·poly(dC) > DNA > poly(dA)·poly(dT).[125]

In order to elucidate binding kinetics of proflavine to poly(dA-dT)·poly(dA-dT) and its brominated analogue poly[d(A-br5U)], Corin and Jovin[128] have utilized the delayed fluorescence decay behavior of the dye at room temperature in addition to conventional kinetic methods such as temperature-jump techniques. They found that the external binding process is much stronger and the intercalation is much weaker in the excited state. This study suggests that the delayed emission spectroscopy may be a useful tool for probing dynamics of DNA-ligand interactions.

7. Phosphorescence Anisotropy Decay

Fast motions such as twisting and bending of the DNA double helix can be seen by time-resolved fluorescence anisotropy methods, using as a probe ethidium bromide intercalated in DNA. In order to measure slower motions, such as the end-over-end tumbling and spinning rotation motions, phosphorescence anisotropy techniques have been developed.[125,128-132]

Hogan et al.[129] have applied phosphorescence anisotropy decay techniques to measure the internal flexibility and overall rotational motions of DNA, covering a time range from 15 ns to 200 μs. Methylene blue has been used as a probe, which binds to DNA by intercalating into the helix and has a high phosphorescence yield. Since the phosphorescence lifetime of methylene blue is about 100 μs, one can monitor motions over a time scale 1000 times larger than fluorescence methods allow. In their experiments, the excitation source was a rhodamine 640 dye laser (λ_{ex} = 604 nm) pumped by a frequency-doubled Nd:YAG pulsed laser (FWHM of 15 ns). The phosphorescence anisotropy decay of intercalated methylene blue is found to be three exponential. For short DNA fragments ($\leq$165 base pairs), the slowest component of anisotropy decay monitors end-over-end tumbling of the helix, which is shown to be as predicted for a rigid rod-like helix. For a longer DNA fragment (600 base pairs), the anisotropy data are fitted quite well by the Barkley-Zimm model,[97] indicating that long DNA helices experience slow segmental motion. Based on the rigidity of the long helix axis, it is concluded that the fast initial decay of methylene blue anisotropy is due to torsional motions of the DNA helix. Although the time range accessible to fluorescence measurements is limited, fluorescence and phosphorescence anisotropy decay curves are shown to be nearly identical, suggesting that the two techniques monitor the same molecular motions.

Hogan et al.[125] have also applied triplet anisotropy decay techniques to study the flexibility of synthetic DNA fragments with different base compositions. It is found that the torsional and bending stiffness of poly(dG)·poly(dC) is much larger than that of poly(dA)·poly(dT) or poly(dA-dC)·poly(dT-dG). This implies that the torsional and bending stiffness may be strongly dependent on base composition and that GC-rich DNA sequences could be sites of weakened nucleosome structure while AT-rich regions may be sites where nucleosome formation is energetically favored. It is also suggested that this rigidity could have important implications for base-pair recognition by proteins.

Wang et al.[130] have used phosphorescence anisotropy techniques to measure the motion of dyes (methylene blue and rhodamine 123) intercalated into nucleosome-complexed DNA. It is found that DNA, when bound to the histone core, experiences large and fast local motion. This motion could be quite accurately fitted by the Barkley-Zimm model[97] for DNA motions, allowing only local twisting motions of the helix and the overall tumbling of the nucleosome. The data are best fitted with a torsional rigidity of $1.8 \pm 1.0 \times 10^{-19}$ erg cm, a value nearly identical to that measured for free DNA.[95,101] Such similarity suggests that large, fast twisting motions of the DNA helix persist, nearly unaltered, when DNA is wrapped to form a nucleosome.

Hogan et al.[132] have used spectroscopic methods to investigate calcium binding to the nucleosome core particle. The results show that the calcium ion binds tightly to the particles and methylene blue, a DNA-specific intercalator, also binds tightly to nucleosomes by intercalation. Data of both phosphorescence anisotropy decay and phosphorescence quenching show that, in the Ca^{2+}-nucleosome complex, methylene blue is capable of wobbling over a substantial angular range at its binding site. To interpret these data, it is proposed that Ca^{2+} binding to nucleosomes causes DNA to fold by means of a series of sharp kinks and methylene blue binds tightly to such kinked sites in the nucleosome. By the photochemical methods which depend on the sensitization of singlet oxygen production and the rate of oxygen diffusion into DNA base pairs, Hogan et al.[133] have suggested that the DNA in the nucleosome is bent or kinked at two sites, which may attract the binding of intercalating dyes such as methylene blue.

Austin et al.[134] have examined the complexes of *E. coli* RNA polymerase and DNA, using rose bengal as a phosphorescence probe. The free enzyme exhibited anisotropic rotations with rotational correlation times of 2 and about 0.7 μs, corresponding to the dimeric and monomeric species. Binding of the polymerase to poly(dA-dT)·poly(dA-dT) or calf thymus DNA led to a slower, more complex anisotropy decay with at least two components, one about 0.5 and the other 10 to 20 μs. The slower component was interpreted to arise from the overall tumbling of the complex which was dominated by the DNA, whereas the faster was proposed to arise from the spinning motions about the long helix of the DNA.

A combination of fluorescence anisotropy and phosphorescence anisotropy decay methods can now utilize to analyze the internal motions of macromolecules that bends, twists, and wobbles over a very wide time range from micro- to picoseconds. It is also possible to extend these techniques to more complex systems which are of biological interest, such as DNA-protein complexes and chromatin.

III. LASER RAMAN SPECTROSCOPY

A. INTRODUCTION

Laser Raman spectroscopy has developed into a useful complement to X-ray diffraction and NMR for the examination of the structural geometry and interactions of nucleic acids. Since Raman spectroscopy is applicable to samples of widely varying morphology, it is particularly well adapted to comparative structural studies of nucleic acids in different supramolecular environments.

Prior to the advent of laser spectroscopy, the usefulness of this technique in the study of biopolymers was limited by the strong elastic Rayleigh scattering encountered and by the significant spectral overlap of the Rayleigh and Raman bands because of the finite range of wavelengths spanned by the incident beam. The greatly improved spectral purity of laser sources has minimized this problem.

Basically, Raman spectroscopy provides an alternative to conventional IR spectroscopy as a means of obtaining information about the vibrational states of molecules. While a detailed theoretical treatment is beyond the scope of this review, it may be useful to present an abbreviated description at the classical level.[135]

If a sample is irradiated with a beam of light of frequency ν, an oscillating dipole of this same frequency will be induced in the molecules. The magnitude of the time-dependent dipole $\bar{\mu}(t)$ is given by

$$\bar{\mu}(t) = \alpha(\nu)\bar{E}_0 \cos 2\pi\nu t \tag{6}$$

where $\alpha(\nu)$ is the polarizability of the molecule and $\bar{E}_0$ is the maximum value of the electric field vector of the incident light. If the molecule possesses vibrational modes of motion, its polarizability will not in general be independent of time, as the vibrational movement of the nuclei pulls the associated electrons along with them, so that the polarizability of the molecule depends upon the nuclear positions and hence upon time:

$$\alpha(\nu) = \alpha_0(\nu) + \alpha'(\nu)\cos 2\pi\nu' t \tag{7}$$

Here α_0 is the polarizability of the molecule in its equilibrium state and ν' is a vibrational frequency. The quantity $\alpha'(\nu)$ governs the change in polarization with nuclear motion.

Combination of the above equations yields:

$$\bar{\mu}(t) = \bar{E}_0\alpha_0(\nu)\cos2\pi\nu t + \bar{E}_0\alpha'(\nu)\cos2\pi\nu' t\cos2\pi\nu t \qquad (8)$$
$$= \bar{\mu}(t) + \bar{\mu}'(t)$$

The initial term corresponds to the elastic interaction of light with the molecules and to the usual induced oscillating dipole of frequency ν. This term gives rise to ordinary Rayleigh light scattering at the frequency of the incident beam.

The second term in Equation 8 corresponds to inelastic light scattering and to the Raman spectrum. For molecules which are large and asymmetric, such as the common biopolymers, the Raman spectrum yields bands which are basically equivalent to those observed by IR spectroscopy. The primary advantage of Raman spectroscopy is the practical one that, because of the weak Raman spectrum of water, biological materials can be readily examined in aqueous media.

Raman spectroscopy has proved to be of particular value in the study of nucleic acid-protein complexes. It is usually possible to resolve a large number of distinct bands, even for intact virus particles, and to identify many of these bands with specific peptide or nucleic acid vibrations. In addition to conformation-dependent shifts and line splittings, Raman spectra also display hypochroism.

In the section to follow, the laser Raman spectra of relaxed and supercoiled DNA will first be discussed. This will be followed by a description of some representative studies of nucleic acid-protein complexes.

B. LASER RAMAN SPECTROSCOPY OF NUCLEIC ACIDS

1. β-Type DNA Configurations in Solution

Raman spectroscopy has become one of the more useful physical techniques for the study of DNA molecules in solution. The method is applicable to nucleic acids of any size and can distinguish between the three primary conformational classes of DNA, namely, A, B, and Z. With the aid of comparative studies upon DNA fibers and crystals, as well as model compounds, it has been possible to establish assignments of most of the Raman bands of DNA.[136,137] These include characteristic bands arising from both the purine and pyrimidine bases and the sugar-phosphate backbone.

A comparison of the Raman spectra in solution of a series of DNAs of varying base composition has indicated, as expected, that the Raman bands fall into two classes with respect to their variation in intensity with GC content.[138] Those bands which have been identified with base ring vibrations show a pronounced dependence upon GC content, while those associated with the sugar-phosphate backbone are relatively independent of GC content. The assignment of the primary Raman bands is given in Table 4.

The intensities of the 730 cm^{-1} adenine band and the 750 cm^{-1} thymine band decrease linearly with increasing GC mole fraction, while those of the 782 cm^{-1} cytosine band and the 682 cm^{-1} guanine-deoxyribose band increase. In contrast, the peak at 833 cm^{-1}, which arises from a phosphate-deoxyribose vibration, shows no significant variation in intensity with GC content.[138]

A comparison of natural DNAs with biosynthetic AT and GC DNA polymers indicates that the spectra obtained with the latter are generally consistent with linear extrapolations to 0 to 100% GC content of the natural DNA data, although occasional minor deviations are observed.[138]

Wartell and Harrell have reported that curve fitting analysis reveals the presence of a minor band near 809 cm^{-1} in the spectra of all the natural and biosynthetic DNAs examined.[138] The intensity of this band is relatively independent of GC content. A Raman band at this frequency is characteristic of the A conformation of DNA.[139] It has been attributed

TABLE 4
**Assignment and Characteristics of Raman Bands of Natural and
Biosynthetic DNAs**

Frequency (cm^{-1})	Half-width (cm^{-1})	Assignment	Frequency (cm^{-1})	Half-width (cm^{-1})	Assignment
490	14	Ade/Thy	1115	5	
499	9	Gua,Thy	1143	10	Ade/Thy
534	8	Ade	1178	16	Gua/Cyt
579	7	Gua/Cyt	1192	7	
596	8.5	Cyt	1213	28	bk/mm
641	11	Ade,Cyt	1237	12—23	bk/mm
670	10	Thy	1256	19	Cyt,Ade
682.5	11.5	Gua-Drib	1268	9—21	
729	8	Ade,u	1302	13.5	Ade
750	11.5	Thy	1317	12	Gua
782	8	Cyt,u	1333	14	bk/mm
791	14.5	bk	1342	16	Ade
809	24.5	bk	1361	17	Gua/Cyt
835	22	bk	1375	15	Ade,Gua,Thy
894	20	bk	1420	11	Drib
924	19	bk	1444	6—20	
998	20		1460	10—30	
1013	27	Ade/Thy	1488	13	Gua,Ade
1052	12	bk	1512	7	Ade
1067	26		1575	14	Gua,Ade
1093	16	bk(PO$_2^-$)			

to a vibrational mode involving both the deoxyribose group in a C3′-endo ring pucker and the C5′-O-P(O$^-$) group.[139,140] The band at 809 cm^{-1} coexists with a band at 835 cm^{-1} which is associated with the B-family conformations.

The simultaneous presence of the bands at 809 and 835 cm^{-1} might arise from the dynamic fluctuation of each deoxyribose ring or from an equilibrium distribution between two populations of furanose conformations. The failure of NMR measurements to detect C3′-endo-type sugar rings suggests that the Raman studies may be sensing the dynamic character of all rings.[141] The response time of NMR is presumably too slow to detect the transient existence of these discrete conformations, sensing instead an average conformation.

In the case of calf thymus DNA, the intensity of the 809 cm^{-1} band in solution is about 20% of that for fibers in the A-conformation. This suggests that about 20% of the deoxyribose rings of the DNA in solution are of the C3′-endo-type.

2. Plasmid DNA

The superhelicity of double-stranded DNA may figure in several important biological processes, including replication and transcription, and is also essential for the close-packing of DNA as it occurs in nucleosomes.[142] It is therefore of substantial interest to determine whether superhelicity modified significantly the conformation of DNA. The possibilities include a localized conversion to the A-form or to a random coil.

Circular supercoiled DNA molecules, such as bacterial plasmids, are attractive systems for testing the structural consequences of superhelicity, permitting the direct comparison of the linear and supercoiled forms.[143] High-resolution laser Raman spectroscopy has been employed to examine the influence of supercoiling upon the DNA conformation of the *E. coli* plasmid pBR322. Vasmel has compared the Raman spectra of the linear and circular forms, with particular regard to the effects of supercoiling upon the conformation of the sugar-phosphate group as well as the degrees of base stacking and base pairing.[143]

The Raman spectra in the region 700 to 1720 cm^{-1} were measured for the closed circular doubly stranded form I and the corresponding linear form III. A quantitative measure of the degree of supercoiling of a circular DNA is σ, the superhelical density, which is equal to the number of superhelical turns per helical duplex turn. The value of σ for the pBR322 plasmid was estimated at about -0.069, which would correspond to about 29 negative superhelical turns for the entire plasmid. The most prominent ring vibrational bands of the purines and pyrimidines occur in the region 1200 to 1600 cm^{-1}. It has been shown from melting studies that the intensity of these bands is strongly dependent upon the extent of base overlap.[139,144]

If the Raman spectra of the linear and circular forms of pBR322 were normalized at the 1580-cm^{-1} band, Gaussian deconvolution of the vibrational bands in the above region indicated that band intensities for the two forms were identical within experimental uncertainty. This result suggested that base stacking interactions make essentially equivalent contributions to the stabilization of the doubly helical structure for both plasmid forms. Thus the altered conformation of superhelical DNA did not appear to result in an altered stacking of the bases. In particular, localized destacking did not appear to be a factor. If such destacking occurred, it would be anticipated to occur preferentially at (A-T)-rich sites and result in an increase in intensity of adenine and thymine bands.

A second point of interest is the question of the conformation of the sugar-phosphate chain in the plasmid molecule. For both the linear and superhelical forms, an intense vibrational band was observed at 1094 cm^{-1}, which has been assigned to the O=P=O^{-} symmetric stretch. A band at 835 cm^{-1} indicates that the major part of the deoxyribose groups of both forms exist in the C$'_2$-endo conformation, which is characteristic of B-DNA. The appearance of a band at 792 cm^{-1}, corresponding to the symmetric O-P-O stretching mode is likewise consistent with the B conformation.

The spectra of the I and III forms were compared directly by resolution into Gaussian components. Good fits were obtained assuming that only 792 and 835 cm^{-1} bands were present. No significant difference between the two forms was observed.

The overall conclusion of this study was that supercoiling to the extent present in pBR322 does not significantly modify the DNA conformation. At least 98% of all bases are fully base paired.

In a parallel investigation, Hayashi et al.[145] have examined the Raman spectrum of the closed-circular DNA of a plasmid, pFB100, which was derived from pBR322 by introduction of a fibroin gene. Preparations of the purified plasmid pFB100, which consisted of 5451 base pairs, was found to exist largely (85%) as an interwound right-handed superhelix with a superhelical density of $1/100 \sim 1/300$; the balance is in the nicked and relaxed form.

Prolonged irradiation with the 514.5-nm beam (100 mW at the sample point) of the argon ion laser used for Raman excitation was found to cause a nick, but no cleavage, of the plasmid DNA. Laser irradiation for 4 h sufficed to cause nicking and relaxation of 50% of the sample, while 30-h exposure resulted in complete relaxation. No other structural changes were found to occur.

In this way it was possible to make a direct comparison, employing the same sample, of the Raman spectra of the supercoiled and relaxed forms of the plasmid DNA and hence to access directly the effects of supercoiling. In harmony with the study upon pBR322 the Raman spectra were found to be very similar for the supercoiled and relaxed states. However, one significant difference was identified in that the minor line at 728 to 729 cm^{-1}, which is present in linear DNAs, is weak or absent in the intact supercoiled plasmid, but appears upon relaxation. This line has been attributed to the ring-breathing vibration of the adenine group.[146] It is of interest that the analogous ring-breathing Raman lines of guanine at 682 cm^{-1} and cytosine at 782 to 785 cm^{-1} are essentially unchanged in the supercoiled and relaxed forms. This is consistent with expectations if any local deformation in the superhelical DNA occurs at AT rather than GC pairs.

Hayashi et al. have proposed that the weakness of the 728- to 729-cm^{-1} line in supercoiled pFB100 DNA is associated with a syn conformer of adenosine.[145] This model is supported by the Raman spectra reported by Benevides et al.[147] for single crystals of [d(CGCATGCG)]$_2$ and [d(m^{5}CGTAm^{5}CG)]$_2$, both of which have Z-DNA structures and contain C3′ endo-syn adenosine.

3. Operator Site of λ Phage

The biological control of gene expression depends upon the interaction of protein repressors with the corresponding operator sites of the DNA of the organism. Models have been proposed which describe possible modes of interaction of protein repressors of known structure with a standard helical structure of the B form of DNA. However, the possible role of any localized deviations in DNA structure has remained uncertain.

Prescott et al.,[148] have examined the Raman spectrum of the O_L1 operator of λ bacteriophage. The operator, which consists of an asymmetric sequence of 17 base pairs, was prepared by synthesis and may be represented as follows:

TACCACTGGCGGTGATA
ATGGTGACCGCCACTAT

The most prominent feature of the Raman spectrum between 600 and 850 cm^{-1} is a composite band centered near 780 cm^{-1} which consists of a cytosine ring mode plus a phosphodiester backbone stretching mode which is characteristic of the B form of DNA.

Three partially resolved bands at 807, 825, and 838 cm^{-1} were assigned to the phosphodiester backbone of O_L1. While the latter two are characteristic of the B form of DNA, the band at 807 cm^{-1} has been assigned to a symmetric vibration of the A form and is presumably equivalent to the band at 809 cm^{-1} which has been detected for DNA solutions.[138]

It was concluded that the backbone of O_L1 could not be described by a single geometrical form shared by all nucleotides and that some fraction may assume a geometry of the A type. However, it was recognized that a complicating factor is present in that the possible role of end effects, which may be important for this short oligonucleotide, remains to be assessed.

C. LASER RAMAN SPECTRA OF DNA-PROTEIN COMPLEXES
1. Calf Thymus Chromatin

The basic structural organization of mammalian chromatin appears to involve condensed structures arising from the coiling of the DNA helical duplex around a histone octamer.[149] The latter consists of one pair each of the histones H2A, H2B, H3, and H4.[149] While there is general agreement on the fundamental outline of the structure, there remains much uncertainty as to the details. It is probable that a major factor in the stabilization of the organized chromosomal structure is the electrostatic interaction between the basic histones and the negatively charged DNA phosphates.

A question of central importance, which can be addressed by Raman spectroscopy, is the extent to which the conformation of DNA is modified by incorporation into the chromatin complex.

Savoie et al.,[150] have examined the laser Raman spectrum of calf thymus chromatin and its constituents in aqueous solution. Spectra were compared for chromatin and core chromatin, from which H1 histone and any nonhistone proteins had been extracted, as well as complexes of DNA with the H3 and H4 histones. All of the above spectra were dominated by the characteristic bands of the DNA bases. From the observed protein/DNA ratio it was concluded that the preparations of chromatin utilized here were essentially free of nonhistone protein and could be regarded as equivalent to polynucleosomes and that the core chromatin consisted of polynucleosomes stripped of H1.

A comparison of the spectra of DNA and chromatin indicated, as expected, a major protein contribution in the latter case. The amide I and III bands are prominent, as is also the δ (CH$_2$) region at 1450 cm^{-1}. Several sharp aromatic bands are also observed, including the 1003-cm^{-1} peak of phenylalanine and the 852-cm^{-1} component of the tyrosine doublet.[150]

In order to compare the spectrum of the H1 subunit incorporated into chromatin with that of free H1, the former spectrum was obtained by subtracting the spectrum of core chromatin from that of chromatin after appropriate scaling. The spectrum of incorporated H1 contained a number of features not present for the free subunit. However, some of these are not of protein origin but arise from environmental differences of DNA in chromatin and core chromatin. The most dramatic difference in the two spectra is in the phenylalanine band at 1003 cm^{-1}, which is about four times as intense for combined as for free H1. The intensity per phenylalanine unit is also substantially greater for combined H1 than for any of the other free histones. A significant increase in intensity of the tyrosine doublet at 832 to 854 cm^{-1} also occurs for combined H1.

There is no very obvious explanation for the above observations. Since the core chromatin spectrum was recorded in 0.6 M NaCl, while that of chromatin was measured in the absence of added salt, it is possible that the change arises from an ionic strength effect, perhaps because of a change in the stacking interactions of the aromatic rings.

It is possible to estimate the contribution of the octameric histone core (H2A, H2B, H3, H4)$_2$ to the Raman spectrum of core chromatin by subtracting the spectrum of DNA from that of intact core chromatin. A comparison of the resultant difference spectrum with that of the isolated histone octamer indicates a striking similarity of the spectra, especially in the amide I and amide III regions. This in turn indicates that the structure of the protein octameric core is modified only to a minor degree from that in the isolated complex.

By a similar procedure the spectrum of the DNA incorporated into chromatin was obtained by subtracting the summed spectra of H1 plus the octamer from that of intact chromatin. The resultant difference spectrum corresponded to that of DNA incorporated into chromatin. A comparison with the spectrum of free DNA reveals a remarkable similarity with only relatively minor differences in intensity of the principal bands, which are not greatly in excess of experimental uncertainty. In particular the relative intensity of the 832-cm^{-1} band arising from the stretching mode of the phosphodiester backbone, is the same in chromatin and core chromatin as in free DNA, suggesting that the basic B conformation of DNA is retained in chromatin and core chromatin.

Nevertheless, perceptible differences in the Raman spectra of free and combined DNA exist. Two weak Raman bands at 1015 and 1146 cm^{-1}, which have been attributed to deoxyribose vibrations in B-DNA, are substantially reduced in intensity for core chromatin, indicating that some minor distortion of the sugar units, or a change in their relative orientation, is present.

It is not possible to deduce from these studies the nature of any specific interactions between the DNA bases in the inner histones in nucleosomes. It is clear that any effects of the interaction upon the Raman spectrum are marginal and will require data of the highest precision for accurate assessment. This study is consistent with earlier results which suggested that the B conformation of DNA is retained on nucleosome formation.

2. Laser Raman Spectra of Salmon Sperm DNA

A novel approach to the question of the conformation of DNA within intact biological systems has been adopted by Kubasek et al.,[151] who, in a complementary study to those on mammalian chromatin summarized above, have compared the Raman spectra of living intact salmon sperm with those of salmon sperm DNA and the deoxyoligonucleotide d-(CGCGAATTCGCG)$_2$. The structure of the latter is known with high precision from X-ray diffraction studies upon the crystalline state and provides a reference model for the B-conformation of DNA.[151]

A comparison of the Raman spectra of the oligonucleotide dodecamer with those of purified salmon sperm DNA and the intact sperm head indicated a close general similarity of all three spectra, which clearly corresponded to the B conformation. However, the possibility of a minor (<1%) contribution of an atypical DNA conformation, such as the left-handed Z form, could not be entirely ruled out.

A few minor differences were detected. Weak bands at 896 and 1450 cm^{-1}, which were present in the native sperm, but not in the purified DNA species, were attributed to membrane components. In addition, slight changes in the relative intensities of the Raman bands in the 1257- to 1340-cm^{-1} region were observed. A further difference arose with respect to the minor band at $\sim$ 810 cm^{-1}, which is detectable by band shape analysis in many purified DNAs of varying GC content, as reported by Wartell and Harrell.[138] The intensity of this band for purified salmon sperm DNA is comparable to that reported for other isolated DNAs. In the dodecamer the band is relatively weak, indicating a relatively small contribution of furanose conformations with a C3'-endo ring pucker. The 810-cm^{-1} band is also very weak for the intact salmon sperm head.

Another perceptible difference concerns the bands at 682 and 668 cm^{-1}, which arise from guanine and thymine, respectively. In intact salmon sperm the 682-cm^{-1} peak has the higher intensity, while the ratio of intensities is reversed for its purified DNA. However this may, at least in part, reflect an effect of ionic strength, as in high (6M) salt the ratio of intensities for the purified DNA becomes equivalent to that for native sperm. In the latter the high level of positively charged spermine might produce electrostatic effects similar to those caused by high salt. However, there is no indication in either case that high ionic strength induces any significant transition to the left-handed Z conformation.

As in the case of mammalian chromatin, incorporation of salmon sperm DNA into a condensed supercoiled structure does not appear to result in a drastic structural change. Only a minor perturbation occurs of its conformation which remains essentially of the B type.

3. Nucleosome Core Particles

The basic structural unit of eukaryotic chromatin is the nucleosome core particle, which is currently believed to consist of 146 base pairs of helical DNA complexed with an octamer of histones. The formation of this structural unit requires a substantial geometrical distortion of the DNA chain. While spectral studies generally agree in assigning nucleosome core DNA to the B conformation[152,153] an X-ray diffraction study indicates that the DNA coiling is not entirely uniform, but contains several sharp bends.[154]

Hayashi et al.,[155] have examined the structure of nucleosome core particles using difference Raman spectroscopy. The source was the 514.5-nm beam from an argon ion laser. The nucleosome core particles were isolated from chicken erythrocytes. The Raman spectra of intact core particles and their isolated DNA and histone octamers were compared.

An initial observation was that the Raman spectrum of the nucleosome particles was essentially equivalent to the summed Raman spectra of its constituents. This result indicates that formation of the nucleosome species does not drastically perturb the conformations of either the DNA or the protein components.[155]

Structural information could also be obtained for the protein moiety from both direct observations upon the isolated histones and spectra obtained indirectly by difference. Both Raman spectra show an amide I band at 1655 cm^{-1} and an amide III band with components at 1275 and 1250 cm^{-1}. The strength and positions of these bands provide compelling evidence that the dominant conformation of the protein chains is α-helical, with a significant contribution of irregular or randomly coiled structure. The absence of a strong amide III component below 1240 cm^{-1} renders unlikely the presence of a significant amount of β-structure.

The characteristic tyrosine doublet occurs at 855 and 836 cm^{-1} with a peak intensity ratio of the former to the latter of 2.7. If all tyrosines contribute equally to the doublet intensity, this finding is consistent with the involvement of each tyrosine OH group in a hydrogen bonding interaction.[155]

The Raman spectrum of the DNA portion could be obtained both from direct observation of the isolated DNA and by subtracting the contribution of the protein component from the spectrum of the intact nucleosome particles. Both spectra displayed the characteristic features of the B conformation of DNA, including a P-O stretching line at 836 cm^{-1}, a PO$_2^-$ symmetric stretching line at 1095 cm^{-1}, thymidine lines at 752, 674, and 1376 cm^{-1}, and a guanosine line at 684 cm^{-1}.

Although the Raman spectrum of the nucleosome core particles is approximately equivalent to the summed spectra of its DNA and protein constituents, a refined analysis indicates that the effects of the interaction are not negligible. In particular, the difference spectrum has distinctive features which can be distinguished from noise and which may be attributed to explicit structural features of the protein and DNA components.

Both positive and negative peaks corresponding to protein bands were observed. The former included tyrosine bands at 642, 850, and 1204 cm^{-1}, an amide II band at 1250 cm^{-1}, and a C-S stretching band at 655 cm^{-1}. The negative bands included an amide I band at 1655 cm^{-1}, a carboxylate band at 1406 cm^{-1}, and phenylalanine bands at 624 and 1005 cm^{-1}. These findings permit the conclusion that the microenvironments of the above aromatic groups are perceptibly altered in the core particles from their state in the isolated protein. Thus, the lower intensities at 624 and 1005 cm^{-1} arising from the phenylalanine residues in the nucleosome core may reflect mutual stacking interactions, while the higher intensities of the tyrosine residues at 642, 850, and 1204 cm^{-1} may conversely reflect a reduced overlap. A positive amide band occurs at 1250 cm^{-1} in the difference spectrum and a negative band at 1655 cm^{-1}, suggesting a change in the secondary structure and in particular a gain in the proportion of random coil. A quantitative estimate places the increase in random coil content at about 28%.[155]

Part of the difference spectrum arises from DNA. The positive bands stemming from adenine at 726 cm^{-1}, cytosine at 1293 cm^{-1}, and thymine or guanine at 1363 cm^{-1}, as well as the negative adenine or guanine band at 1326 cm^{-1}, are consistent with significantly different stacking geometries of DNA in the nucleosomes and in the free state. The 1673 cm^{-1} line, which is attributed to the hydrogen-bonded C=O stretching mode of thymine is weaker in the nucleosome core particles than in free DNA, suggesting some localized denaturation in A-T rich sites. Peaks in the difference spectrum arising from the phosphodiester backbone occur at 816, 836, 1091, and 1102 cm^{-1}, indicating small but significant alterations in the phosphodiester conformation.

Qualitatively, the observed difference spectrum is consistent with that expected for a significant B → A conversion. Hayashi et al.[155] accordingly propose that, in the nucleosome core particles, there is a small degree of fluctuation from the predominant B-form to the A-conformation, especially in the zones rich in A-T base pairs.[155] These results thus differ from those cited earlier for thymus chromatin and salmon sperm in providing somewhat more definite evidence for an increased contribution of the A conformation.

D. LASER RAMAN SPECTRA OF VIRUS PARTICLES
1. Belladonna Mottle Virus

Belladonna mottle virus (BDMV), a spherical plant virus of the tymovirus group, consists of a single RNA strand of molecular weight 1.9 × 10^6 plus 180 identical protein subunits of molecular weight 2.03 × 10^4. NMR studies have indicated that the RNA of BDMV undergoes a conformational transition within the capsid upon adjustment of the pH from below pH 6.5 to above 6.8.[156,157] At slightly acid pHs the encapsidated RNA appears to be relatively rigid, while at higher pHs it acquires substantial mobility.

The influence of pH upon Raman spectrum has been monitored for the free RNA and protein capsid, as well as for the intact virus. In 10 mM Ca^{2+}, the virus particles remain intact and no significant loss of RNA occurs at pHs up to 8. Under these conditions a pronounced difference spectrum for free RNA is developed upon altering the pH from 5.0 to 8.0. Both positive and negative peaks are present. A comparison with model compounds indicates that the majority of the peaks in the difference spectrum can be attributed to adenine and cytosine bases. Both bases are protonated to a significant extent at pH 5, but not at pH 8. From the magnitude of the intensity changes of the cytosine bands at 790 and 1260 cm^{-1} it may be computed that about 21% of the cytosines are protonated at pH 5 if it is assumed that none are protonated at pH 8.[158] A similar analysis indicates that about 5 to 10% of the adenines are protonated at pH 5.[158]

In addition to the difference spectrum arising from base protonation, a minor band associated with the RNA phosphodiester backbone appears at 813 cm^{-1}. The change is consistent with a small (~5%) increase in A-helix content at the lower pH. The Raman spectrum at pH 8 indicates that at least 85% of the nucleotide residues exist in the A conformation.

In contrast to the behavior of the free RNA, the Raman spectrum of the RNA-free BDMV capsid was found to be essentially independent of pH. No difference spectrum was developed between pH 5 and pH 8, indicating that no significant change in secondary structure occurred over this pH range. The observed spectrum shows that the protein subunit is rich in β-sheet secondary structure, as has also been found to be the case for the capsid proteins of several other spherical plant viruses.

Information about the hydrogen bonding of the −OH groups of the tyrosine residues may be obtained from the intensity ratio of the tyrosine doublet at 830 and 852 cm^{-1}. The observed ratio ($I_{852}/I_{830} = 1.85$) is consistent with the involvement of two of the four tyrosines per subunit as acceptors of strong hydrogen bonds from positive donor groups and of the remaining two in hydrogen bonding as both donor and acceptor. The Raman lines at 756, 1357, and 1554 cm^{-1} of the single tryptophan per subunit suggest that it exists in a hydrophilic microenvironment.

The pH dependence of the Raman spectrum of the intact virus generally resembles that of the free viral RNA, except that the difference spectrum peaks are more intense for the former, suggesting a greater protonation of cytosine groups for the native virus. If all of the cytosine difference spectrum arises from this cause, about 30% of the cytosines are protonated at pH 5. Intensity changes also occur in the adenine region at 1485 and 1570 cm^{-1}, which are difficult to explain solely in terms of protonation of the adenine ring and may reflect, at least in part, alterations in the degree of adenine stacking.

The Raman difference spectrum of BDMV RNA in the encapsidated and free states was obtained by subtracting the capsid spectrum from that of the intact virus and then subtracting the spectrum derived in this way from that of the free viral RNA. The changes in the Raman spectrum arising from encapsidation at pH 5 are small but significant. Bands at 1255 and 1299 cm^{-1}, which arise from cytosine, are altered in the direction expected for increased protonation upon encapsidation. Lines at 1397, 1481, and 1568 cm^{-1}, which arise primarily from adenine, are changed in the direction corresponding to reduced base stacking for the encapsidated form. The intensity ratio of the phosphate lines at 809 and 1097 is marginally reduced, suggesting a minor loss of A-type RNA structure upon encapsidation.

The contribution of the capsid protein to the difference spectrum at pH 5 is relatively minor. No peaks corresponding to the amide I and amide III lines were present, in harmony with the conclusion cited earlier that the secondary structure of the subunit is not changed by encapsidation. There is some indication of slight changes in the microenvironments of phenylalanine and tyrosine upon encapsidation. For the difference spectrum arising from encapsidation at pH 8.0, the contribution of base protonation is lost. Only the weak phosphate bands remain.

The overall pattern of the above finding is of relatively mild structural effects accompanying the assembly of the BDMV virus from its protein and RNA constituents. The conformations of both the protein and the RNA are altered only to a minor extent by incorporation into the virus. Much of the difference spectrum observed for assembly of the virus at pH 5 arises from increased protonation of the RNA bases. This contribution is lost at pH 8 in the presence of Ca^{2+} and the residual difference spectrum is so slight as to indicate little or no protein-RNA interaction under the latter conditions.

2. Laser Raman Spectrum of Cowpea Chlorotic Mottle Virus

Another example of the application of laser spectroscopy to a virus particle is provided by the investigation of Verduin et al., upon the virion of cowpea chlorotic mottle virus (CCMV), which is a member of the bromovirus group.[159] This class of viruses is characterized by a multicomponent RNA genome and by the capacity to undergo structural transitions as a function of conditions.[160] The CCMV virion is a mixture of three nonidentical ribonucleoprotein particles of slightly different buoyant density. The particle of highest density contains the largest RNA particle, RNA-1; the least dense particle contains RNA-2, while the intermediate particle contains the two smallest RNA units, RNA-3 and RNA-4. All three particles appear to consist of 180 identical subunits arranged in an icosahedral lattice.

Laser Raman spectroscopy has been employed by Verduin et al.[159] to examine the structure of CCMV and to monitor conformational changes of the RNA and coat protein of CCMV as a function of pH and ionic strength in the presence of 10 mM Mg^{2+}. A comparison of the spectra at pH 5 of the intact virus with the summed spectra of the isolated RNA and capsid protein components indicated that the two were basically similar, but that subtle differences were present. The amide I and amide III bands of the capsid were centered at 1666 and 1246 cm^{-1}, respectively, indicating that the secondary structure of the protein subunit is primarily β-sheet. However, the presence of shoulders to the main amide III band at 1266 and 1284 cm^{-1} indicates the occurrence to a significant extent of randomly coiled polypeptide and α-helix. A comparison with the spectra of proteins of known structure yielded an estimate of 40 to 60% for the β-sheet content, with the balance divided roughly equally between the randomly coiled and α-helical conformations.

The tyrosine doublet band occurs in the capsid at 853 and 830 cm^{-1}. The peak intensity ratio I_{853}/I_{830} is 1.4. This is suggestive of the existence of moderate hydrogen bonding of each phenolic hydroxyl group as both donor and acceptor, as expected for tyrosine side chains in contact with aqueous solvent. The three tryptophan residues of each subunit of capsid protein show characteristic Raman lines near 760, 874, 1368, and 1548 cm^{-1}; these values are typical of tryptophan in a hydrophilic environment.

The Raman spectrum at pH 5.0 of the purified RNA of CCMV is typical of single-stranded RNA. The intensity ratio I_{812}/I_{1099} is equal to 1.57; this is characteristic of a high degree of stacking of the bases. It is of interest that no features indicative of protonated adenine or cytosine were detected for protein-free RNA from CCMV at pH 5.0.

Environmental effects upon the capsid protein were examined by comparing the observed spectrum of the purified capsid protein with that computed by subtracting the spectrum of purified RNA from that of the virus. A difference spectrum of the two was then generated, which was diagnostic of the residues differing in the virion and capsid states. The results indicated substantial differences in the environments of phenylalanine and tryptophan side chains in the virion and free capsid states. The band intensities at 757 and 1004 cm^{-1} are less intense in the intact virus than in the free capsid; the implied increased hypochromism in the virion suggests enhanced stacking interactions of phenylalanine and tryptophan in this structure.

The Raman intensities of the amide I line at 1662 cm^{-1} and the amide III line at 1228 cm^{-1} are greater in the virus than in the free capsid protein, suggesting a perceptible loss of β-structure upon separation from RNA. The Raman difference spectra thus are indicative of significant conformational differences between the free capsid protein and that incorporated into the virus particle.

Incorporation into the virion also results in significant changes in the Raman spectrum of the RNA of CCMV. The spectrum of encapsidated RNA was computed by subtraction of the spectrum of the capsid protein from that of the intact virus and compared with that of purified RNA from CCMV. A difference spectrum was also constructed for the two. It corresponded closely, as expected, with the difference spectrum described above, which was obtained by subtracting the computed from the observed capsid spectrum; both are equivalent to the spectral differences between the virion and the sum of its protein and RNA constituents.

The adenine-guanine lines at 1483 and 1572 cm^{-1} are stronger for the encapsidated RNA than for the isolated RNA molecule, perhaps because of a reduction in purine-purine base stacking in the former. This, as well as minor spectral changes associated with the backbone, is consistent with a slight loss of ordered secondary structure upon incorporation into the virion.[159]

An increase of pH from 5.0 to 7.7 results in an increase in effective hydrodynamic volume of CCMV.[159] Computation of a difference spectrum for the virus at these two pHs reveals that only a few Raman lines are significantly altered by the pH-induced swelling. Somewhat surprisingly, no negative band at 1400 to 1420 cm^{-1} (corresponding to COO^- groups) or positive band at 1700 to 1750 cm^{-1} (corresponding to COOH groups) appeared upon subtracting the pH 7.7 from the pH 5.0 spectrum, indicating that no titration of the carboxyl groups of aspartic and glutamic acids occurs over this pH range. Intense negative bands appear at 1255 and 1297 cm^{-1}. The former is in the amide III region and suggests that some loss of ordered structure of the capsid protein occurs at the high pH. The latter band has been attributed to the deprotonation of a minor fraction of the RNA adenines at pH 7.7. Thus, in contrast to the protein-free RNA, significant protonation of encapsidated RNA bases occurs at pH 5.0.

As in the case of the BDMV, the general pattern of CCMV is of relatively mild conformational changes accompanying the combination of RNA and capsid protein.

ACKNOWLEDGMENTS

Equipment acquisition and development was made possible by NSF and NIH facility grants. Ian Davidson performed delicate operations on the Tennelec CFD. Minyung Lee and Yong Rok Kim have made important contributions to the experimental methods. Philip Anfenrud has independently solved many of the problems mentioned here, and has shared solutions with us. Finally, Robin Hochstrasser has enthusiastically supported the construction of this apparatus and has generated a number of fresh ideas to be investigated with it.

REFERENCES

1. **Lewis, C., Ware, W. R., Doemeny, L. J., and Nemzek, T. L.,** The measurement of short-lived fluorescence decay using the single photon counting methods, *Rev. Sci. Instrum.*, 44, 107, 1973.
2. **Leskovar, B., Lo, C. C., Hartig, P. R., and Sauer, K.,** Photon counting system for subnanosecond fluorescence lifetime measurements, *Rev. Sci. Instrum.*, 47, 1113, 1976.
3. **Spears, K. G., Cramer, L. E., and Hoffland, L. D.,** Subnanosecond time-correlated photon counting with tunable lasers, *Rev. Sci. Instrum.*, 49, 255, 1978.
4. **Gratton, E., Jameson, D. M., Rosato, N., and Weber, G.,** Multifrequency cross-correlation phase fluorimeter using synchrotron radiation, *Rev. Sci. Instrum.*, 55, 486, 1984.

5. **Gratton, E., Jameson, D. M., and Hall, R. D.,** Multifrequency phase and modulation fluorometry, *Annu. Rev. Biophys. Bioeng.,* 13, 105, 1984.
6. **Lakowicz, J. R., Laczko, G., Cherek, H., Gratton, E., and Limkeman, M.,** Analysis of fluorescence decay kinetics from variable-frequency phase shift and modulation data, *Biophys. J.,* 46, 463, 1984.
7. **Lempert, R. A., Chester, L. A., Phillips, D., O'Connor, D. V., Roberts, A. J., and Meech, S. R.,** Standards for nanosecond fluorescence decay time measurements, *Anal. Chem.,* 55, 68, 1983.
8. **Kinoshita, S., Ohta, H., and Kushida, T.,** Subnanosecond fluorescence-lifetime measuring system using single photon counting method with mode-locked laser excitation, *Rev. Sci. Instrum.,* 52, 572, 1981.
9. **Chang, M. C., Courtney, S. H., Cross, A. J., Gulotty, R. J., Petrich, J. W., and Fleming, G. R.,** Time-correlated single photon counting with microchannel plate detectors, *Anal. Instrum.,* 14, 433, 1985.
10. **Yamazaki, I., Tamai, N., Kume, H., Tsuchiya, H., and Oka, K.,** Microchannel-plate photomultiplier applicability to the time-correlated photon-counting method, *Rev. Sci. Instrum.,* 56, 1187, 1985.
11. **Bebelaar, D.,** Time response of various types of photomultipliers and its wavelength dependence in time-correlated single-photon counting with an ultimate resolution of 47 ps FWHM, *Rev. Sci. Instrum.,* 57, 1116, 1986.
12. **Holtom, G.,** to be published.
13. **Lakowicz, J. R.,** A review of photon-counting and phase-modulation measurements of fluorescence decay kinetics, in *Applications of Fluorescence in the Biomedical Sciences,* Taylor, D. L., Waggoner, A. S., Murphy, R. F., Lanni, F., and Birge, R. R., Eds., Alan R. Liss, New York, 1986, 29.
14. **Barkley, M. D., Kowalczyk, A. A., and Brand, L.,** Fluorescence decay studies of anisotropic rotations of small molecules, *J. Chem. Phys.,* 75, 3581, 1981.
15. **Millar, D. P., Robbins, R. J., and Zewail, A. H.,** Torsion and bending of nucleic acids studied by subnanosecond time-resolved fluorescence depolarization of intercalated dyes, *J. Chem. Phys.,* 76, 2080, 1982.
16. **Cross, A. J., Waldock, D. H., and Fleming, G. R.,** Time resolved polarization spectroscopy: level kinetics and rotational diffusion, *J. Chem. Phys.,* 78, 6455, 1983.
17. **Bebelaar, D.,** Compensator for the time dispersion in a monochromator, *Rev. Sci. Instrum.,* 57, 1686, 1986.
18. **Coyne, B.,** Tennelec Corp., Oak Ridge, TN, private communication, 1989.
19. **Anfinrud, P.,** Ph.D. thesis, Department of Chemistry, Iowa State University, 1987.
20. **O'Connor, D. V., Ware, W. R., and Andre, J. C.,** Deconvolution of fluorescence decay curves. A critical comparison of techniques, *J. Phys. Chem.,* 83, 1333, 1979.
21. **Grinvald, A. and Steinberg, I. Z.,** On the analysis of fluorescence decay kinetics by the method of least-squares, *Anat. Biochem.,* 59, 583, 1974.
22. **Beechem, J. M. and Brand, L.,** Global analysis of fluorescence decay: applications to some unusual experimental and theoretical studies, *Photochem. Photobiol.,* 44, 323, 1986.
23. **Wahl, P.,** Analysis of fluorescence anisotropy decays by a least square method, *Biophys. Chem.,* 10, 91, 1979.
24. **Baumann, J. and Fayer, M. D.,** Excitation transfer in disordered two-dimensional and anisotropic three-dimensional systems: effects of spatial geometry on time-resolved observables, *J. Chem. Phys.,* 85, 4087, 1986.
25. **Förster, T.,** Zwischenmolekulare Energiewanderung und Fluoreszenz, *Ann. Phys. Leipzig,* 2, 55, 1948.
26. **Barkley, M. D. and Zimm, B. H.,** Theory of twisting and bending of chain macromolecules: analysis of the fluorescence depolarization of DNA, *J. Chem. Phys.,* 70, 2991, 1979.
27. **Szabo, A.,** Theory of fluorescence depolarization in macromolecules and membranes, *J. Chem. Phys.,* 81, 150, 1984.
28. **Guéron, M., Eisinger, J., and Lamola, A. A.,** Excited states of nucleic acids, in *Basic Principles in Nucleic Acid Chemistry,* Vol. 1, Ts'o, P. O. P., Ed., Academic Press, New York, 1974, chap. 4.
29. **Vigny, P. and Duquesne, M.,** On the fluorescence properties of nucleotides and polynucleotides at room temperature, in *Excited States of Biological Molecules,* Birks, J. B., Ed., John Wiley & Sons, New York, 1976, 167.
30. **Daniels, M.,** Excited states of the nucleic acids: bases, mononucleosides, and mononucleotides, in *Photochemistry and Photobiology of Nucleic Acids,* Vol. 1, Wang, S. Y., Ed., Academic Press, New York, 1976, chap. 2.
31. **Hauswirth, W. W. and Daniels, M.,** Excited states of the nucleic acids: polymeric forms, in *Photochemistry and Photobiology of Nucleic Acids,* Vol. 1, Wang, S. Y., Ed., Academic Press, New York, 1976, chap. 3.
32. **Vigny, P. and Ballini, J. P.,** Excited states of nucleic acids 300 K and electronic energy transfer, in *Excited States in Organic Chemistry and Biochemistry,* Pullman, B. and Goldblum, N., Eds., D. Reidel, Dordrecht, Netherlands, 1977, 1.
33. **Callis, P. R.,** Electronic states and luminescence of nucleic acid systems, *Ann. Rev. Phys. Chem.,* 34, 329, 1983.

34. **Callis, P. R.,** Polarized fluorescence and estimated lifetimes of the DNA bases at room temperature, *Chem. Phys. Lett.,* 61, 563, 1979.

35. **Ballini, J. P., Daniels, M., and Vigny, P.,** Wavelength-resolved lifetime measurements of emissions from DNA components and poly rA at room temperature excited with synchrotron radiation, *J. Lumin.,* 27, 389, 1982.

36. **Ballini, J. P., Vigny, P., and Daniels, M.,** Synchrotron excitation of DNA fluorescence decay time evidence for excimer emission at room temperature, *Biophys. Chem.,* 18, 61, 1983.

37. **Rigler, R., Claesens, F., and Lomakka, G.,** Picosecond single photon fluorescence spectroscopy of nucleic acids, in *Ultrafast Phenomena, Part IV* (Springer Series in Chemical Physics, Vol. 38), Auston, D. H. and Eisenthal, K. B., Eds., Springer-Verlag, New York, 1984, 472.

38. **Rigler, R., Claesens, F., and Kristensen, O.,** Picosecond fluorescence spectroscopy in the analysis of structure and motion of biopolymers, *Anal. Instrum.,* 14, 525, 1985.

39. **Bevington, P. B.,** *Data Reduction and Error Analysis for the Physical Sciences,* McGraw-Hill, New York, 1969.

40. **Kobayashi, S., Yamashita, M., Sato, T., and Muramatsu, S.,** A single-photon sensitive synchroscan streak camera for room temperature picosecond emission dynamics of adenine and polyadenylic acid, *IEEE J. Quantum Electron.,* QE-20, 1383, 1984.

41. **Kobayashi, S. and Yamashita, M.,** Quenching of polyadenylic acid emission by divalent metal ions at room temperature, *Nucleic Acids Res. Symp. Ser.,* 17, 215, 1986.

42. **Yamashita, M., Kobayashi, S., Torizuka, K., and Sato, T.,** Observation of diffusion-free intermolecular excimer of 9-methyl adenine aqueous solution by picosecond time-resolved spectroscopy, *Chem. Phys. Lett.,* 137, 578, 1987.

43. **Birks, J. B.,** *Photophysics of Aromatic Molecules,* Wiley-Interscience, New York, 1970, chap. 7.

44. **Saenger, W.,** *Principles of Nucleic Acid Structure,* Springer-Verlag, New York, 1984, chap. 6.

45. **Morgan, J. P. and Daniels, M.,** Excited states of DNA and its components at room temperature. III. Spectral, polarization and quantum yields of emissions from ApA and poly rA, *Photochem. Photobiol.,* 31, 101, 1980.

46. **Morgan, J. P. and Daniels, M.,** Excited states of DNA and its components at room temperature. IV. Spectral, polarization and quantum yield studies of emissions from CpC and poly rC, *Photochem. Photobiol.,* 31, 207, 1980.

47. **Shaar, C. S., Morgan, J. P., and Daniels, M.,** Excited states of DNA and its components at room temperature. V. Spectral, polarization and quantum yield studies of cytidylyl-(3',5')-adenosine, *Photochem. Photobiol.,* 39, 747, 1984.

48. **Anders, A.,** DNA fluorescence at room temperature excited by means of a dye laser, *Chem. Phys. Lett.,* 81, 270, 1981.

49. **Aoki, T. I. and Callis, P. R.,** The fluorescence of native DNA at room temperature, *Chem. Phys. Lett.,* 92, 327, 1982.

50. **Georghiou, S., Nordlund, T. M., and Saim, A. M.,** Picosecond fluorescence decay time measurements of nucleic acids at room temperature in aqueous solution, *Photochem. Photobiol.,* 41, 209, 1985.

51. **Georghiou, S. and Saim, A. M.,** Excited-state properties of DNA methylated at the N-7 position of guanine and its free fluorophore at room temperature, *Photochem. Photobiol.,* 44, 733, 1986.

52. **Pegg, A. E.,** Formation and metabolism of alkylated nucleosides: possible role in carcinogenesis by nitroso compounds and alkylating agents, *Adv. Cancer Res.,* 25, 195, 1977.

53. **Leonard, N. J.,** Etheno-substituted nucleotides and coenzymes: fluorescence and biological activity, *CRC Critical Rev. Biochem.,* 15, 125, 1984.

54. **Barrio, J. R., Secrist, J. A., III, and Leonard, N. J.,** Fluorescent adenosine and cytidine derivatives, *Biochem. Biophys. Res. Commun.,* 46, 597, 1972.

55. **Kubota, Y., Motoda, Y., and Nakamura, H.,** Fluorescence studies of the interaction between $1,N^6$-ethenoadenosine monophosphates and nucleotides, *Biophys. Chem.,* 9, 105, 1979.

56. **Baker, B. M., Vanderkooi, J., and Kallenbach, N. R.,** Base stacking in a fluorescent dinucleoside monophosphate: ϵApϵA, *Biopolymers,* 17, 1361, 1978.

57. **Kubota, Y., Motoda, Y., Fujisaki, Y., and Steiner, R. F.,** Fluorescence decay studies of modified dinucleoside monophosphates containing $1,N^6$-ethenoadenosine, *Biophys. Chem.,* 18, 225, 1983.

58. **Lee, C. H. and Tinoco, I., Jr.,** Studies of the conformation of modified dinucleoside phosphates containing $1,N^6$-ethenoadenosine and 2'-*O*-methylcytidine by 360-MHz ^{1}H nuclear magnetic resonance spectroscopy. Investigation of the solution conformations of dinucleoside phosphates, *Biochemistry,* 16, 5403, 1977.

59. **Reich, C. and Tinoco, I., Jr.,** Fluorescence-detected circular dichroism of dinucleoside phosphates. A study of solution conformations and the two-state model, *Biopolymers,* 19, 833, 1980.

60. **Kubota, Y., Sanjoh, A., Fujisaki, Y., and Steiner, R. F.,** Fluorescence decay studies of poly(riboadenylic acid) containing $1,N^6$-ethenoadenosine, *Biophys. Chem.,* 18, 233, 1983.

61. **Kubota, Y., Fujisaki, Y., and Steiner, R. F.,** Fluorescence studies on the complex formation between poly(rA) containing $1,N^6$-ethenoadenosine and poly(rU), *Nucleic Acids Res. Symp. Ser.,* 16, 21, 1985.

62. **Rigler, R. and Wintermeyer, W.,** Dynamics of tRNA, *Annu. Rev. Biophys. Bioeng.,* 12, 475, 1983.
63. **Wells, B. D.,** The conformation of the tRNA^Phe anticodon loop monitored by fluorescence, *Nucleic Acids Res.,* 12, 2157, 1984.
64. **Claesens, F. and Rigler, R.,** Conformational dynamics of the anticodon loop in yeast tRNA^Phe as sensed by the fluorescence wyebutine, *Eur. Biophys. J.,* 13, 331, 1986.
65. **Wells, B. D. and Lakowicz, J. R.,** Intensity and anisotropy decays of the Wye base of yeast tRNA^Phe as measured by frequency-domain fluorometry, *Biophys. Chem.,* 26, 39, 1987.
66. **Quigley, G. J., Teeter, M. M., and Rich, A.,** Structural analysis of spermine and magnesium ion binding to yeast phenylalanine transfer RNA, *Proc. Natl. Acad. Sci. U.S.A.,* 75, 64, 1978.
67. **Lakowicz, J. R.,** *Principles of Fluorescence Spectroscopy,* Plenum Press, New York, 1983, chap. 3.
68. **Lakowicz, J. R. and Maliwal, B. P.,** Construction and performance of a variable-frequency phase-modulation fluorometer, *Biophys. Chem.,* 21, 61, 1985.
69. **Lakowicz, J. R.,** Fluorescence studies of structural fluctuations in macromolecules as observed by fluorescence spectroscopy in the time, lifetime, and frequency domains, *Methods Enzymol.,* 131, 518, 1986.
70. **Peacocke, A. R.,** The interaction of acridines with nucleic acids, in *The Chemistry of Heterocyclic Compounds: Acridines,* Vol. 9, 2nd ed., Acheson, R. M., Ed., Interscience, New York, 1973, chap. XIV.
71. **Lochmann, E. R. and Micheler, A.,** Binding of organic dyes to nucleic acids and the photodynamic effect, in *Physicochemical Properties of Nucleic Acids,* Vol. 1, Duchesne, J., Ed., Academic Press, New York, 1973, chap. 8.
72. **Löber, G.,** The fluorescence of dye-nucleic acid complexes, *J. Lumin.,* 22, 221, 1981.
73. **Steiner, R. F. and Kubota, Y.,** Fluorescent dye-nucleic acid complexes, in *Excited States of Biopolymers,* Steiner, R. F., Plenum Press, New York, 1983, chap. 6.
74. **Lerman, L. S.,** Structural considerations in the interaction of DNA and acridines, *J. Mol. Biol.,* 3, 18, 1961.
75. **Fuller, W. and Waring, M. J.,** A molecular model for the interaction of ethidium bromide with deoxyribonucleic acid, *Ber. Bunsenges. Phys. Chem.,* 68, 805, 1964.
76. **Shapiro, S. L.,** Ultrafast techniques applied to DNA studies, in *Biological Events Probed by Ultrafast Laser Spectroscopy,* Alfano, R. R., Ed., Academic Press, New York, 1982, chap. 16.
77. **Kubota, Y. and Motoda, Y.,** Nanosecond fluorescence decay studies of the deoxyribonucleic acid-9-aminoacridine and deoxyribonucleic acid-9-amino-10-methylacridinium complexes, *Biochemistry,* 19, 4189, 1980.
78. **Kubota, Y., Motoda, Y., Kuromi, Y., and Fujisaki, Y.,** Fluorescence decay studies of the DNA-3,6-diaminoacridine complexes, *Biophys. Chem.,* 19, 25, 1984.
79. **Olmsted, J., III and Kearns, D. R.,** Mechanism of ethidium fluorescence enhancement on binding to nucleic acids, *Biochemistry,* 16, 3647, 1977.
80. **Atherton, S. J. and Beaumont, P. C.,** Ethidium bromide as a fluorescent probe of the accessibility of water to the interior of DNA, *Photobiochem. Photobiophys.,* 8, 103, 1984.
81. **Sommer, J. H., Nordlund, T. M., McGuire, M., and McLendon, G.,** Picosecond time-resolved fluorescence spectra of ethidium bromide: evidence for a nonactivated reaction, *J. Phys. Chem.,* 90, 5173, 1986.
82. **Sobell, H. M.,** Structural and dynamic aspects of drug intercalation into DNA and RNA, in *Nucleic Acid Geometry and Dynamics,* Sarma, R. H., Ed., Pergamon, New York, 1980, 289.
83. **Zimmer, C. and Wähnert, U.,** Non-intercalating DNA-binding ligands: specificity of the interaction and their use as tools in biophysical, biochemical and biological investigations of the genetic material, *Prog. Biophys. Mol. Biol.,* 47, 31, 1986.
84. **Commings, D. E.,** Mechanisms of chromosome banding. VIII. Hoechst 33258-DNA interactions, *Chromosoma,* 52, 229, 1975.
85. **Bontemps, J., Houssier, C., and Fredericq, E.,** Physicochemical study of the complexes of '33258 Hoechst' with DNA and nucleohistone, *Nucleic Acids Res.,* 2, 971, 1975.
86. **Steiner, R. F. and Sternberg, H.,** The interaction of Hoechst 33258 with natural and biosynthetic nucleic acids, *Arch. Biochem. Biophys.,* 197, 580, 1979.
87. **Mikhailov, M. V., Zasedatelov, A. S., Krylov, A. S., and Gurskii, G. V.,** Mechanism of the "recognition" of at pairs in DNA by molecules of the dye Hoechst 33258, *Mol. Biol.,* 15, 541, 1981.
88. **Kubota, Y., Murashige, S., and Steiner, R. F.,** unpublished data, 1987.
89. **Pjura, P. E., Grzeskowiak, K., and Dickerson, R. E.,** Binding of Hoechst 33258 to the minor groove of B-DNA, *J. Mol. Biol.,* 197, 257, 1987.
90. **Pullman, A. and Pullman, B.,** Molecular electrostatic potential of the nucleic acids, *Q. Rev. Biophys.,* 14, 289, 1981.
91. **Kapuscinski, J. and Szer, W.,** Interaction of 4',6-diamidine-2-phenylindole with synthetic polynucleotides, *Nucleic Acids Res.,* 6, 3519, 1979.
92. **Cavatorta, P., Masotti, L., and Szabo, A. G.,** A time-resolved fluorescence study of 4',6'-diamidine-2-phenylindole dihydrochloride binding to polynucleotides, *Biophys. Chem.,* 22, 11, 1985.

93. **Szabo, A. G., Krajcarski, D. T., Cavatorta, P., Masotti, L., and Barcellona, M. L.,** Excited state pK$_a$ behaviour of DAPI. A rationalization of the fluorescence enhancement of DAPI in DAPI-nucleic acid complexes, *Photochem. Photobiol.,* 44, 143, 1986.

94. **Kubista, M., Akerman, B., and Nordén, B.,** Characterization of interaction between DNA and 4′,6-diamino-2-phenylindole by optical spectroscopy, *Biochemistry,* 26, 4545, 1987.

95. **Millar, D. P., Robbins, R. J., and Zewail, A. H.,** Direct observation of the torsional dynamics of DNA and RNA by picosecond spectroscopy, *Proc. Natl. Acad. Sci. U.S.A.,* 77, 5593, 1980.

96. **Millar, D. P., Robbins, R. J., and Zewail, A. H.,** Time-resolved spectroscopy of macromolecules: effect of helical structure on the torsional dynamics of DNA and RNA, *J. Chem. Phys.,* 74, 4200, 1981.

97. **Barkley, M. D. and Zimm, B. H.,** Theory of twisting and bending of chain macromolecules; analysis of the fluorescence depolarization of DNA, *J. Chem. Phys.,* 70, 2991, 1979.

98. **Allison, S. A. and Schurr, J. M.,** Torsion dynamics and depolarization of fluorescence of linear macromolecules. I. Theory and application to DNA, *Chem. Phys.,* 41, 35, 1979.

99. **Schurr, J. M.,** Rotational diffusion of deformable macromolecules with mean local cylindrical symmetry, *Chem. Phys.,* 84, 71, 1984.

100. **Thomas, J. C., Allison, S. A., Appellof, C. J., and Schurr, J. M.,** Torsional dynamics and depolarization of fluorescence of linear macromolecules. II. Fluorescence polarization anisotropy measurements on a clean viral ϕ29 DNA, *Biophys. Chem.,* 12, 177, 1980.

101. **Millar, D. P., Robbins, R. J., and Zewail, A. H.,** Torsion and bending of nucleic acids studied by subnanosecond time-resolved fluorescence depolarization of intercalated dyes, *J. Chem. Phys.,* 76, 2080, 1982.

102. **Magde, D., Zappala, M., Knox, W. H., and Nordlund, T. M.,** Picosecond fluorescence anisotropy decay in the ethidium/DNA complex, *J. Phys. Chem.,* 87, 3286, 1983.

103. **Fujimoto, B. S., Shibata, J. H., Schurr, R. L., and Schurr, J. M.,** Torsional dynamics and rigidity of fractionated poly(dGdC), *Biopolymers,* 24, 1009, 1985.

104. **Shibata, J. H., Fujimoto, B. S. and Schurr, J. M.,** Rotational dynamics of DNA from 10^{-10} to 10^{-5} seconds: comparison of theory with optical experiments, *Biopolymers,* 24, 1909, 1985.

105. **Wu, P., Fujimoto, B. S., and Schurr, J. M.,** Time-resolved fluorescence polarization anisotropy of short restriction fragments: the friction factor for rotation of DNA about its symmetry axis, *Biopolymers,* 26, 1463, 1987.

106. **Ashikawa, I., Kinoshita, K., Jr., Ikegami, A., Nishimura, Y., Tsuboi, M., Watanabe, K., Iso, K., and Nakano, T.,** Internal motion of deoxyribonucleic acid in chromatin. Nanosecond fluorescence studies of intercalated ethidium, *Biochemistry,* 22, 6018, 1983.

107. **Ashikawa, I., Kinoshita, K., Jr., and Ikegami, A.,** Dynamics of Z-form DNA, *Biochim. Biophys. Acta,* 782, 87, 1984.

108. **Langowski, J., Benight, A. S., Fujimoto, B. S., and Schurr, J. M.,** Change of conformation and internal dynamics of supercoiled DNA upon binding of *Escherichia coli* single-strand binding protein, *Biochemistry,* 24, 4022, 1985.

109. **Genest, D., Mirau, P., and Kearns, D. R.,** Investigation of DNA dynamics and drug-DNA interaction by steady state fluorescence anisotropy, *Nucleic Acids Res.,* 13, 2603, 1985.

110. **Härd, T. and Kearns, D. R.,** Anisotropic overall and internal motion of short DNA fragments, *Nucleic Acids Res.,* 14, 3945, 1986.

111. **Härd, T. and Kearns, D. R.,** Association of short DNA fragments: steady state fluorescence polarization study, *Biopolymers,* 25, 1519, 1986.

112. **Härd, T. and Kearns, D. R.,** Anisotropic motions in intercalative DNA-dye complexes, *J. Phys. Chem.,* 90, 3437, 1986.

113. **Härd, T.,** Out-of-plane mobility in the ethidium/DNA complex, *Biopolymers,* 26, 613, 1987.

114. **Fujimoto, B. S. and Schurr, J. M.,** An analysis of steady-state fluorescence polarization anisotropy measurements on dyes intercalated in DNA, *J. Phys. Chem.,* 91, 1947, 1987.

115. **Barkley, M. D., Kowalczyk, A. A., and Brand, L.,** Fluorescence decay studies of anisotropic rotations of small molecules, *J. Chem. Phys.,* 75, 3581, 1981.

116. **Szabo, A.,** Theory of fluorescence depolarization in macromolecules and membranes, *J. Chem. Phys.,* 81, 150, 1984.

117. **Anders, A.,** Energy transfer in nucleic acid-dye complexes, *Opt. Commun.,* 26, 339, 1978.

118. **Anders, A.,** Models of DNA-dye-complexes: energy transfer and molecular structures as revealed by laser excitation, *Appl. Phys.,* 18, 333, 1979.

119. **Andreoni, A., Cubeddu, R., de Silvestri, S., Laporta, P., and Svelto, O.,** Time-resolved two-step selective laser photodamage of dye-biomolecule complexes, *Phys. Rev. Lett.,* 45, 431, 1980.

120. **Atherton, S. J. and Beaumont, P. C.,** Quenching of the fluorescence of DNA-intercalated ethidium bromide by some transition-metal ions, *J. Phys. Chem.,* 90, 2252, 1986.

121. **Atherton, S. J. and Beaumont, P. C.,** Laser flash photolysis studies of DNA-complexed ethidium bromide, *Photochem. Photobiol.,* 44, 103, 1986.

122. **Baguley, B. C. and Le Bret, M.,** Quenching of DNA-ethidium fluorescence by amsacrine and other antitumor agents: a possible electron-transfer effect, *Biochemistry,* 23, 937, 1984.
123. **Atherton, S. J. and Beaumont, P. C.,** Laser flash photolysis of DNA-intercalated ethidium bromide in the presence of methylviologen, *J. Phys. Chem.,* 91, 3993, 1987.
124. **Berkoff, B., Hogan, M., Legrange, J., and Austin, R.,** Dependence of oxygen quenching of intercalated methylene blue triplet lifetime on DNA-base-pair composition, *Biopolymers,* 25, 307, 1986.
125. **Hogan, M., LeGrange, J., and Austin, B.,** Dependence of DNA helix flexibility on base composition, *Nature,* 304, 752, 1983.
126. **Kelly, J. M., van der Putten, W. J. M., and McConnell, D. J.,** Laser flash spectroscopy of methylene blue with nucleic acids, *Photochem. Photobiol.,* 45, 167, 1987.
127. **Milton, J. G. and Galley, W. C.,** Evidence for heterogeneity in DNA-associated solvent mobility from acridine phosphorescence spectra, *Biopolymers,* 25, 1673, 1986.
128. **Corin, A. F. and Jovin, T. M.,** Proflavine binding to poly[d(A-T)] and poly[d(A-br⁵U)]: triplet state and temperature-jump kinetics, *Biochemistry,* 25, 3995, 1986.
129. **Hogan, M., Wang, J., Austin, R. H., Monitto, C. L., and Hershkowitz, S.,** Molecular motion of DNA as a measured by triplet anisotropy decay, *Proc. Natl. Acad. Sci. U.S.A.,* 79, 3518, 1982.
130. **Wang, J., Hogan, M., and Austin, R. H.,** DNA motions in the nucleosome core particle, *Proc. Natl. Acad. Sci. U.S.A.,* 79, 5896, 1982.
131. **Corbin, A. F., Matayoshi, E. D., and Jovin, T. M.,** Triplet-state spectroscopy for investigating diffusion and chemical kinetics, in *Spectroscopy and the Dynamics of Molecular Biological Systems,* Bayley, P. M. and Dale, R. E., Eds., Academic Press, New York, 1985, 53.
132. **Hogan, M. E., Hayes, B., and Wang, N. C.,** Ion-induced DNA structure change in nucleosomes, *Biochemistry,* 25, 5070, 1986.
133. **Hogan, M. E., Rooney, T. F., and Austin, R. H.,** Evidence for kinks in DNA folding in the nucleosome, *Nature,* 328, 554, 1987.
134. **Austin, R. H., Karohl, J., and Jovin, T. H.,** Rotational diffusion of *Escherichia coli* RNA polymerase free and bound to deoxyribonucleic acid in nonspecific complexes, *Biochemistry,* 22, 3082, 1983.
135. **Cantor, C. R. and Schimmel, P. R.,** *Biophysical Chemistry,* Vol. 2, Freeman, San Francisco, 1980, 472.
136. **Thomas, G. A. and Peticolas, W. L.,** Flexibility of nucleic acid conformations. I. Comparison of the intensities of the Raman-active backbone vibrations in double-helical nucleic acids and model double-helical dinucleotide crystals, *J. Am. Chem. Soc.,* 105, 986, 1983.
137. **Martin, J. C. and Wartell, R. M.,** Changes in Raman vibrational bands of calf thymus DNA during the B-to-A transition, *Biopolymers,* 21, 499, 1982.
138. **Wartell, R. M. and Harrell, J. T.,** Characteristics and variations of B-type DNA conformations in solution, *Biochemistry,* 25, 2664, 1986.
139. **Erfurth, S. C., Bond, P. J., and Peticolas, W. L.,** Characteristics of the A-B transition of DNA in fibers and gels by laser Raman spectroscopy, *Biopolymers,* 14, 1245, 1975.
140. **Lu, K. C., Prohofsky, E. W., and Van Zandt, L. L.,** Vibrational modes of A-DNA, B-DNA, and A-RNA backbones: an application of a green-function refinement procedure, *Biopolymers,* 16, 2491, 1977.
141. **Assa-Munt, N. and Kearns, D. R.,** Poly (dA-dT) has a right-handed B conformation in solution: a two-dimensional NMR study, *Biochemistry,* 23, 791, 1984.
142. **Cozzarelli, N. R.,** DNA gyrase and the supercoiling of DNA, *Science,* 207, 953, 1980.
143. **Vasmel, H.,** Influence of supercoiling on DNA structure, *Biopolymers,* 24, 1001, 1985.
144. **Painter, P. C. and Loenig, J. L.,** The solution conformation of poly (LLysine). A Raman and infrared spectroscopy study, *Biopolymers,* 15, 241, 1976.
145. **Hayashi, H., Nishimura, Y., Tsuboi, M., Sekimizu, K., and Natori, S.,** Raman spectrum of a closed-circular DNA, *Biopolymers,* 24, 1107, 1985.
146. **Tsuboi, M., Takahashi, S., and Harada, T.,** in *Physicochemical Properties of Nucleic Acids,* Vol. 2, Duchesne, J., Ed., Academic Press, New York, 1973.
147. **Benevides, J. M., Wang, A. H., van der Marel, G. A., van Boom, J. H., Rich, A., and Thomas, G. J., Jr.,** The Raman spectra of left-handed DNA oligomers incorporating adenine-thymine base pairs, *Nucleic Acids Research,* 12, 5913, 1984.
148. **Prescott, B., Benevides, J. M., Weiss, M. A., and Thomas, G. J., Jr.,** Raman spectrum and secondary structure of the phage operator site O_L1, *Spectrochim. Acta,* 42A, 223, 1986.
149. **McGhee, J. D. and Felsenfeld, G.,** Nucleosome structure, *Annu. Rev. Biochem.,* 49, 1115, 1980.
150. **Savoie, R,. Jutier, J. J., Alex, S., Nadeau, P., and Lewis, P. N.,** Laser Raman spectra of calf thymus chromatin and its constituents, *Biophys. J.,* 47, 451, 1985.
151. **Kubasek, W. L., Wang, Y., Thomas, G. A., Patapoff, T. W., Schoenwalder, K. H., Van der Sande, J. H., and Peticolas, W. L.,** Raman spectra of the model B-DNA oligomer d(CGCGAATTCGCG)₂ and of the DNA in living salmon sperm show that both have very similar B-type conformations, *Biochemistry,* 25, 7440, 1986.

152. **Thomas, G. J., Jr., Prescott, B., and Olins, D. E.,** Secondary structure of histones and DNA in chromatin, *Science,* 197, 385, 1977.

153. **Goodwin, D. C. and Brahms, J.,** Form of DNA and the nature of interaction with protein in chromatin, *Nucleic Acids Res.,* 5, 835, 1978.

154. **Richmond, T. J., Finch, J. T., Rushton, B., Rhodes, D., and Klug, A.,** Structure of the nucleosome core particle at 7 Å resolution, *Nature,* 311, 532, 1984.

155. **Hayashi, H., Nishimura, Y., Katahira, M., and Tsuboi, M.,** The structure of nucleosome core particles as revealed by difference Raman spectroscopy, *Nucleic Acids Res.,* 14, 2583, 1986.

156. **Virudachalam, R., Sitaraman, K., Heuss, K. L., Markley, J. L., and Argos, P.,** Evidence for pH-induced release of RNA from Belladonna Mottle virus and the stabilizing effect of polyamines and cations, *Virology,* 130, 351, 1983.

157. **Virudachalam, R., Stiaraman, K., Heuss, K. L., Markley, J. L., and Argos, P.,** Carbon-13 and proton nuclear magnetic resonance spectroscopy of plant viruses: evidence for protein-nucleic acid interactions in Belladonna Mottle virus and detection of polyamines in Turnip Yellow Mosaic virus, *Virology,* 130, 300, 1983.

158. **Prescott, B., Sitaraman, K., Argos, P., and Thomas, G. J., Jr.,** Protein-RNA interactions in Belladonna Mottle virus investigated by laser Raman spectroscopy, *Biochemistry,* 24, 1226, 1985.

159. **Verduin, B. J. M., Prescott, B., and Thomas, G. J., Jr.,** RNA-protein interactions and secondary structures of cowpea chlorotic Mottle virus for *in vitro* assembly, *Biochemistry,* 23, 4301, 1984.

160. **Kaper, J. M.,** *The Chemical Basis of Virus Structure, Dissociation, and Reassembly,* North-Holland, Amsterdam, 1975.

Chapter 6

PICOSECOND LASER SPECTROSCOPY AND OPTICALLY DETECTED MAGNETIC RESONANCE ON MODEL PHOTOSYNTHETIC SYSTEMS IN BIOPOLYMERS*

G. F. W. Searle, T. J. Schaafsma, and A. van Hoek

TABLE OF CONTENTS

---

* Dedicated to R. Avarmaa (1941-1987).

I. GENERAL INTRODUCTION

There has been a trend the last 10 to 20 years to study biological systems of increasing size and complexity with physical techniques. A first prerequisite for such studies is naturally a well-characterized biological preparation and a second is a good understanding of the technique in order to interpret the results correctly. A combination of biological and physical expertise is not always available, however in the field of photosynthesis, which is characteristically interdisciplinary by nature, there is an exceptionally high number of such research groups.

The primary photophysical processes of plant or bacterial photosynthesis, such as excitation energy transfer and charge separation (trapping), occur on the femtosecond-nanosecond timescale.[1] These processes are of interest from a biophysical and physicochemical point of view and will be discussed by focusing on some representative models of the photosynthetic system.

Although picosecond laser spectroscopy has a relatively short history,[2] it has been applied extensively to photosynthetic systems, and more recently to their model systems, because of the ease of exciting such systems with picosecond light pulses, followed by optical detection with fluorescence or absorption.

Optically detected magnetic resonance (ODMR) can be regarded as a zero magnetic field version of the well known electron spin resonance (ESR) technique.[3]

It is not necessary to perform ODMR experiments on a picosecond timescale, as changes in triplet state populations take place in microseconds or longer.[3] Therefore, this technique cannot be used to study primary photosynthetic processes directly, but it has been shown to give information indirectly on energy transfer between pigments via molecular triplet states.[4]

This review is primarily concerned with lasers and their applications. Lasers are used in order to achieve the time resolution needed to observe photophysical processes, very similar to those occurring in photosynthesis, in real time. For example, the singlet state kinetics of chlorophylls and porphyrins are on a sub-10 ns timescale. Lasers are also used as high-intensity excitation sources in order to generate a significant population of triplets in (bio)polymers, probing the microenvironment of the excited molecule.

The central question, which we address in this review, is how the use of picosecond laser spectroscopic studies combined with the use of ODMR techniques can yield information on the transfer and storage of energy in model photosynthetic systems. These systems can be isolated from the *in vivo* apparatus biochemically and consist of chlorophyll and carotenoid molecules bound noncovalently to protein,[5] or synthesized chemically using porphyrins with various functional groups attached.[6]

The basic system is a collection of specialized pigment molecules

$$^1(\text{antenna})_n^* \rightleftarrows {}^1\text{P*I} \rightleftarrows {}^1[\text{P}^+\text{I}^-] <> {}^3[\text{P}^+\text{I}^-]$$
$$\longrightarrow {}^3\text{PI} \longleftarrow$$

exhibiting singlet excitation energy transfer and energy trapping, for example, by charge separation and recombination in complexes of molecules; one type may act as an electron donor (P), the other as an electron acceptor (I) or intersystem crossing within the donor molecule.

A. ENERGY TRANSFER

How important is the function of the biopolymer in inducing and controlling energy transfer in the antenna pigments through pigment-biopolymer interactions? What are the physical mechanisms involved in exerting this function?

Förster-type resonance energy transfer is inversely proportional to the sixth power of the distance between a pair of molecules,[7] and it is therefore critically important for the overall efficiency of a multistep transfer that the antenna pigments are closely spaced. However, in order to prevent strong interactions between pigments, which can stimulate internal conversion, this spacing must be held at about 1 nm. Förster transfer also depends on the relative orientation of the pigments,[7] but in an unordered system containing many molecules this factor is less critical. In chlorophyll protein complexes, the protein and biopolymer play an important role in the structural organization of the pigments and determine, to a great extent, the efficiency of the Förster transfer.

The three-dimensional organization of the pigments in the biopolymer appears at first sight to be a rather random affair.[8] There is a striking absence of any regularity in the organization, and this noncrystalline structure determines the coherence time for the energy transfer.[9] This time is short, leading to energy transfer with a diffusive and incoherent character.

Another role for the biopolymer is that of inducing transfer through energy (optical spectroscopic shifts). The pigments are bound noncovalently to the biopolymer[5] and it is highly unlikely that the rather weak chlorophyll-protein interactions would lead to a significant distortion of the rigid chlorophyll ring system itself. However, interactions through ligation to the central Mg atom, and through hydrogen-bonding to the carbonyl (aldehyde and ketone) side groups, lead to small shifts in the optical spectra.[10] In this way the protein can influence Förster transfer, because this depends on the overlap integral[7] representing the degree of resonance between donor and acceptor molecules in the energy transfer chain.

The heterogeneity of the pigments introduced by binding to the biopolymer can give rise to an important characteristic of photosynthetic chlorophyll protein complexes, namely that of directionality or channeling of the excitation energy. This will occur if there is a correlation of the structural organization of chlorophylls with their energy. In most mathematical models of energy transfer,[7] this element is ignored, which seriously reduces their relevance to the study of *in vivo* systems. The overall rate for energy transfer from the site of absorption of the photon to the reaction center is considerably increased by this directionality. This would otherwise be a random diffusion of excitation in three dimensions, but it can become a pseudo-unidirectional diffusion in specialized antenna complexes.[11] It is striking that in these complexes a threefold rotational symmetry often exists, but the significance of this function has so far remained a challenging puzzle.[12]

The channeling of excitation energy can be seen as a partial localization. A more complete localization is achieved when the excitation reaches the pigments surrounding the trapping site, which is most pronounced at low temperatures. In this way the efficiency of trapping can be considerably increased as the excitation can visit the trapping site many times before it is transferred away or lost as fluorescence.[7]

Pigment heterogeneity within a single antenna complex molecule coupled to structural organization is thus a useful characteristic of *in vivo* systems, not only for its function, but also for the study of energy transfer on a picosecond timescale.

The range of interactions possible with the functional groups of the protein, both in strength and in nature, also leads to an inhomogeneous broadening of the pigment absorption and fluorescence spectral bands compared to spectra for the monomeric pigments in dilute solution.[13] This allows the overlap integral for Förster transfer for endothermic (reverse or uphill) transfer steps to take nonzero and significant values.

The possibility for reverse transfer can help in optimizing the overall efficiency of energy transfer and trapping because it allows the diffusion of excitation away from a trapping site which is not able to carry out electron transfer (a closed trap), to one which is active (an open trap). Besides these static roles for the biopolymer, there could be more dynamic roles. For example, we might question whether the vibrational levels of the biopolymer couple to the pigment vibrational states during electronic state interconversion. This is in turn dependent upon the relative rates of energy transfer and of relaxation of the higher vibrational states of the biopolymer.[9] In a Förster transfer, which is best described by a hopping of a localized excitation, i.e., transfer between pigment molecules slow compared to vibrational relaxation, there might be little coupling, although even here the timescales for excitation energy transfer and vibrational relaxation are not very different (0.1 to 1.0 ps).

B. ENERGY TRAPPING

Let us now consider energy trapping in the basic system. How important are pigment-biopolymer interactions in optimizing the efficiency of charge separation and inhibiting the recombination of the radical ion pair? In a recent review, Miller[14] has discussed this question along with the factors which control electron transfer. In this review we shall outline briefly the principles involved, insofar as they relate to our discussion of pigment-biopolymer interactions in energy transfer.

As for energy transfer, the biopolymer has a ''static'' role in the control of electron transfer by determining the distance and orientation of the pigments involved and by influencing the redox potential (energy) of the pigments. Further effects can be local electrical fields (induced by point charges of the protein), on the transfer of electrons, or on the (de)stabilization of charge transfer states, also, a quantum mechanical tunneling of electrons through space involving H-bonds or through biopolymer sigma-bonds. Neither of these latter effects have a counterpart in energy transfer.

We should not expect any effects of coupling of vibrational levels of pigment electronic states to biopolymer vibrational states in the system at times >10 ps after electron transfer, as this is an expected upper limit for vibrational relaxation times in these systems. At times <10 ps this coupling can in principle be present. In the photosynthetic electron-transport chain any such effects would only be significant for electron transfer from the primary donor (P) to the primary acceptor (I), and from I to the secondary acceptor (A) because electron donation to P^+ (the oxidized primary donor) occurs on a timescale of 1 ns or longer, whilst electron transfer from A takes about 200 ps.[15] Therefore, during the primary steps the electronic states may not reach vibrational equilibrium and the effects on electron transfer may be significant.

II. INSTRUMENTAL METHODS

Many spectroscopic techniques are applied to the study of energy migration in photosynthetic model systems. This treatment is limited to picosecond time-correlated photon counting and ODMR.

A. PICOSECOND TIME-CORRELATED PHOTON COUNTING

Many molecular decay processes in photopigments can be studied by picosecond fluorescence spectroscopy.[16,17] In order to obtain reliable kinetic data it is important to be aware of several kinds of annihilation processes which may occur during excitation, e.g., singlet-singlet, triplet-triplet, and exitonic annihilation.[18-20] Time-correlated single-photon counting uses pulse energies down to the subnanojoule level. In Figure 1 a block diagram of a typical setup is shown and discussed below.

Excitation from a mode-locked continuous wave laser is used, supplying optical pulses of about 100 ps full width at half maximum (FWHM) in the case of the Ar ion laser and less than 5 ps FWHM in the case of dye lasers.[21] The energy of the pulses is too low to induce singlet-singlet annihilation with an unfocused laser beam. The repetition rate of the pulses of 76 MHz for Ar ion lasers is, however, sufficiently high to initiate triplet-triplet annihilation in, for instance, chlorophyll-protein complexes. For this reason, and also to

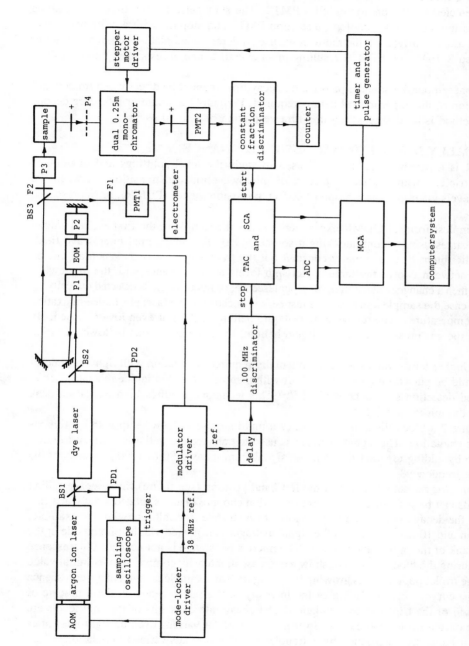

FIGURE 1. Fluorescence decay spectrometer. ADC = analog to digital converter; AOM = acoustooptic mode locker, BS1, BS2, BS3 = beamsplitters; EOM = electrooptic modulator; F1, F2 = neutral density filters; MCA = multichannel analyzer; P1, P2, P3 = optical crystal polarizers; P4 = sheet type polarizer under magic angle; PD1, PD2 = fast photodiodes; PMT1 = side window monitor photomultiplier; PMT2 = microchannel plate photomultiplier R1645UO1; TAC and SCA = time to amplitude converter with amplitude window discriminator.

increase the experimental time window, the repetition rate of optical pulses is reduced using an electrooptic modulator.[22]

The wavelength of the observed fluorescence is selected using optical filters, however, for recording time-resolved fluorescence spectra a monochromator can be used.[23] Fluorescence is detected by the photomultiplier PMT2. The start pulse for the time-to-amplitude converter is initiated with a photon pulse form PMT2 and stopped with a reference signal from the modulator driver. At the moment we use a microchannel plate-type photomultiplier (Hamamatsu R 1645U) for PMT2 resulting in an overall temporal response of about 140 ps FWHM.

With that temporal response, picosecond decay times cannot be directly determined and deconvolution of the experimental data is required. Preferably a mimic or reference deconvolution method is used,[24] resulting in a time resolution of 1 to 10 ps.[25]

B. OPTICALLY DETECTED MAGNETIC RESONANCE

ODMR is a family of techniques[26] used to study the level splittings and kinetics of molecular triplets, using optical detection of microwave-induced transitions between the triplet sublevels via optical absorption (ADMR), phosphorescence (PDMR), or fluorescence (FDMR).

The simplest form of ODMR experiment excites the sample to the first excited singlet and triplet states using a continuous light source and applies an external microwave field, of which the intensity is nominally constant but the frequency is varied. When an energy difference between two of the three triplet spin levels is in resonance with the microwave frequency, then a change in absorption, phosphorescence, or fluorescence is detected optically.

In this case the sample temperature must be low because of the fast spin-lattice relaxation at higher temperatures. The temperature of liquid helium (4.2 K) or even lower is used. By decreasing the gas pressure above the liquid helium, the temperature can be lowered to 1.2 K.

When higher temperatures are experimentally required, excitation with light and microwaves should be pulsed on a timescale short with respect to the spin lattice relaxation time, and a gated detection should be applied.[27-29] This technique requires high excitation peak powers of the microwave field.

In Figure 2 a block diagram is shown of a multipurpose ODMR setup with continuous wave laser excitation. The signal averager is used to enhance the initial low signal-to-noise (S/N) ratio by adding together the sampled signal from many (100 to 10000) sweeps of the microwave frequency.

With the trigger select switch in the left hand position (see figure) fluorescence fading experiments can be performed. The light excitation chopper blade is rotated slowly at a rate limited by the decay rate constants of the lowest triplet state (for chlorophyll *a*, for example, 10 ms open and 10 ms closed) and the signal averager is triggered via the photodiode at the light-on flank of the optical block pulse. The microwave source is not needed. Fluorescence detected during the block pulse will show a decrease in intensity compared to its initial value because the triplet population is growing in. In Figure 3 an example is given of a fluorescence fading (FF) curve, i.e., the fading of the intensity of the fluorescence on the timescale of the ingrowth of the triplet state population. The decay rate constants of the spin levels are dependent on the triplet population. So from FF curves at various exciting light intensities the rate constants are determined by extrapolating to zero relative triplet population.

With the trigger-select switch in the middle position, an FDMR experiment can be carried out to determine the molecular triplet spin level energy splittings. The light is no longer modulated by the chopper, but now the microwave field is applied and its frequency is repeatedly varied (swept). The signal averager is triggered each time at the beginning of the frequency sweep. A photomultiplier detects the fluorescence change when a certain

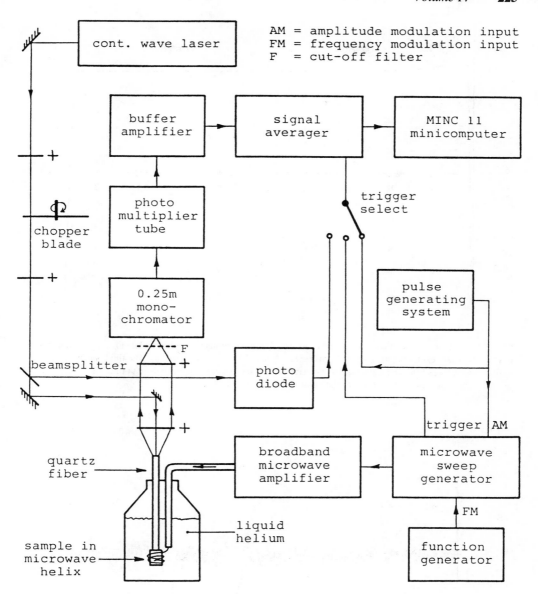

FIGURE 2.   Multipurpose FDMR, FF, and MIRF apparatus.

frequency out of the sweep ''fits'' between two of the three triplet spin levels. These changes in the fluorescence intensity are repeatedly registered by the averager. In Figure 4 an example is presented of an FDMR spectrum.

When the trigger-select switch is in the right hand position, so-called microwave induced response of fluorescence (MIRF) experiments can be performed (Figure 5). In this case the light excitation is also unmodulated, the microwave source is fixed at the resonance frequency for coupling of two of the three triplet spin levels, and the microwave intensity is gated by the pulse-generating system. The microwave source is slightly modulated in frequency by the function generator controlling the FM input of the sweep generator to avoid too selective an excitation. Here again the very fast process of fluorescence is used as a monitor signal; in this case, to study the kinetic response of the molecular triplet sublevel populations upon short circuiting the levels by resonant external microwave radiation. This time-dependent

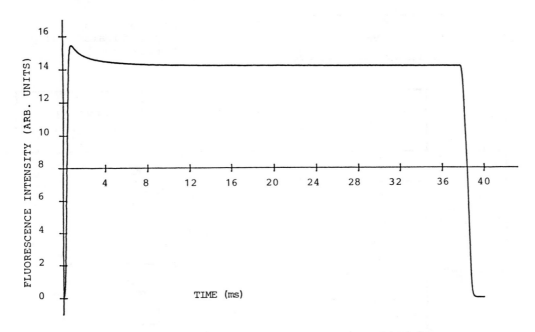

FIGURE 3.   Fluorescence Fading curve of a natural chlorophyll-protein complex, isolated from plant photosystem I, at 4.2 K. Laser excitation at 458 nm, 40 mW. Detection at 718 nm, 10000 sweeps. The final percentage decrease in fluorescence was 9%. The decrease was described by a sum of three exponentials with rate constants (s^{-1}) 2133 ± 300, 522 ± 17, and 93 ± 6, and relative amplitudes 0.11, 0.71, and 0.18.

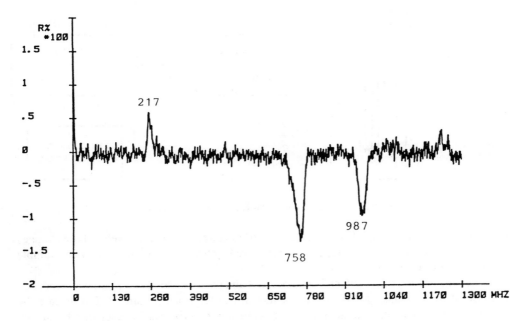

FIGURE 4.   FDMR spectrum of a natural light-harvesting chlorophyll-protein complex, isolated from plants, at 4.2 K. Laser excitation at 476.5 nm, 80 mW. Detection at 683 nm, bandwidth 5 nm.

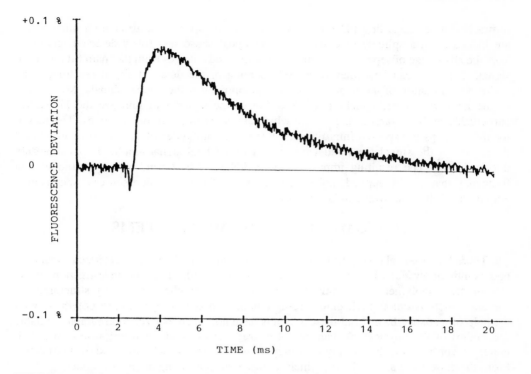

FIGURE 5. MIRF curve of a natural chlorophyll-protein complex, isolated from plant photosystem I, at 4.2 K. Microwave pulse duration 200 μs, microwave frequency in resonance with the D-E transition, and microwave power nonsaturating.

FDMR technique provides in principle the same information on triplet state decay rate constants as the FF method and both methods have their benefits and disadvantages (see Reference 30).

ODMR is experimentally, thus far, a rather simple technique. The most critical part of the setup is possibly the microwave antenna. A microwave helix is used for recording ODMR spectra because of its broad band characteristic.[31] However, a slotted tube or split ring resonator can be used when the microwave frequency is kept constant, for example, when optical spectra are recorded during microwave excitation or in the case of kinetic experiments.[32-35] The latter type of antenna is essential when a polarized H_1 field is required.[36] These tuned resonators also have the added advantage of a more efficient H_1 field generation.

The most direct way of measuring parameters of the molecular triplet state is to observe phosphorescence (or triplet-triplet absorbance). If phosphorescence cannot be detected, only absorbance and fluorescence techniques are left. In practice these are complementary techniques, either the one or the other being the most suitable for a particular system. Observing microwave-induced changes in the fluorescence intensity is an indirect detection method but has the advantage of high sensitivity resulting in a detection limit of as low as 10^{10} to 10^{11} molecules under favorable, but not exceptional, conditions.

Noise in ODMR experiments originates primarily from the light source if a tungsten lamp or stabilized laser is not used. The large dimensions of the ODMR setup can also lead to noise due to mechanical vibrations of the optical components, and this is mainly in the low (0.01 to 100 Hz) frequency range similar to the noise introduced by the use of arc lamps. In general, no improvement in the S/N ratio should be expected from modulating the excitation, coupled to phase sensitive detection, for those situations where the detected signal contains noise originating predominantly from one of the excitation sources. However, in practice when recording FDMR, PDMR, or ADMR spectra the irradiating microwaves are indeed often amplitude modulated and the signal phase sensitively detected, giving at least the advantage of separating the microwave-induced changes from the main fluorescence signal. This also can be achieved by AC coupling of the detected signal resulting in an easier determination of the relative deviation amplitudes of the optical signals. Modulation of the microwaves and synchronous detection of the signal will improve the S/N ratio, particularly in those parts of the spectra which are between the resonance lines. The reason for this perhaps unexpected improvement is that the amplitudes of the microwave-induced deviations of the signal are small with respect to the total signal amplitude (usually only 0.1 to 1.0%). The resulting improvement in S/N ratio is dependent on the modulation frequency that can be applied, noise characteristics of the excitation sources, and relative amplitude of the microwave-induced changes.

III. PIGMENT-BIOPOLYMER SYSTEMS

There has been relatively little success in the synthesis of well-characterized models of photosynthetic systems based on biopolymers which would allow an investigation of the role of the biopolymer. There have been reports of the synthesis of poorly structured absorbates, such as chlorophyll onto serum albumin,[37] and chlorophyll onto polythene with the help of a coabsorbate,[38] but these attempts have not given experimental samples suitable for studies of the physical mechanisms of energy transfer and electron transfer. We shall therefore confine ourselves to pigment-biopolymer systems which have a known structure. From these we have selected two, which represent an improving degree of modeling of the *in vivo* photosynthetic system in two stages: (1) model pigment — model biopolymer and (2) model pigment — natural biopolymer.

A. Model Pigment — Model Biopolymer

The group of Vacek has developed a system consisting of pheophorbide *a* bound covalently, via the C-17 carboxyl group, to some of the epsilon amino groups of the lysine in a poly-tripeptide(L-lys-L-ala-L-ala), see Figure 6.

Pheophorbide is a chemically stable derivative of the natural pigment chlorophyll. The central Mg atom is replaced by protons and the long hydrophobic phytyl alcohol side chain is removed, giving a model pigment which has spectroscopic properties very similar to those of chlorophyll itself. There is, however, no possibility for ligation, so that variation of the energy through pigment-biopolymer interaction is confined to that via the ring V carbonyl group, and energy level shifts due to pigment-pigment interaction will probably dominate. The phytyl chain, which *in vivo* helps to anchor the chlorophyll molecule to protein, has its role taken over by the covalent amide bond.

The polytripeptide has been chosen to contain a hydrophilic, charged amino acid, and a hydrophobic amino acid in approximately the same ratio as found for the natural chlorophyll-binding proteins.[39] The lysine also provides the pigment anchoring site. The modeling is, of course, not perfect; the main deficiency being that *in vivo* the hydrophobic and hydrophilic regions of the protein are 20 to 30 amino acids long — the hydrophobic alpha-helical stretches are known to span the lipid membrane.[11] The polytripeptide can be brought into a folded configuration depending on the solvent so that the pheophorbide molecules can interact with each other, resulting in dimer-like low-energy trapping sites serving as models for reaction centers.

The linking of the pheophorbide to the polytripeptide introduces an unfortunate uncertainty into the model because complete substitution of the lysines is not possible and the distribution of the pheophorbides along the polypeptide is not known.

In 1981 Fidler et al.[40] reported that the pheophorbide *a* polytripeptide showed a two-

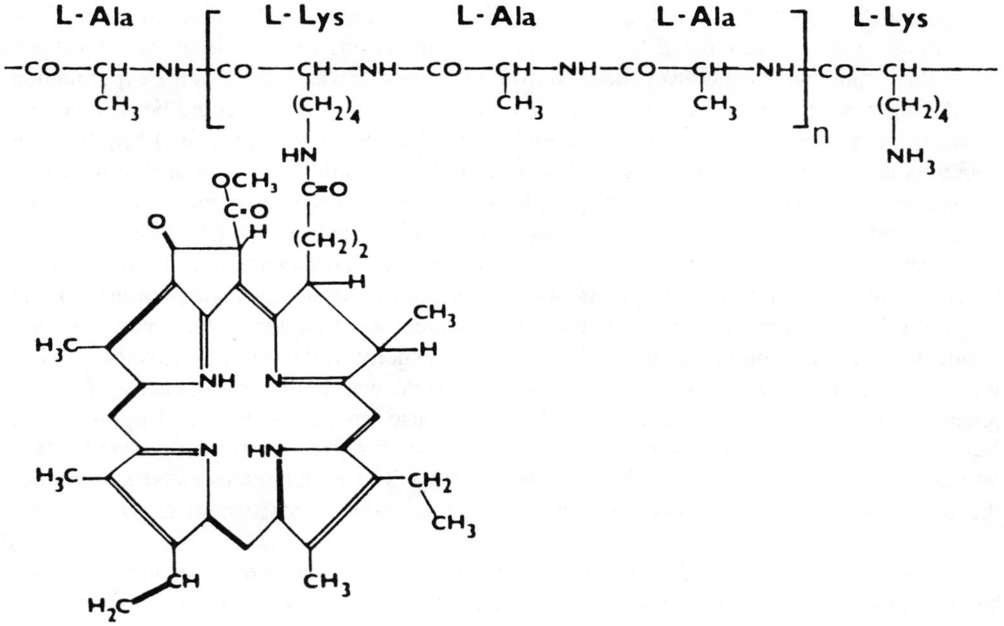

FIGURE 6. Structure of the photosynthetic model system consisting of pheophorbide *a* bound covalently to a synthetic polytripeptide; average molecular weight is 10000. Solvent: dimethylformamide. (From Hala, J., Searle, G. F. W., Schaafsma, T. J., van Hoek, A., Pancoska, P., Blaha, K., and Vacek, K., *Photochem. Photobiol.*, 44, 527, 1986. With permission.)

component fluorescence decay kinetics, using picosecond laser excitation and single photon counting detection. The longer component of 3.2 ns could be attributed to monomer pigment, while the shorter component was supposed to arise from interactions of two or more pigments which were brought into close proximity with a plane to plane distance of less than 0.2 nm.

In 1985 the Prague group reported further studies on this interesting system[41] using excitation from a dye laser, which was pumped by a nitrogen laser, in order to excite with a narrow bandwidth. They could measure fine structure in the fluorescence emission spectrum of pheophorbide a polytripeptide at 10 K in solution in dimethylformamide, and the frequencies of normal vibrations in the first excited singlet state could be compared to those for isolated (unbound) pheophorbide a. The fluorescence spectra displayed the sharp vibronic bands at different wave numbers for different dye laser excitation wavelengths, superimposed on a broad band background. It was concluded that the covalent binding of the pheophorbide-a molecule produced only slight changes in the frequencies of normal vibrations. However, significant broadening and shifts to lower energies in the site distribution function were seen, which was due to the pigment-polypeptide interaction.

In 1986 the first FDMR experiments on this system were reported jointly by the Prague group and ourselves.[42] The FDMR spectra of isolated (free) and polypeptide-bound pheophorbide a in solution in dimethylformamide were detected in the emission peak at 667 to 669 nm. There was a difference not only in microwave frequency as might be expected on interaction of the ring V carbonyl of pheophorbide a with the polypeptide — the value of the zero field splitting parameter D increased from 0.0337 in the isolated pigment to 0.0359 cm^{-1} in the bound pigment — but also in the amplitude of the transitions. The D − E transition which was negative for free pigment became positive for the bound form. The inversion of the sign was not so clear for the D + E transition, probably due to additional effects due to hydrogen bonding to the carbonyl, which are known from studies with pheophytin[43,62] to affect the amplitude of the D + E transition. The signs of the FDMR transitions were also calculated from the relative populations and the decay rate constants of the three spin levels which were derived from FF curves. It was found that the experimental FDMR signs were opposite to those calculated from FF. This conflict could be resolved by evoking the mechanism of inversion by energy transfer, which was presented briefly in an appendix and is discussed in more detail in Section IV. It should be noted that sign inversion has also been noted in several pigment-protein complexes isolated from *in vivo* photosynthetic systems and appears to be a general phenomenon in plants and bacteria.[44]

In the same paper, further results on the picosecond fluorescence decay kinetics were reported using excitation from a synchronously pumped dye laser and single photon counting detection. If the fluorescence was monitored at different wavelengths, using a narrow band-width, then it was found possible to detect a heterogeneity in the emitting species. At 730 nm, a 50 ps component having 66% of the amplitude was found in the decay of bound pheophorbide *a*, but this was absent at 680 nm, and also absent at both wavelengths for the free pigment. This component was attributed to interacting pheophorbide-*a* molecules (aggregates) having an emission shifted to lower energy. These aggregates acted as traps for the excitation energy, probably through an increased rate of intersystem crossing to the triplet state.

It was proposed that the triplet state detected by FDMR in the main monomer emission band of bound pheophorbide *a* was that of the aggregate (trap) with the spectrum inverted by energy transfer.

These studies with a model pigment — model biopolymer system demonstrate the importance of the biopolymer in directing the interactions of the pigments and allowing transfer of excitation energy between them. In this model system, the folding of the polypeptide in the polar solvent dimethylformamide results in the formation of a number of pigment aggregates, acting as traps for the excitation energy.

B. MODEL PIGMENT — NATURAL BIOPOLYMER

The group of Boxer has been primarily responsible for the development of a system in which pyrochlorophyllides are bound noncovalently to myo- and hemoglobins. These globular proteins normally bind heme, an Fe-containing porphyrin molecule, with a precisely known stoichiometry; one per protein in myoglobin and four per protein in hemoglobin. The structures of the crystallized heme proteins are known from X-ray diffraction studies.[45]

The heme moieties can be removed from the protein, giving the so-called apoprotein, to quantitatively add back pyrochlorophyllide.[46] The determination of the crystal structure of the resulting complexes is still in progress, but it is a reasonable assumption that the structure is relatively unaltered as the ring systems of heme and pyrochlorophyllide are quite analogous. Therefore these model complexes have a known structure, with regard to both the pigment and the biopolymer. The pyrochlorophyllides used are stable Mg or Zn derivatives, differing from chlorophyll only in the absence of the phytyl chain and in other minor ways.

In 1979 the group of Boxer introduced the chlorophyllide *a* myoglobin system,[47] and in 1981 reported briefly on the fluorescence decay kinetics after synchrotron excitation and single photon counting detection. In addition, high field ESR measurements of a series of chlorophyllide-apomyglobin complexes in solution was reported.[48] There was apparently very little effect of the binding to the apomyoglobin on the fluorescence lifetime (single component) of the chlorophyllide derivatives tested, in most cases a slight, about 20%, lengthening was seen. The ESR-derived zero field splitting (ZFS) parameters of the bound pigments also showed little change from the values for the free pigments in organic glasses. In 1982 the same group described the absorption spectroscopic properties of pyrochlorophyllide *a*-substituted apomyoglobin in crystalline form.[45]

In 1982 the first ODMR studies of this system were carried out by the same group in cooperation with Clarke — the FDMR spectrum of the complex at 2 K was compared to that for the free pigment in an organic matrix.[49] The fluorescence emission spectrum at 2 K showed two peaks at 671 and 685 nm instead of the single broad peak at 670 to 680 nm of the free pigment, and this was interpreted as arising from two sites for the pigment. Independent evidence for two sites had previously been obtained from proton nuclear magnetic resonance (NMR) studies of these systems in solution,[48,50] and these were ascribed to different configurations of the pigment-binding site of the protein induced by the presence of the pigment. The triplet states detected by FDMR in the two resolved fluorescence bands could be distinguished by their ZFS parameters (D = 0.0297, E = 0.0037 cm^{-1} and D = 0.0284, E = 0.0033 cm^{-1}, respectively) and a small difference in their decay rates. Both triplets were distinct from that of free pyrochlorophyllide *a*.

The authors came to the conclusion that the protein (biopolymer) was capable of dynamical interconversion between two conformations at room temperature, as suggested by the earlier NMR work, and that at the cryogenic temperatures of the FDMR measurements the protein complex is frozen into a statistical distribution of both conformations. The slightly different electrostatic protein environment about the pyrochlorophyllide ring results in two triplet states with different ZFS parameters. There is no difference however in the binding (ligation) of the pigment to the biopolymer in the two conformations.

This result illustrates the usefulness of ODMR in studying pigment-biopolymer interactions by using the triplet state of the incorporated pigment as an internal probe, but also introduces an element of uncertainty in the supposition that the structure of the complex is identical to that already determined for myoglobin.

In 1983 the group of Boxer extended the system to contain more than one pigment molecule per protein molecule in order to study the role of pigment-biopolymer interactions in transfer of excitation energy between the pigments.[46] They used the tetrameric apo-hemoglobin and complexed it with zinc pyrochlorophyllide *a*. Hemoglobin hybrids in which the heme was only partially substituted by chlorophyllide could also be synthesized. Evidence for energy transfer was obtained from steady state fluorescence anisotropy. In 1984 the picosecond fluorescence kinetics were reported[51] and energy transfer could be confirmed to occur between the two chlorophyllides in a hybrid chlorophyllide/heme apohemoglobin by following the decrease in the fluorescence anisotropy on a nanosecond timescale; the bimolecular rate constant was calculated to be 4 to 7 × 10^8 s$_{-1}$. The fluorescence lifetime is not affected by energy transfer between chlorophyllides, and 3.73 ns was measured in fully chlorophyllide-substituted apohemoglobin compared to 3.9 ns in the chlorophyllide-apomyoglobin complex. However in chlorophyllide/heme hybrids, the lifetime was found to be shortened by exothermic energy transfer from chlorophyllide to heme (in the deoxy or cyanomet form). The authors came to the conclusion that the observed migration of excitation energy could be completely and quantitatively explained by the use of Förster's resonant excitation transfer theory so that this system should represent a suitable model to test out the effects of pigment-biopolymer interactions on energy transfer.

In a review in 1985,[52] Boxer discussed the results on these myo- and hemoglobin systems and suggested future experiments in which molecular biological techniques would be used to systematically modify amino acid residues in the vicinity of the heme pocket. He reported that very large (gram) quantities of synthetic myoglobin had already been produced using such techniques.

IV. THEORY

A. FLUORESCENCE-DETECTED MAGNETIC RESONANCE
1. Principles

Triplet state FDMR spectroscopy makes use of the fact that the threefold degeneracy of

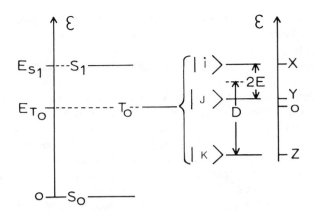

FIGURE 7.   Optical pumping cycle involving the ground state (S_0), the
first excited singlet state (S_1), and the lowest triplet state (T_0). The energies
(ϵ) of the nondegenerate spin sublevels of T_0 (X, Y, and Z) are related to
the zero field splitting parameters D and E as indicated. The zero field
spin levels are labelled $|i>$, $|j>$, and $|k>$.

the molecular triplet state is lifted by spin-spin interaction, even in the absence of an external
magnetic field. For photosynthetic pigments, the resulting zero-field splittings (ZFSs) are
~10^{-2} cm $^{-1}$. The energies of these nondegenerate triplet spin levels are denoted X, Y,
and Z and are in the order X > Y > Z, representing eigenvalues of the spin-Hamiltonian[53]
of the triplet state, see Figure 7.

Thus, the distances between the spin levels (ZFSs) are given by X-Y, Y-Z, and X-Z in
energy units. Conventionally, these splittings are expressed in parameters D and E, related
to X, Y, and Z by[53]

$$D = -\frac{3}{2}Z \tag{1}$$

$$E = -\frac{1}{2}(X - Y) \tag{2}$$

The creation of a molecule in its lowest triplet state T_0 is usually obtained after optical
excitation into an excited singlet state S_n, followed by intersystem crossing $S_n \rightarrow T_0$. Resonance transitions between the spinlevels of T_0 (X-Y, Y-Z, and X-Z) can then in principle
be detected as a change in the intensity of the phosphorescence of T_0, due to microwave
absorption at any of the three resonance frequencies. At low temperatures (<4 K) relaxation
between T_0 spin levels can be neglected, which is a necessary condition for observing FDMR
spectra.

For photosynthetic pigments, phosphorescence is too weak[54-57] to be suitable for observing zero field magnetic resonance. However, the fluorescence intensity arising from S_1
$\rightarrow S_0$ emission can be expressed as[58]

$$I_f = A\left[N - \sum_{i=1}^{3} n_i\right] \tag{3}$$

where A is an instrumental constant, N is the total number of pigment molecules, and
$\sum_i n_i$ represents the number of molecules in the spin levels of the lowest triplet state (T_0).

It is assumed that the number of molecules in excited singlet states can be neglected w.r.t.
that in S_0 and T_0 under steady-state illumination of the sample.

Absorption of microwaves between two spin levels $|i>$ and $|j>$ gives rise to a change of fluorescence ΔI_f given by[59-62]

$$\Delta I_f = A(k_i - k_j)(k_i + k_j)^{-1}(n_i^0 - n_j^0) \tag{4}$$

where $i \neq j = 1,2,3$ and $n_{i,j}^0$ represents the steady-state population of the T_0 spin levels $|i>$ or $|j>$ under continuous illumination and in the absence of microwaves; $k_{i,j}$ are the decay rate constants of these spin levels. This equation shows that resonances may be observed only when $n_i^0 \neq n_j^0$ and $k_i \neq k_j$.

Using microwave pulses saturating the $|i> \leftrightarrow |j>$ transition, the resulting time dependence of the fluorescence contains the triplet state kinetics. Following the onset of the microwave pulse, the change of fluorescence is given by[61]

$$\Delta I_f(t) = A(k_i - k_j)(k_i + k_j)^{-1}(n_i^0 - n_j^0)\left\{\exp\left[-\frac{1}{2}(k_i + k_j)t\right] - 1\right\} \tag{5}$$

whereas

$$\Delta I_f(t) = Ak_ik_j(k_i + k_j)^{-1}(n_i^0 - n_j^0)\{k_i^{-1}\exp(-k_it) - k_j^{-1}\exp(-k_jt)\} \tag{6}$$

after switching off the microwaves.

The time dependence of the fluorescence $\Delta I_f(t)$ can be analyzed, yielding the decay rate constants k_λ and steady-state populations n_λ^0 of the triplet state, provided that the duration of the microwave pulse is much shorter than the smallest value of the lifetime of spin level λ. For a distribution of spin level decay rate constants, as may occur for a pigment in different microenvironments of a (bio)polymer, there is no easy check whether this condition is fulfilled. For a number of sites, the microwave pulsewidth may in fact be long, compared to the lifetime of one or more of the triplet spin levels. Under these conditions, the microwave pulse acts as a microwave step. Then, in order to obtain reliable kinetic data, the microwave step must be fully saturating the zero field transition between a pair of triplet spin levels.[63]

2. FDMR in the Presence of Energy Transfer

Figure 8 represents the lowest energy levels S_0, S_1, and T_0 of a molecule (t), acting as a trap for singlet excitation energy, transferred to this trap from a collection of excited molecules, denoted as antenna (a). Energies of the S_1^a and S_1^t, or T_0^a and T_0^t states as well as intermolecular distances are assumed to allow energy transfer to occur. We assume both the number of antenna molecules, $N_a >> N_t$, the number of trap molecules, both per unit of volume. Therefore, only the antenna is considered to be excited. Neglecting diffusional processes, singlet energy transfer can be described by the bimolecular reaction:

$$S_1^a + S_0^t \underset{k_1}{\overset{k_2}{\rightleftharpoons}} S_0^a + S_1^t \tag{7}$$

followed by

$$S_1^a \xrightarrow[k_{ST}^a]{} T_0^a \tag{8}$$

and/or

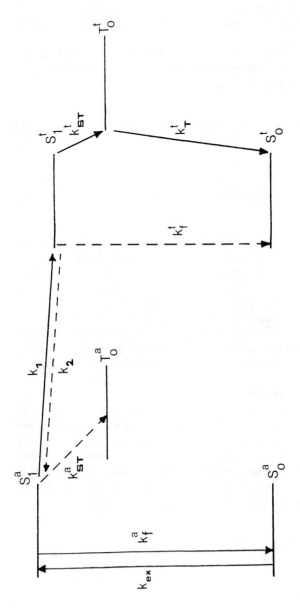

FIGURE 8.   Kinetic scheme for singlet energy transfer from (a) antenna to (t) trap, and for intersystem crossing $S_1^t \rightarrow T_0$, in covalently bound pheophorbide-*a*. Kinetic constants are defined in the text.

$$S_1^t \xrightarrow[k_{ST}^t]{} T_0^t \tag{9}$$

Furthermore

$$S_1^a + S_0^a + T_0^a \equiv N_a \tag{10}$$

and

$$S_1^t + S_0^t + T_0^t \equiv N_t \tag{11}$$

where S_0, S_1, and T_0 denote the number of antenna and trap molecules per unit volume in the various electronic states.

Combining the rate equations

$$\dot{S}_1^a = S_0^a(k_{ex} + k_2 S_1^t) - S_1^a(k_f^a + k_{ST}^a + k_1 S_0^t) \tag{12}$$

$$\dot{S}_1^t = k_1 S_0^t S_1^a - S_1^t(k_f^t + k_{ST}^t + k_2 S_0^a) \tag{13}$$

$$\dot{S}_0^a = k_1 S_1^a S_0^t - k_2 S_0^a S_1^t + k_T^a T_0^a \tag{14}$$

$$\dot{S}_0^t = -k_1 S_1^a S_0^t + k_2 S_0^a S_1^t + k_T^t T_0^t \tag{15}$$

with Equations 10 and 11 and neglecting S_1^a and S_1^t w.r.t. S_0^a and S_0^t, respectively, yields relations between S_1^a and S_1^t on one hand and T_0^a and T_0^t on the other hand, under steady-state conditions, i.e., under continuous illumination.

Saturating any of the three transitions between the spin levels $| i>$, $| j>$ and $| k>$ of T_0^a or T_0^t results in a change of the total population of the triplet levels ΔT_0^a or ΔT_0^t.

By straightforward but tedious algebra, one finds for the resulting change of the steady-state fluorescence of antenna and trap, respectively:[42]

$$\Delta I_f^{a,t} = k_f^a \Delta S_1^a = k_{ex}\phi_f^a\phi_E(1 - \phi_E)(1 - \phi_{E'})(1 - \phi_E\phi_{E'})^{-2}\left(\frac{N_a}{N_t}\right) \cdot \Delta T_0^t \tag{16}$$

$$\Delta I_f^{a,a} = k_f^a \Delta S_1^a = -k_{ex}\phi_f^a(1 - \phi_E)(1 - \phi_E\phi_{E'}^2)(1 - \phi_E\phi_{E'})^{-2}\Delta T_0^a \tag{17}$$

$$\Delta I_f^{t,a} = k_f^t \Delta S_1^t = -k_{ex}\phi_f^t\phi_E(1 - \phi_{E'})^2(1 - \phi_E\phi_{E'})^{-2}\Delta T_0^a \tag{18}$$

$$\Delta I_f^{t,t} = k_f^t \Delta S_1^t = -k_{ex}\phi_f^t\phi_E(1 - \phi_E)(1 - \phi_{E'})(1 - \phi_E\phi_{E'})^{-2}\left(\frac{N_a}{N_t}\right)\Delta T_0^t \tag{19}$$

with

$$\phi_f^a \equiv \frac{k_f^a}{k_f^a + k_{ST}^a} \tag{20}$$

$$\phi_f^t \equiv \frac{k_f^t}{k_f^t + k_{ST}^t} \tag{21}$$

$$\phi_E \equiv \frac{k_1 N_t}{k_f^a + k_{ST}^a + k_1 N_t} \tag{22}$$

$$\phi_{E'} \equiv \frac{k_2 N_a}{k_f^t + k_{ST}^t + k_2 N_a} \tag{23}$$

ϕ_f^a and ϕ_f^t represent the fluorescence yields of antenna and trap molecules, respectively, in the absence of energy transfer, whereas ϕ_E and $\phi_{E'}$ denote the yields for forward and backward singlet energy transfer.

The equations (16 to 19) represent the four possible combinations of detecting a change in fluorescence intensity of either antenna or trap, resulting from a change in triplet population in either of these.

Several conclusions can be drawn from Equations 16 to 19:

1.   If $\phi_{E'} \cong 1$ (complete back transfer), no triplets can be formed in the trap molecules (Equations 16, 18, and 19) and only $\Delta I_f^{a,a}/\Delta T_0^a \neq 0$, so that only antenna triplets can be detected when monitoring the antenna fluorescence. On the other hand, if $\phi_E \cong 1$, only $\Delta I_f^{t,a}/\Delta_0^a \neq 0$. However, no antenna triplets are formed (Equation 23) and there is no FDMR transition observed at all.

2.   In the absence of energy transfer, i.e., if $\phi_E = 0$, only $\Delta I_f^a/\Delta T_0^a \neq 0$, implying that FDMR spectra only contain transitions due to triplets in the antenna. These transitions have opposite sign w.r.t. those due to triplets of trap molecules observed via antenna fluorescence, i.e., in the presence of singlet energy transfer.

3.   For intermediate values of ϕ_E and neglecting back transfer ($\phi_{E'} = 0$) triplets of antenna — as well as trapmolecules — can in principle be detected by monitoring the fluorescence of both types of molecules. Then the Equations 16 to 19 simplify to:

$$\Delta I_f^{a,t} = k_{ex}\phi_f^a\phi_E(1 - \phi_E)\left(\frac{N_a}{N_t}\right)\Delta T_0^t \tag{24}$$

$$\Delta I_f^{a,a} = -k_{ex}\phi_f^a(1 - \phi_E)\Delta T_0^a \tag{25}$$

$$\Delta I_f^{t,a} = -k_{ex}\phi_f^t\phi_E\Delta T_0^a \tag{26}$$

$$\Delta I_f^{t,t} = -k_{ex}\phi_f^t\phi_E(1 - \phi_E)\left(\frac{N_a}{N_t}\right)\Delta T_0^t \tag{27}$$

4.   Equations 24 and 27 contain the ratio N_a/N_t, which may be a large number, thus amplifying the fluorescence change, resulting from a change in triplet population. In fact, both expressions have a different sign and differ by the ratio ϕ_f^a/ϕ_f^t. Since the trap is at lower energy than the antenna, radiationless processes are expected to be more efficient for the trap, resulting in $\phi_f^t < \phi_f^a$. Often, there are secondary processes in the trap, e.g., charge-separation, if the trap is a photochemically active photosynthetic reaction center, which results in $\phi_f^t << \phi_f^a$ and no trap fluorescence is observable.

Equations 16 to 19 and 23 to 27 can be applied to both model and *in vivo* systems.

For photosynthetic pigments embedded in a protein (constituting a photochemically active reaction center surrounded by antenna) $\phi_{E'} = 0$ and $\phi_f^t \approx 0$, and triplets formed in the reaction center (see Reference 3) can only be observed via antenna-fluorescence (Equation 16). If part of the antenna-protein is denatured during preparation, the embedded antenna-pigments cannot transfer their singlet energy to the trap and may form antenna triplets. Usually, denaturing the protein causes a blue-shift of the fluorescence emission of the embedded pigments. This blue-shifted fluorescence then can be used to obtain the FDMR spectrum of the antenna triplets (Equation 25), whereas the native fluorescence emission gives rise to the FDMR spectrum of the triplet state of the reaction center. Both spectra are expected to be very similar, but of opposite sign,[42,44,64] as is evident by comparing Equations 24 and 25.

For model systems, modification of the (bio)polymer structure, leading to unfolding of the polymer-chain, is predicted to result in $\phi_E = 0$, thus eliminating inverted FDMR transitions due to trap molecules in the triplet state.

The triplet state of the pigment can therefore be used as a probe, reflecting the ordering of the (bio)polymer, provided the distribution of intermolecular distances between the pigment molecules allows singlet energy transfer to occur. The spatial resolution is expected to be in the range of tens of nanometers, the distance over which Förster singlet energy transfer can occur.

The results using Equations 24 to 27 can also be rationalized on a more simple basis. Inspecting Figure 8 again, a positive change ΔI_f^a, resulting from a positive change ΔT_0^t can be understood as follows: increasing T_0^t leads to a decrease of S_0^t and less deactivation of S_1^a via singlet energy transfer, and fluorescence from S_1^a therefore increases. On the other hand, for isolated pigment molecules, i.e., in the absence of energy transfer, a positive change ΔT_0^a results in a decrease of S_0^a, and thus in a decrease of the steady-state fluorescence from S_1^a.

By a similar reasoning it is straightforward to see that for triplet-triplet ($T_0^a \rightarrow T_0^t$) energy transfer, no sign inversion of the FDMR transitions of the triplet state of trap molecules is observed. For T-T energy transfer, a positive change ΔT_0^t again results in a decrease of S_0^t and an increase of T_0^a under steady-state excitation of the antenna part, and this leads to a decrease of S_1^a. Similarly, $\Delta I_f^a/\Delta T_0^a$, $\Delta I_f^t/\Delta T_0^a$, and $\Delta I_f^t/\Delta T_0^t$ are all <0.

B. FLUORESCENCE FADING

Contrary to what its name suggests, FF involves only fully reversible photophysical processes and does not include irreversible photochemical changes of the investigated compound.[60,65,66] The method employs the response of the fluorescence to an optical excitation step from the singlet-ground state S_0 to an excited singlet state S_n. Assuming an instantaneous response of the S_1 population to this excitation step, the subsequent change of the fluorescence intensity $I_f(t)$ monitors the time-dependent change of the S_1 population due to transfer of molecules to the lowest triplet state T_0 by intersystem crossing. $I_f(t)$ contains the kinetics of the T_0 state. For an isolated pigment molecule, embedded in a (bio)polymer we consider the states S_0, S_1, and T_0, forming an optical pumping cycle.

The rate equations for the S_0 and S_1 states are

$$\dot{S}_0 = k_{ex}S_0 + k_fS_1 + k_TT_0 \qquad (28)$$

$$\dot{S}_1 = k_{ex}S_0 - (k_f + k_{ISC})S_1 \qquad (29)$$

The kinetic constants are defined in Figure 9A. The total number of molecules (N) per unit of volume is given by

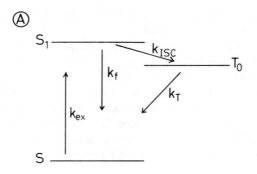

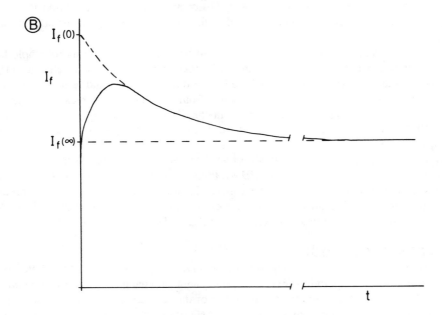

FIGURE 9. (A) Kinetic scheme underlying an FF experiment. The rate constants are optical excitation $S_0 \rightarrow S_1$, k_{ex}; radiative decay $S_1 \rightarrow S_0$, k_f; intersystem crossing, k_{isc}; and decay (radiative + nonradiative) of T_0, k_T. (B) Definition of $I_f(0)$, $I_f(\infty)$, and $I_f(t)$ following an optical excitation step starting at the origin (t = 0).

$$N = S_0 + S_1 + T_0 \tag{30}$$

where S_0, T_1, and T_0 denote the number of molecules in the corresponding states.

The response of the fluorescence intensity to an excitation step can be obtained by solving the rate equations (28 and 29) and using the relation (30). This results in:

$$S_1(t) - S_1(\infty) = A\exp(r_1 t) + B\exp(r_2 t) \tag{31}$$

$$r_1 = -k - k_{ex}(1 - \phi_T) \tag{32}$$

$$r_2 = k_T - k_{ex}\phi_T \tag{33}$$

$$k = k_f + k_{ISC} \tag{34}$$

$$A = (r_1 - r_2)^{-1}[k_{ex}N + r_2S_1(\infty)] \tag{35}$$

$$B = -(r_1 - r_2)^{-1}[k_{ex}N + r_1S_1(\infty)] \tag{36}$$

The roots r_1 and r_2 represent approximate solutions, retaining only terms linear in the excitation rate k_{ex}, which is assumed to be small w.r.t. $k_f + k_{ISC}$; the triplet quantum yield $\phi_T \doteq k_{ISC}/k$. The constants A and B are obtained from the boundary condition $S_1(0) = 0$ and $\dot{S}_1(0) = k_{ex}N$. Noting that in general $r_1 \gg r_2$, Equation 29 can be transformed into:

$$S_1(t) = k_{ex}Nk^{-1}[1 - \exp(r_1t) - k_{ex}k_{ISC}k^{-1}k_T^{-1}\{1 - \exp(r_2t)\}] \tag{37}$$

In steady state it follows that:

$$T_0(\infty) = S_1(\infty)k_{ISC}k_T^{-1} \tag{38}$$

$$S_1(\infty) = k_{ex}Nk^{-1} \tag{39}$$

Converting $S_1(t)$ into $I_f(t)$ — the fluorescence intensity — by $I_f = k_fS_1 = \phi_fkS_1$, where ϕ_f is the fluorescence quantum yield, Equation 37 takes the form:

$$I_f(t) = k_eN\phi_f\left[1 - \exp(r_1t) - \frac{T_0(\infty)}{N}\{1 - \exp(r_2t)\}\right] \tag{40}$$

Equation 38 reveals the presence of two different processes: (1) a rapid rise of the fluorescence to a steady state level, due to populating of S_1 immediately after the onset of excitation, i.e., in a time interval $k^{-1} < t < k_T^{-1}$ and (2) a much slower decrease of $I_f(t)$ due to populating T_0 — referred to as FF. Assuming $k \gg k_T$, we choose to ignore the first process for $I_f(t)$, for instance, i.e., we choose the time interval such that it only refers to process (2), implying that effectively $t = 0$ somewhere in the interval $k^{-1} \ll t < k_T^{-1}$. Then, Equation 40 can be written as

$$I_f(t) = I_f(0)\left[1 - \frac{T_0(\infty)}{N}\{1 - \exp(r_2t)\}\right] \tag{41}$$

where $I_f(0)$ is the fluorescence intensity, found by extrapolating the fluorescence fading curve to $t = 0$. From Equation 41 we obtain:

$$I_f(t) - I_f(\infty) = [I_f(0) - I_f(\infty)]\exp(r_2t) \tag{42}$$

$$\frac{I_f(0) - I_f(\infty)}{I_f(0)} = T_0(\infty)/N \tag{43}$$

These results are a simplified version of those obtained by Avarmaa.[66] The shape of $I_f(t)$, expressed by Equation 42 is presented in Figure 9B.

By combining Equations 33 and 43, it can be easily seen that r_2 in Equation 33 can also be written as

$$r_2 = -k_T\left[1 - \frac{T_0(\infty)}{N}\right] \tag{44}$$

at low excitation rates. Thus, the true molecular triplet decay rate k_T can be simply obtained from the experimental FF curve by subtracting a fraction $\frac{T_0(\infty)}{N}$ from the measured value of r_2. This fraction also represents the experimental fractional decrease of the fluorescence intensity from its initial to its steady state value, due to transfer of molecules to T_0 during excitation.

A similar treatment as given above applies to the case where the presence of three spin levels $|i\rangle$, $|j\rangle$, and $|k\rangle$ in T_0 is taken into account. Now the $I_f(t)$ curve contains the roots $k_\lambda + \phi_T^\lambda k_e$ ($\lambda = i,j,k$), analogously to Equation 42, where k_λ is the decay rate constant of spin level $|\lambda\rangle$. Then a three-exponential analysis of $I_f(t)$, e.g., using Provencher's methods,[67] yields the three decay rate constants of the T_0 spin levels after proper extrapolation to zero optical excitation rate. The amplitudes of these three roots in $I_f(t)$ are a direct measure of the corresponding steady-state populations $n_\lambda(\infty)$.

By combining $n_\lambda(\infty)$ and k_λ for $\lambda = i,j,k$, the relative populating rates $p\lambda$ of the corresponding spin levels can be determined from $p_\lambda = n_\lambda \cdot k_\lambda$. Thus, the populating and decay kinetics of T_0 can be fully resolved without making use of transitions between the T_0 sublevels, as in FDMR. There may occur complications, however, e.g., when one of the spin levels carries no steady-state population, e.g., due to a rigid selection rule, resulting in zero populating rate. Then, of course, the decay rate constant of this level cannot be determined and one finds two, instead of three, decay components. This situation cannot be distinguished from that with two accidentally equal decay rate constants. The FF method has been employed to study the solvation of several photosynthetic pigments in noncrystalline solid solutions at low temperatures (e.g., see References 10, 65, 66, and 68).

The FF method can also be applied to systems containing both isolated pigments and traps for excitation energy, such as described by Hala et al.[42] Energy transfer may occur between the isolated pigments acting as an antenna for optical excitation and the traps. At low to moderate pigment concentration, singlet energy transfer is more likely to occur than triplet energy transfer and we will limit the following treatment to the former type of energy transfer.

Singlet energy transfer is described by the bimolecular forward and backward reactions as treated in Section IV.A2. It is assumed that the singlet energy transfer from antenna to trap is sufficiently efficient to neglect the formation of antenna triplets, i.e., no T_0^a states are considered. Neglecting back transfer ($k_2 S_1^t \ll k_{ex}$), equating S_0^a to N_a (no formation of antenna triplets), Equation 12 simplifies to

$$\dot{S}_1^a = k_{ex}N_a - (k_f^a + k_{ST}^a)S_1^a - k_1 S_0^t S_1^a \tag{45}$$

Noting that

$$\dot{S}_0^t = -k_1 S_1^a S_0^t + k_2 S_0^a S_1^t + k_T^a T_0^a \approx k_1 S_1^a S_0^t \tag{46}$$

Equation 45 can be rewritten as

$$\dot{S}_1^a = k_{ex}N_a - (k_f^a + k_{ST}^a)S_1^a + \dot{S}_0^t \qquad (47)$$

The last term in Equation 47 is dominant at $t \gg (k_f^a + k_{ST}^a + k_{ex})^{-1}$, so that

$$\dot{S}_1^a = \dot{S}_0^t \qquad (48)$$

Noting that $T_0^t(0) = 0$

$$S_1(t) - S_1(0) = -T_0^t(t) = T_0^t(\infty)(e^{-k_T^t t} - 1) \qquad (49)$$

and the fluorescence fading $I_f(t)$ is given by

$$I_f^a(t) - I_f^a(0) = k_f^a[S_1^a(t) - S_1^a(0)] = k_f^a T_0^t(\infty)[e^{-k_T^t t} - 1] \qquad (50)$$

As for an isolated molecule, the antenna fluorescence is a continuously decreasing function of time. A log plot of $\dfrac{I_f^a(t) - I_f^a(\infty)}{I_f^a(0) - I_f^a(\infty)}$ vs. t yields k_T^t.

At sufficiently low temperature, usually ≤4 K, from such a plot the three separate decay rate constants of the spin levels of T_0^t can be obtained. Similarly to the analysis of FF curves for isolated molecules, the steady-state populations of the three T_0^t spin levels can be derived from the amplitudes of the three exponential decays, constituting the total FF curve.

In conclusion, the full sublevel kinetics of the lowest triplet state of a trap, embedded in a collection of isolated pigment molecules transferring singlet-excitation energy to the trap, can be determined using FF curves.

C. FDMR VS. FF

By inserting the kinetic data for the T_0^t state of the trap into Equation 4, the sign of the FDMR transitions of this state, detected via the antenna fluorescence, can be predicted. In the presence of singlet to singlet energy transfer from antenna to trap, the sign of all FDMR transitions is predicted to be opposite to those calculated from the values of k_λ and n_λ^0 obtained from the FF kinetics. For T-T and other types of energy transfer, such sign inversion of FDMR spectra is not predicted, as can be easily concluded from a qualitative analysis of the biomolecular energy transfer kinetics in an optical pumping cycle, including energy transfer between T_0^a and T_0^t states. For photosynthetic pigments in nonordered (bio)polymers or organic glasses, often the triplet state kinetics cannot be determined in a reliable manner. Presumably, this is due to the spread in transition frequencies resulting from the distribution of microenvironments of the pigment in such media. Then, the microwave field is not fully saturating for each site and the response of the fluorescence, following a microwave pulse or step, cannot be fitted by three exponentials. In addition, there may be more fundamental reasons, such as increased spin-lattice relaxation between the triplet sublevels in unordered media, preventing a reliable analysis of the microwave-induced fluorescence response. On the other hand, FF curves do not suffer from this phenomenon and yield a set of kinetic data for each of the triplet sublevels, averaged over the site distribution.

V. CONCLUSIONS AND PERSPECTIVES

The model systems which we have discussed are a collection of photosynthetic pigments embedded in a biopolymer and acting as an antenna for optical excitation of a trap. They are representative of well-characterized models of complexes of chlorophylls with protein, which show efficient excitation energy transfer and charge separation in plants and photosynthetic bacteria.

In these man-made model pigment-polymer systems, a distinction can be made between those molecules with an antenna function on one hand and traps for excitation energy on the other. The antenna and trap contain several chemically different pigments or pigments in different environments.

In photosynthetic systems at room temperature, traps may consist of reaction centers or at low temperature, impurity traps in the antenna. Higher plants have intricate antenna systems consisting of complexes of protein and various forms of chlorophylls: the light-harvesting complex (LHC), and several forms of antannae associated with each of the two reaction centers. Photosynthetic bacteria have similar antennae, organized in a similar way to plants, but simplified by the presence of only one type of reaction center.

The use of ODMR techniques, in particular the inversion of the sign of FDMR transitions, has given useful information on the trapping of excitation energy in pigment-biopolymer model systems. Both ADMR and FDMR can give structural information on the pigment molecule, which acts as the trap in the complex.

Picosecond laser spectroscopy is a useful technique for studying energy transfer between pigments bound to a biopolymer. However, the results to date are not able to tell us very much more than that the Förster description is probably the correct one and that this is dictated by the "static" interaction of pigment and biopolymer.

It will be necessary to investigate even more complicated models containing many pigment molecules in order to test such concepts as the channeling of energy and its localization in certain subpopulations of pigments. The synthesis of such models is difficult, however it may be unnecessary if current attempts to crystallize the isolated chlorophyll-protein complexes from plants and photosynthetic bacteria prove successful. The determination of the structure of these crystalline samples to a sufficient resolution to obtain the atomic coordinates of the antenna and reaction-center pigments will give us model systems containing both the natural pigments and the natural biopolymer, and moreover with a precisely known three-dimensional structure.

Knowledge of the atomic coordinates of the protein biopolymer (the polypeptide backbone, and especially the amino acid side groups) may allow detailed studies of the dynamic role of the biopolymer in electron transfer and possibly in energy transfer, which could be tested by subpicosecond kinetic experiments.

REFERENCES

1. **Breton, J., Martin, J.-L., Migus, A., Antonetti, A., and Orszag, A.,** Femtosecond spectroscopy of excitation energy transfer and initial charge separation in the reaction center of the photosynthetic bacterium *Rhodopseudomonas viridis, Proc. Natl. Acad. Sci. U.S.A.,* 83, 5121, 1986.
2. **Pellegrino, F.,** Ultrafast energy transfer processes in photosynthetic systems probed by picosecond fluorescence spectroscopy, *Opt. Eng.,* 22, 508, 1983.
3. **Schaafsma, T. J.,** ODMR spectroscopy in photosynthesis I, in *Triplet State ODMR Spectroscopy: Techniques and Applications to Biophysical Systems,* Clarke, R. H., Ed., Wiley-Interscience, New York, 1982, chap. 8.
4. **Hoff, A. J. and Gorter de Vries, H.,** Energy transfer at 1.5 K in some photosynthetic bacteria monitored by microwave-induced fluorescence (MIF) spectra, *Isr. J. Chem.,* 21, 277, 1981.
5. **Markwell, J. P., Thornber, J. P., and Boggs, R. T.,** Higher plant chloroplasts: evidence that all the chlorophyll exists as chlorophyll-protein complexes, *Proc. Natl. Acad. Sci. U.S.A.,* 76, 1233, 1979.
6. **Boxer, S. G.,** Model reactions in photosynthesis, *Biochim. Biophys. Acta,* 726, 265, 1983.
7. **Pearlstein, R. M.,** Chlorophyll singlet excitons, in *Photosynthesis: Energy Conversion by Plants and Bacteria,* Vol. 1, Govindjee, Ed., Academic Press, New York, 1982, 293.

8. **Matthews, B. W. and Fenna, R. E.,** Structure of a green bacteriochlorophyll protein, *Acc. Chem. Res.,* 13, 309, 1980.
9. **Kenkre, V. M.,** Theoretical methods of energy transfer in molecular systems, in *Proc. 4th Int. Seminar on Energy Transfer in Condensed Matter,* Pantoflicek, J. and Zachoval, L., Eds., Society of Czechoslovak Mathematicians and Physicists, Prague, 1981, 54.
10. **Mauring, K., Renge, I., Sarv, P., and Avarmaa, R.,** Fluorescence detected triplet kinetic study of the specifically solvated chlorophyll a and protochlorophyll in frozen solutions, *Spectrochim. Acta,* 43A, 507, 1987.
11. **Zuber, H.,** Structure of light-harvesting antenna complexes of photosynthetic bacteria, cyanobacteria and red algae, *Trends Biochem. Sci.,* 11, 414, 1986.
12. **Deisenhofer, J., Michel, H., and Huber, R.,** The structural basis of photosynthetic light reactions in bacteria, *Trends Biochem. Sci.,* 10, 243, 1985.
13. **Rebane, K. K. and Avarmaa, R. A.,** Sharp line vibronic spectra of chlorophyll and its derivatives in solid solutions, *Chem. Phys.,* 68, 191, 1982.
14. **Miller, J. R.,** Controlling charge separation through effects of energy distance and molecular structure on electron transfer rates, *N. J. Chem.,* 11, 83, 1987.
15. **Parson, W. W.,** Photosynthetic bacterial reaction centers: interactions among the bacteriochlorophylls and bacteriopheophytins, *Annu. Rev. Biophys. Bioenerg.,* 11, 57, 1982.
16. **Gulotty, R. J., Mets, L., Alberte, R. S., and Fleming, G. R.,** Picosecond fluorescence study of photosynthetic mutants of *Chlamydomonas reinhardii:* origin of the fluorescence decay kinetics of chloroplasts, *Photochem. Photobiol.,* 41, 487, 1985.
17. **Van Hoek, A., Vos, K., and Visser, A. J. W. G.,** Ultrasensitive time-resolved polarized fluorescence spectroscopy as a tool in biology and medicine, *IEEE J. Quantum Electron.,* QE-23, 1812, 1987.
18. **Tredwell, C. J., Synowiec, J. A., Searle, G. F. W., Porter, G., and Barber, J.,** Picosecond time resolved fluorescence of chlorophyll *in vivo, Photochem. Photobiol.,* 28, 1013, 1978.
19. **Geacintov, N. E., Husiak, T., and Kolubayev, T.,** Excitation annihilation versus excited state absorption and stimulated emission effects in laser studies of fluorescence quenching of chlorophyll *in vitro* and *in vivo, Chem. Phys. Lett.,* 66, 154, 1979.
20. **Kolubayev, T., Geacintov, N. E., Paillotin, G., and Breton, J.,** Domain sizes in chloroplasts and chlorophyll-protein complexes probed by fluorescence yield quenching induced by singlet-triplet exciton annihilation, *Biochim. Biophys. Acta,* 808, 66, 1985.
21. **Van Hoek, A. and Visser, A. J. W. G.,** Artefact and distortion sources in time correlated single photon counting, *Anal. Instrum.,* 14, 359, 1985.
22. **Van Hoek, A. and Visser, A. J. W. G.,** Pulse selection system with electro-optic modulators applied to mode-locked cw lasers and time-resolved single photon counting, *Rev. Sci. Instrum.,* 52, 1199, 1981.
23. **Van Hoek, A., Vervoort, J., and Visser, A. J. W. G.,** A subnanosecond resolving spectrofluorimeter for the analysis of protein fluorescence kinetics, *J. Biochem. Biophys. Meth.,* 7, 243, 1983.
24. **Vos, K., van Hoek, A., and Visser, A. J. W. G.,** Application of a reference deconvolution method to tryptophan fluorescence in proteins. A refined description of rotational dynamics, *Eur. J. Biochem.,* 165, 55, 1987.
25. **Van Hoek, A. and Visser, A. J. W. G.,** Modelocked and synchronously pumped cw lasers and time-resolved single photon counting: a time resolution study, *Proc. Int. Conf. Lasers '87,* Lake Tahoe, NV, Society for Optical and Quantum Electronics, McLean, VA, 1988, 1083.
26. **Hoff, A. J.,** ODMR spectroscopy in photosynthesis II. The reaction center triplet in bacterial photosynthesis, in *Triplet State ODMR Spectroscopy,* Clarke, R. H., Ed., John Wiley & Sons, New York, 1982, 367.
27. **Chan, I. Y., Sandroff, C. F., and Goldenberg, B. L.,** A high temperature ODMR study of the two crystalline phases of benzil, *Chem. Phys. Lett.,* 61, 465, 1979.
28. **Gillies, R., Spendel, W. U., and Ponte Goncalves, A. M.,** Nanosecond-resolved optically detected magnetic resonance: application to the study of triplet-state spin-lattice relaxation in benzil crystals, *Chem. Phys. Lett.,* 66, 121, 1979.
29. **Trifunac, A. D. and Smith, J. P.,** Optically detected time resolved EPR of radical ion pairs in pulse radiolysis of liquids, *Chem. Phys. Lett.,* 73, 94, 1980.
30. **Benthem, L.,** Tetraphenylporphyrin Dimers: An Optical and Magnetic Resonance Study, Thesis Wageningen Agricultural University, Netherlands, 1984.
31. **Clarke, R. H.,** Zero-field ODMR techniques: nonphosphorescence detection, in *Triplet State ODMR Spectroscopy,* Clarke, R. H., Ed., John Wiley & Sons, New York, 1982, 25.
32. **Momo, F., Sotgiu, A., and Zonta, R.,** On the design of a split ring resonator for ESR spectroscopy between 1 and 4 GHz, *J. Phys. E:,* 16, 43, 1983.
33. **Hardy, W. N. and Whitehead, L. A.,** Split-ring resonator for use in magnetic resonance from 200-2000 MHz, *Rev. Sci. Instrum.,* 52, 213, 1981.
34. **Mehring, M. and Freysoldt, F.,** A slotted tube resonator STR for pulsed ESR and ODMR experiments, *J. Phys. E:,* 13, 894, 1980.

35. **Frenois, C.,** Broadband tunable cavities with helical microstrip lines, *J. Phys. E:,* 17, 35, 1984.
36. **Den Blanken, H. J., Meiburg, R. F., and Hoff, A. J.,** Polarised triplet-minus-singlet absorbance difference spectra measured by absorbance-detected resonance. An application to photosynthetic reaction centers, *Chem. Phys. Lett.,* 105, 336, 1984.
37. **Szalay, L., Tombácz, E., Várkonyi, Z., and Faludi-Dániel, Á.,** Detergent effects on an albumin-chlorophyll complex model of photosynthetic protein-pigment complexes, *Acta Phys. Acad. Sci. Hung.,* 53, 225, 1982.
38. **Kusumoto, Y., Senthilathipan, V., and Seely, G. R.,** Association of chlorophyll with amides on platicized polyethylene particles. III. Unusual spectra of chlorophyll a with N-methylmyristamide, *Photochem. Photobiol.,* 37, 571, 1983.
39. **Breton, J. and Nabedryk, E.,** Transmembrane orientation of α-helixes and the organization of chlorophylls in photosynthetic pigment-protein complexes, *FEBS Lett.,* 176, 355, 1984.
40. **Fidler, V., Pancoska, P., Porter, G., Vacek, K., and Blaha, K.,** Pheophorbide a covalently linked to polypeptide: a model system for studying pigment-protein interactions, in *Proc. 4th Int. Seminar on Energy Transfer in Condensed Matter,* Pantoflicek, J. and Zachoval, L., Eds., Society of Czechoslovak Mathematicians and Physicists, Prague, 1981, 172.
41. **Hala, J., Pelant, I., Ambroz, M., Pancoska, P., and Vacek, K.,** Site-selection and Shpolskii spectroscopy of model photosynthetic systems, *Photochem. Photobiol.,* 41, 643, 1985.
42. **Hala, J., Searle, G. F. W., Schaafsma, T. J., van Hoek, A., Pancoska, P., Blaha, K., and Vacek, K.,** Picosecond laser spectroscopy and optically detected magnetic resonance on a model photosynthetic system, *Photochem. Photobiol.,* 44, 527, 1986.
43. **Kooyman, R. P. H.,** Complexes and Aggregates of Chlorophylls, Thesis, Wageningen Agricultural University, Netherlands, 1980.
44. **Hoff, A. J. and Gorter de Vries, H.,** Energy transfer at 1.5 K in some photosynthetic bacteria monitored by microwave-induced fluorescence (MIF) spectra, *Isr. J. Chem.,* 21, 277, 1981.
45. **Boxer, S. G., Kuki, A., Wright, K. A., Katz, B. A., and Xuong, N. H.,** Oriented properties of the chlorophylls: electronic absorption spectroscopu of orthorhombic pyrochlorophyllide a-apomyoglobin single crystals, *Proc. Natl. Acad. Sci. U.S.A.,* 79, 1121, 1982.
46. **Kuki, A. and Boxer, S. G.,** Chlorophyllide-substituted hemoglobin tetramers and hybrids: preparation, characterization, and energy transfer, *Biochemistry,* 22, 2923, 1983.
47. **Boxer, S. G. and Wright, K. A.,** Preparation and properties of a chlorophyllide-apomyoglobin complex, *J. Am. Chem. Soc.,* 101, 6791, 1979.
48. **Wright, K. A. and Boxer, S. G.,** Solution properties of synthetic chlorophyllide-and bacteriochlorophyllide-apomyoglobin complexes, *Biochemistry,* 20, 7546, 1981.
49. **Clarke, R. H., Hanlon, E. B., and Boxer, S. G.,** Investigation of the lowest triplet state of the pyrochlorophyllide a apomyoglobin complex by zero-field optically detected magnetic resonance spectroscopy, *Chem. Phys. Lett.,* 89, 41, 1982.
50. **Boxer, S. G. and Bucks, R. R.,** Chlorophyll-amino acid interactions in synthetic models, *Isr. J. Chem.,* 21, 259, 1981.
51. **Moog, R. S., Kuki, A., Fayer, M. D., and Boxer, S. G.,** Excitation transport and trapping in a synthetic chlorophyllide substituted hemoglobin: orientation of the chlorophyll S_1 transition dipole, *Biochemistry,* 23, 1564, 1984.
52. **Boxer, S. G.,** Structure and energetics in reaction centers and semi-synthetic chlorophyll protein complexes, in *Antennas and Reaction Centers of Photosynthetic Bacteria, Structure, Interactions, and Dynamics,* Michel-Beyerle, M. E., Ed., Springer-Verlag, Berlin, 1985, 306.
53. **van der Waals, J. H. and de Groot, M. S.,** Magnetic interactions related to phosphorescence, in *The Triplet State,* Zahlan, A. B., Ed., University Press, Cambridge, 1967, 101.
54. **Krasnovskii, A. A.,** Delayed fluorescence and phosphorescence of plant pigments, *Photochem. Photobiol.,* 36, 733, 1982.
55. **Krasnovski, A. A., Lebedev, N. N., and Litvin, F. F.,** Phosphorescence and delayed fluorescence of chlorophyll and its precursors in solutions, leaves and chloroplasts at 77 K, *Stud. Biophys.,* 65, 81, 1977.
56. **Naus, J.,** Phosphorescence and Delayed Light Emission of Chlorophyll Pigments, Thesis, Charles University, Prague, 1977.
57. **Kleibeuker, J. F., Platenkamp, R. J., and Schaafsma, T. J.,** The triplet state of photosynthetic pigments. I. Pheophytins, *Chem. Phys.,* 27, 51, 1978.
58. **van der Bent, S. J., de Jager, P. A., and Schaafsma, T. J.,** Optical detection and electronic simulation of magnetic resonance in zero magnetic field of dihydroporphin free base, *Rev. Sci. Instrum.,* 47, 117, 1976.
59. **Clarke, R. H. and Hofeldt, R. H.,** Optically detected zero field magnetic resonance studies of photoexcited triplet states of chlorophyll a and b, *J. Chem. Phys.,* 61, 4582, 1974.

60. **van Dorp, W. G., Schaafsma, T. J., Soma, M., and van der Waals, J. H.,** Investigation of the lowest triplet state of free base porphin; microwave induced changes in its fluorescence, *Chem. Phys. Lett.,* 21, 221, 1973.

61. **van Dorp, W. G., Schoemaker, W. H., Soma, M., and van der Waals, J. H.,** The lowest triplet state of free base porphin: determination of its kinetics of populating and depopulating from microwave induced transients in the fluorescence intensity, *Mol. Phys.,* 30, 1701, 1975.

62. **van der Bent, S. J. and Schaafsma, T. J.,** Interaction of light and microwaves with pheophytins, *Chem. Phys. Lett.,* 35, 45, 1975.

63. **Hoff, A. J. and Cornelissen, B.,** Microwave power dependence of triplet state kinetics as measured with fluorescence detected magnetic resonance in zero field. An application to the reaction center bacteriochlorophyll triplet in bacterial photosynthesis, *Mol. Phys.,* 45, 413, 1982.

64. **Beck, J., Kaiser, G. H., von Schütz, I. U., and Wolf, H. C.,** Optically excited triplet states in the bacteria *Rhodopseudomonas sphaeroides* ''wild type'' detected by magnetic resonance in zero field, *Biochim. Biophys. Acta,* 634, 165, 1981.

65. **Avarmaa, R.,** Fluorescence detection of triplet state kinetics of chlorophyll, *Chem. Phys. Lett.,* 46, 279, 1977.

66. **Avarmaa, R.,** Fluorescence detection study of molecular triplet states in chlorophyll and related compounds, *Mol. Phys.,* 37, 441, 1979.

67. **Provencher, S. W.,** A Fourier method for the analysis of exponential decay curves, *Biophys. J.,* 16, 27, 1976.

68. **Avarmaa, R. and Schaafsma, T. J.,** Site-selected fluorescence detection of magnetic resonance of protochlorophyll and related chlorophylls, *Chem. Phys. Lett.,* 71, 339, 1980.

INDEX

time-resolved, 108, 111—117
Liquid crystalline polymers, see Mesomorphic
  polymers
Lower critical solution temperature (LCST), 109
Low-intensity therapy, 162—163
Lyotropic polymer, 36

M

Macrostructures, 116
Main-chain mesomorphic polymers, 36
Malto-oligosaccharides, 12
Mass spectrometers, types, 2—4
Mass spectrometry, application to polymer analysis,
  1—31
Maxwell equations, 39—40, 47
MCA, see Multichannel analyzer
Medicine, laser spectroscopy, 124
Mesophases, 102
Mesomorphic polymers, 34—37, 82—87
Methylated DNA, emission spectroscopy, 187—188
Microbeam systems, 134—135, 163
Microchannel plate-type photomultiplier, 224
Microwave induced response of fluorescence (MIRF),
  225, 227
Mie scattering techniques, 163
Mie theory, 47—50, 52, 54, 101, 106
MIRF, see Microwave induced response of
  fluorescence
Model photosynthetic systems, 219—245
  aggregates as traps, 230
  chlorophyllide *a* myoglobin system, 230
  FDMR spectra, 229, 231
  low-energy trapping sites, 228
  pheophorbide-*a*-polytripeptide, 228—229
  pigment-biopolymer systems, 228—231
  site distribution function, 229
Model pigment-model biopolymer, 228—230
Model pigment-natural biopolymer, 230—231
Modulation metods, 172
Molecular dynamics, 34
Molecular triplet states, 220
Monochromaticity, 107, 124—125
Monochromator dispersion, 179—181
Morphology, 101, 104—108, 118
Morphology formation, multicomponent polymer
  systems, 109—117
Motion, 155, 156, 165, 197
MPI, see Multiphoton ionization
Multichannel analyzer (MCA), 181—182
Multicomponent polymer systems
  morphology formation in, 109—117
  morphology of, as studied by light scattering, 104—
    108
  optical properties of, 105—108
  structure of, 105—108
Multiphoton ionization (MPI), 2, 26—27
Myoglobin, 125, 157—158, 160—161

N

Nd:YAG lasers, 126, 133—134, 144, 148, 173, 176

clinical medicine, 162
  nanosecond absorption spectra, 159
  Raman scattering, 153
Nd:YLF laser, 173
Negative ions, 9
Nematic polymers, 34—36, 82, 84, 87
Neurosurgery, 162
Neutron scattering, 34, 37
NG, see Nucleation and growth
NMR, see Nuclear magnetic resonance
Nonlaser ionization sources, 4—7
Nuclear magnetic resonance (NMR), 189
Nucleation and growth (NG) mechanism, 110—112,
  114—115, 118
Nucleic acid bases, emission spectroscopy, 184—190
Nucleic acid-dye complexes, emission spectroscopy,
  190—201, see also Emission spectroscopy
Nucleic acids, 125, 128, 147—157, 165
  conformation, 155—157
  emission spectroscopy, 184—190
  energy transfer, 152—155
  excited states, 151—153
  excitons, 154—155
  fluorescence, 151—153, 157
  motion, 155—157
  Raman spectroscopy, 202—205
  structural investigations, 148—151
Nucleosome core particles, Raman spectra, 207—208
Nucleus, 110

O

ODMR, see Optically detected magnetic resonance
Ophthalmology, 162
Optically detected magnetic resonance (ODMR),
  224—228
  chlorophyllide-a-myoglobin systems, 231
  multipurpose setup, 224—225
  noise, 227
  pigment-biopolymer interactions, 231
  pulsed, 224
Optical multichannel analyzer, 58—59
Optical pumping cycle, 232, 237
Optical rotation, 45, 82, 84—85
Orientational disorder, 76
Oriented crystallization, 101

P

Particle bombardment, 5—7
PDMR, see Phosphorescence detected magnetic
  resonance
PEG, see Poly(ethylene glycol)
PEI, see Poly(ethylene imine)
PEG methyl ether, 23
Phase separation, 108—110, 115
Phosphorescence, 199—201
Phosphorescence detected magnetic resonance
  (PDMR), 224, 228
Photoactivation of enzymes, 125
Photobiological reaction, 129
Photobiology, 125, 136—161, see also specific topics